CAMBRIDGE LIBRARY COLLECTION

Books of enduring scholarly value

Physical Sciences

From ancient times, humans have tried to understand the workings of the world around them. The roots of modern physical science go back to the very earliest mechanical devices such as levers and rollers, the mixing of paints and dyes, and the importance of the heavenly bodies in early religious observance and navigation. The physical sciences as we know them today began to emerge as independent academic subjects during the early modern period, in the work of Newton and other 'natural philosophers', and numerous sub-disciplines developed during the centuries that followed. This part of the Cambridge Library Collection is devoted to landmark publications in this area which will be of interest to historians of science concerned with individual scientists, particular discoveries, and advances in scientific method, or with the establishment and development of scientific institutions around the world.

The System of the Stars

Agnes Mary Clerke (1842-1907) published *The System of the Stars* in 1890 when she was a well-established popular science writer. The volume was intended to bring the educated public up to date with the progress made during the nineteenth century in the field of sidereal astronomy. The work was one of the first publications to be illustrated with astrophotography: it contains five astronomical photographs of nebulae. Such photographs had significant impact on the reception and popular acceptance of astrophotography as scientific data. In *The System of the Stars*, Clerke used the photographs to argue that the natural beauty and symmetry of the universe, displayed by astrophotography, proved the existence of a creator. The work is an important piece of popular Victorian scientific literature, and remains significant today in the context of the nineteenth-century intellectual debates on the relationship between the sciences and religious belief.

Cambridge University Press has long been a pioneer in the reissuing of out-of-print titles from its own backlist, producing digital reprints of books that are still sought after by scholars and students but could not be reprinted economically using traditional technology. The Cambridge Library Collection extends this activity to a wider range of books which are still of importance to researchers and professionals, either for the source material they contain, or as landmarks in the history of their academic discipline.

Drawing from the world-renowned collections in the Cambridge University Library, and guided by the advice of experts in each subject area, Cambridge University Press is using state-of-the-art scanning machines in its own Printing House to capture the content of each book selected for inclusion. The files are processed to give a consistently clear, crisp image, and the books finished to the high quality standard for which the Press is recognised around the world. The latest print-on-demand technology ensures that the books will remain available indefinitely, and that orders for single or multiple copies can quickly be supplied.

The Cambridge Library Collection will bring back to life books of enduring scholarly value (including out-of-copyright works originally issued by other publishers) across a wide range of disciplines in the humanities and social sciences and in science and technology.

The System of the Stars

Agnes Mary Clerke

CAMBRIDGE UNIVERSITY PRESS

Cambridge, New York, Melbourne, Madrid, Cape Town, Singapore,
São Paolo, Delhi, Dubai, Tokyo

Published in the United States of America by Cambridge University Press, New York

www.cambridge.org
Information on this title: www.cambridge.org/9781108014168

This edition first published 1890
This digitally printed version 2010

ISBN 978-1-108-01416-8 Paperback

THE

SYSTEM OF THE STARS

Photographic Chart of the PLEIADES by M. M. Paul and Prosp

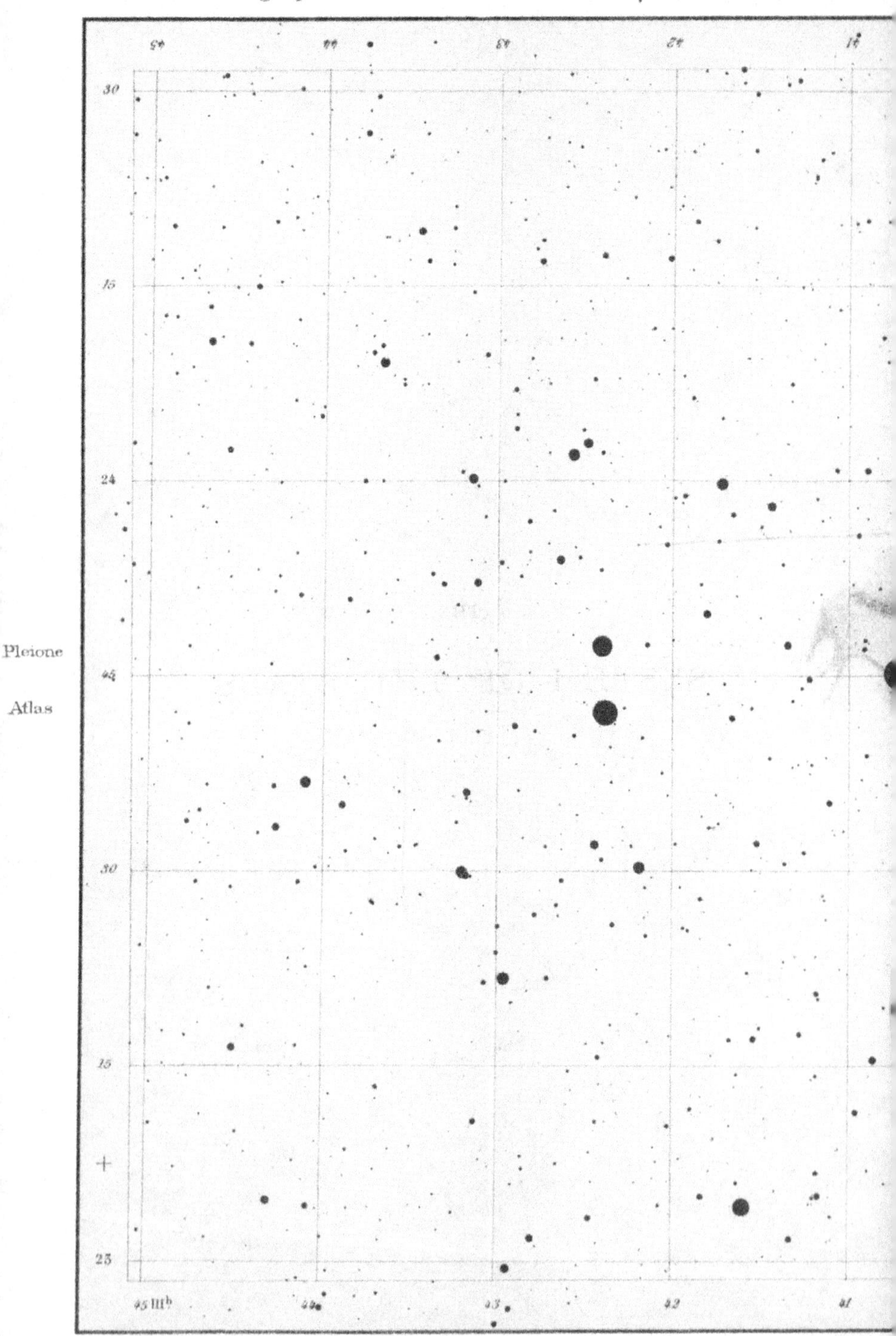

F. S. Weller lithe.

Longmans & Co., London

Henry containing 2326 Stars with nebulae intermixed.

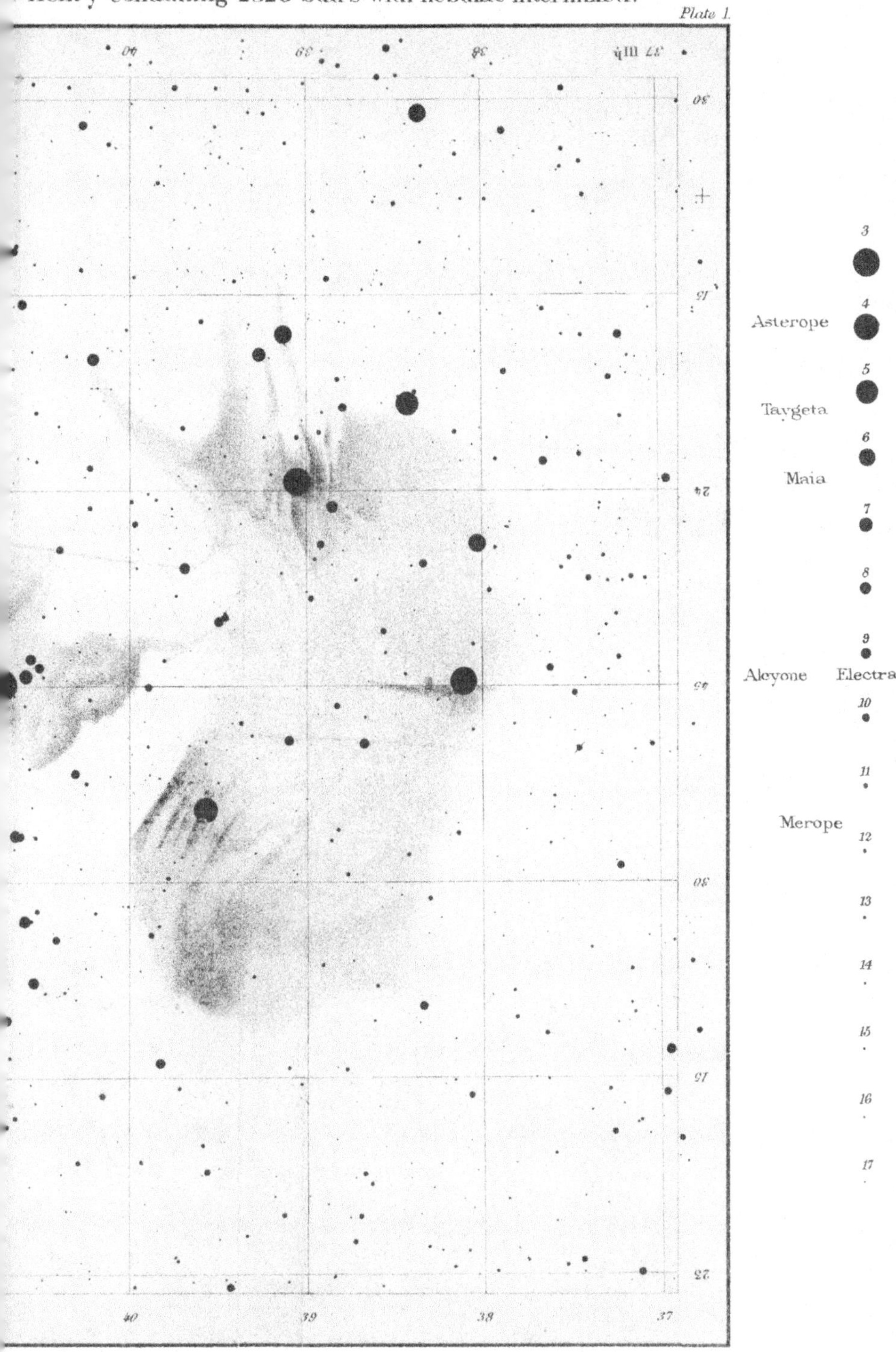

New York.

Gravé par Lamblin.

THE

SYSTEM OF THE STARS

BY

AGNES M. CLERKE

AUTHOR OF 'A POPULAR HISTORY OF ASTRONOMY
DURING THE NINETEENTH CENTURY'

*'Io vidi delle cose belle
Che porta 'l ciel'*

LONDON
LONGMANS, GREEN, AND CO.
AND NEW YORK: 15 EAST 16th STREET
1890

TO THE MEMORY OF

MY FATHER

JOHN WILLIAM CLERKE

WHO DIED IN LONDON

FEBRUARY 24, 1890

PREFACE

SIDEREAL science has a great future before it. The prospects of its advance are incalculable; the possibilities of its development virtually infinite. No other branch of knowledge attracts efforts for its promotion, at once so wide-spread, so varied, and so enthusiastic; and in no other is anticipation so continually outrun by the brilliant significance of the results achieved.

For the due appreciation, however, of these results, some preliminary knowledge is required, and is possessed by few. To bring it within the reach of many is the object aimed at in the publication of the present volume. Astronomy is essentially a popular science. The general public has an indefeasible right of access to its lofty halls, which it is the more important to keep cleared of unnecessary technical impediments, since the natural tendency of all sciences is to become specialised as they advance. But literary treatment is the foe of specialisation, and helps to secure, accordingly, the topics it is applied to, against being secluded from the interest and understanding of ordinarily educated men and women. Now, in the whole astonishing history of the human intellect, there is no more astonishing chapter than that concerned with the sidereal researches of the last quarter of a century. Nor can the resources of thought be more effectually widened, or its principles be more surely ennobled through the vision of a Higher Wisdom, than by rendering it, so far as possible, intelligible to all.

The following pages then embody an attempt to combine, in a general survey, some definite particulars of knowledge regarding our sidereal surroundings. The plan pursued has been to instruct by illustrative examples, to select typical instances from each class of phenomena, dwelling upon them with sufficient detail to awaken interest and assist realisation, while avoiding the tediousness inseparable from exhaustive treatment. The statement of facts has been kept primarily in view; but the more important efforts to interpret them have been noticed, and the difficulties attending rival theories impartially pointed out. In developing the subject it seemed best to proceed from the particular to the general; to start with describing the physical constitution of individual bodies, and, ascending by degrees through continually added complexities of mutual relationships, reach at last the crowning problem of the Construction of the Heavens.

The writer gratefully acknowledges the assistance derived in the preparation of the present work, from the kindness of Dr. Gill, H.M. Astronomer at the Cape, first and chiefly, in affording her an opportunity of observing the southern skies; secondly, in reading over several of its chapters in manuscript. Her thanks are also due to Professor E. S. Holden, Director of the Lick Observatory, to Messrs. Burnham, Keeler, and Barnard of the same establishment, to Professor E. C. Pickering, Director of Harvard College observatory, to Dr. and Mrs. Huggins, Professors Lockyer, Vogel, Schönfeld, and others, for communications of great interest and value. To the generous assistance of Mr. Isaac Roberts, of Crowborough, the author is indebted for the photographs of nebulæ and clusters reproduced in the Plates; others have been kindly furnished by Mr. Barnard; while permission to use photographs and drawings already published has, with the utmost courtesy, been accorded to her by Admiral Mouchez and the MM. Henry, by Professors Holden and Pickering, Dr. and Mrs. Huggins, Professor Vogel and Dr. Scheiner of Potsdam, and by Dr. Dunér, Director of the Observatory of Upsala. In designing and selecting the illustrations, elucidation, rather than ornament, was aimed at; how far the end has been attained, must be left to the public to decide.

London: *October*, 1890.

CONTENTS

CHAPTER I.

THE TASK OF SIDEREAL ASTRONOMY.

CHAPTER II.

THE METHODS OF SIDEREAL RESEARCH.

CHAPTER III.

SIRIAN AND SOLAR STARS.

CHAPTER IV.

STARS WITH BANDED SPECTRA.

CHAPTER XIV.

MULTIPLE STARS.

CHAPTER XV.

THE PLEIADES.

CHAPTER XVI.

STAR CLUSTERS.

CHAPTER XVII.

THE FORMS OF NEBULÆ.

CHAPTER XVIII.

THE GREAT NEBULÆ.

CHAPTER XXIII.

THE MILKY WAY.

CHAPTER XXIV.

STATUS OF THE NEBULÆ.

CHAPTER XXV.

THE CONSTRUCTION OF THE HEAVENS.

ILLUSTRATIONS

PLATES

WOODCUTS IN TEXT

ADDENDA

Supply at p. 3, *note*, line 2 from bottom:

Spica = α Virginis.

Add to list of Algol-variables at p. 136:

Name	Discoverer	Period	Amount and duration of change
12 Antliæ	Paul, 1888	7h. 47m.	6·7 to 7·3 mag. in 4h. 50 m. (?)

Note to page 284.

The spectrum of the trifid nebula, as observed by Mr. Keeler at Lick in 1890, is continuous in the green and blue, with a brightening near the middle.

Note to page 290.

Mr. Keeler has lately succeeded in measuring, with the great Lick refractor, the movements in line of sight of several planetary nebulæ. See *Publications Astr. Pacific Society*, No. 11, 1890.

Note to page 342.

The velocities of recession ascribed to β and ε Orionis in 1888 can only be regarded as provisionally determined. With improved apparatus, evidence has been obtained at Potsdam of variability in the movement of Rigel, due doubtless to orbital revolution.

THE SYSTEM OF THE STARS.

CHAPTER I.

THE TASK OF SIDEREAL ASTRONOMY.

When all the stars blaze out on a clear, moonless night, it seems as if it would be impossible to count them; and yet it is seldom that more than 2,000 are visible together to the unaided eye. The number, however, depends very much upon climate and sharpness of sight. Argelander enumerated at Bonn, where rather more than eight-tenths of the sphere come successively into view, 3,237 stars.[1] But of these no more than 2,000 could be, at any one time, above the horizon, and so many would not be *visibly* above it, owing to the quenching power of the air in its neighbourhood. Heis, at Münster, saw 1,445 stars more than Argelander at Bonn;[2] Houzeau recorded 5,719 at Jamaica;[3] Gould 10,649 at Cordoba in South America.[4] The discrepancies of these figures, setting aside the comparatively slight effect of the increased area of the heavens displayed in low latitudes, are due to the multitude of small stars always, it might be said, hovering on the verge of visibility. If, indeed, the atmosphere could be wholly withdrawn, fully 25,000 stars would, according to a trustworthy estimate, become apparent to moderately good eyes.[5]

Our system of designating the stars has come down to us

[1] *Uranometria Nova*, 1843.

[2] Heis, *De Magnitudine Numeroque Stellarum*, p. 16, 1852.

[3] *Uranométrie Générale*, Annales de l'Observatoire de Bruxelles, t. i. 1878.

[4] *Uranometria Argentina*, 1879.

[5] Backhouse, *Journal Liverpool Astr. Society*, vol. vii. p. 226.

from a hoar antiquity. It is a highly incommodious one. 'The constellations,' Sir John Herschel remarks,[1] 'seem to have been almost purposely named and delineated to cause as much confusion and inconvenience as possible. Innumerable snakes twine through long and contorted areas of the heavens where no memory can follow them; bears, lions, and fishes, large and small, northern and southern, confuse all nomenclature.' And yet we could ill afford to dispense with the picturesque associations of a menagerie largely stocked from the banks of the Euphrates. The signs of the Zodiac, which are undoubtedly of Chaldean origin, embody legendary cycles of thought already, some four thousand years ago, the worse for wear and dilapidated by time. Homer and Hesiod were familiar with the Bear, Arcturus, and the Dog-star, with 'the Hyades, and the Pleiades, and the strength of Orion.' The Little Bear was introduced from Phœnicia, when the Pole-star became the mariner's 'cynosure.' Finally, a number of individual stars have Arabic appellations, dating from the epoch of Saracen supremacy over science. Thus 'Vega,' the current name of the brightest star in the Greek constellation of the Lyre, is the remnant of an Arabic phrase signifying the 'Falling Eagle,' while 'Altair' stands for the 'Flying Eagle;' 'Deneb' means the Tail of the Swan; 'Fomalhaut,' the 'Mouth of the Fish;' 'Rigel' in Orion is the Foot, 'Betelgeux,' the 'Shoulder of the Giant,' and so on.

The constellations[2] now generally recognised are sixty-seven in number, of which forty-eight are found in Ptolemy's 'Almagest.' From Ptolemy, too, is derived the method of classifying the stars by 'magnitudes.' This is a most inappropriate term, since none of the stars have any perceptible dimensions. They are literally what Shelley calls them, 'atoms of intensest light'—globes shrunken by distance to the semblance of mere shining needle-points. Our own sun, removed to the place of the nearest fixed star, would be in

[1] *Treatise on Astronomy*, p. 163, *note*.

[2] For an easy method of identifying the chief northern stars, see Sir Robert Ball's *Story of the Heavens*, p. 372; also the 'Uranography' in Young's *Elements of Astronomy*, 1890.

the same condition; its diameter of $\frac{1}{143}''$ would be utterly inappreciable with the largest telescope. It is true that the telescopic images of the stars appear to be of measurable size; but this is so purely an optical effect that the 'spurious discs' shown by them actually grow smaller instead of larger as the power of the instrument is increased.

Thus 'magnitude' has nothing to do with apparent size, but refers entirely to apparent lustre, which depends upon distance and intensity of shining, as well as upon real dimensions. The faintest stars have the highest numerical magnitudes; and it has been found that the gap between each successive order, as represented by the stars traditionally belonging to it, corresponds to a falling-off of light in the proportion of about 2½ to 1. The arrangement by magnitudes is, of course, entirely arbitrary; natural gradations are not by a flight of steps, but along an inclined plane. Stars classed as of the first magnitude (of which there are ten in each hemisphere) [1] differ accordingly very much among themselves. Sirius exceeds Fomalhaut no less than twelve times; Vega is nearly thrice as brilliant as *α* Crucis. Arcturus outshines every other northern star, but three southern luminaries—Sirius, Canopus, and *α* Centauri—are superior to it. Of second magnitude are the seven stars grouped to form 'Charles's Wain,' the Pole-star, and the most vivid gems in Perseus, Cassiopeia, and the Swan. Stars of the sixth magnitude are the faintest ordinarily visible to the naked eye; but those of the seventh can be seen under advantageous circumstances. The plan, introduced by Bayer in 1603, of naming the stars of the several constellations roughly in order of brightness by the letters of the Greek alphabet, established for each a kind of light-sequence, useful though far from exact. The smaller stars are usually distin-

[1] The ten brightest stars north of the equator are: Arcturus = *α* Boötis, Vega = *α* Lyræ, Capella = *α* Aurigæ, Procyon = *α* Canis Minoris, Betelgeux = *α* Orionis, Aldebaran = *α* Tauri, Altair = *α* Aquilæ, Pollux = *β* Geminorum, Regulus = *α* Leonis, and Deneb = *α* Cygni. The ten brightest southern stars are: Sirius = *α* Canis Majoris, Canopus = *α* Argûs, *α* Centauri, Rigel = *β* Orionis, Achernar = *α* Eridani, *β* Centauri, *α* Crucis, Antares = *α* Scorpii, and Fomalhaut = *α* Piscis Austrini.

guished by the numbers attached to them in various catalogues.

One of the most notorious circumstances about the stars is their 'twinkling.' They undergo, especially when near the horizon, extremely rapid changes of lustre, attended sometimes by the glinting of prismatic colours. Nor do all stand, in this respect, on the same level. White stars twinkle more than red ones. Even early and untutored observers noticed how

> The fiery Sirius alters hue,
> And bickers into red and emerald.

whence it was called by Aratus ποικίλος, the 'many-coloured;' and chromatic unsteadiness was a marked peculiarity of the 'new stars' of 1572 and 1604.

It is easy to see that this effect is in some way due to the atmosphere. Like refraction, it vanishes at the zenith; it varies in intensity with weather and climate. The first rational conjecture as to its cause was made in 1667 by Robert Hooke, who attributed it to irregular refractions in the various air-strata. More exact inquiries on the subject have, in recent times, led to some curious results.

The impressions of light on the retina last, according to Plateau's careful determination, 0·34—say one-third—of a second. This is the limit of their individual perceptibility. With more frequent recurrence, they become merged indistinguishably together. But the changes visually integrated as scintillation succeed each other much more rapidly than three times in a second. Hence the need of some means of separating and analysing them.

These are provided by M. Montigny's 'scintillometer,'[1] in which the principle of employing the sensibility of different parts of the retina for the registration of a swift succession of impressions is skilfully turned to account. By the rotation of a glass-plate obliquely inserted in front of the eye-piece of a refracting telescope, the image of a star viewed with it is

[1] Described in *Bulletin de l'Acad. des Sciences*, Bruxelles, t. xvii. p. 261, 2nd ser.; *Monthly Notices*, vol. xxxvii. p. 203; *Ciel et Terre* (Fievez), t. i. p. 369.

made to describe a perfect circle in the field. The line of light thus traced out is, in the absence of scintillation, continuous and of a uniform hue, but breaks up, under its influence, into vividly tinted arcs, at times into prismatic 'pearls.' The addition of a pair of crossed wires facilitates the reckoning of the colour-fluctuations thus rendered separately visible; and they are found to occur, on an average, in white stars standing thirty degrees above the horizon, seventy-eight times in a second, in yellow and red stars similarly placed, sixty-eight and fifty-six times respectively.[1]

The explanation of these appearances is evidently to be sought in the refractive power, combined with the disturbed condition of our atmosphere. For a different path through its strata is necessarily pursued by each of the differently refrangible beams united to form the image of a star. The violet enters them higher up, since it is more bent in transit than the red; and so proportionately of the rest. Each then is liable to encounter different vicissitudes on the way, betrayed to our sight by rapid flashes of colour. Each is affected by innumerable small deviations and momentary caprices of refraction; so that the bundle of rays picturing a star at a given instant is, as it were, a fortuitous and eminently unstable combination. It is dissolved, and a new one constituted, sixty or seventy times in a second; and the elements temporarily missing determine the resulting colour.

We can now understand that white stars twinkle more than red, because the sheaf of their beams being fuller, interceptions of them are more frequent. But planets which are radiating *discs*, and not merely *points*, scarcely show the effect at all, because the absence of rays from one part is compensated by the arrival of rays from other parts of their surfaces. That stars do not scintillate in large telescopes is due to the neutralisation of each casual stoppage by the great number of the rays collected together. Instead of a twinkling image, however, a blurred and distended one is formed, and observation gains nothing by the exchange. And since the degree

[1] *Bull. de l'Acad.* Bruxelles, t. xxxvii. p. 185, 2nd ser.

to which this phenomenon is present varies very much with locality, the sites for powerful instruments should certainly not be chosen, as M. Exner has remarked,[1] regardless of its prevalence. At Vienna it is particularly troublesome, and Dr. Pernter's experiments go to prove that even an ascent to considerable altitudes affords no security against it. He found Sirius, in fact, actually to scintillate *more* at the summit than at the foot of the Sonnblick (10,000 feet high).[2]

Scintillation, like astronomical refraction, augments as the thermometer falls and as the barometer rises. This is inevitable, since the first condition of its occurrence is differential refractive action on the various light-rays.[3] But it has other less obviously accountable meteorological relations, established by M. Montigny during nearly forty years of observation. With the quantity of moisture in the air the twinkling of the stars increases so markedly as to serve for a useful prognostic of rain. Hurricanes and cyclones are still more emphatically announced by it, and it is extremely sensitive to magnetic disturbances.[4] Ussher was struck in the last century, with the surprising vividness of scintillation during auroræ; M. Montigny finds that the coincidence extends to magnetic commotions perceptible only instrumentally. Moreover, Weber remarked at Peckeloh in 1880, that stars situated near the magnetic meridian twinkled more than elsewhere in the sky;[5] and an element of change in this respect, depending upon the points of the compass, has since been fully ascertained. So that it is far from improbable that scintillation may turn out to be one of the many terrestrial phenomena associated with vicissitudes in the physical condition of the sun.

The world of stars thrown open by the telescope may fairly be called boundless. Using a glass only two and a half inches across, Argelander registered 324,189 down to 9½ magnitude, all in the northern hemisphere, enlarged by only a very narrow zone (one degree wide) of the southern. The work

[1] *Astr. Nach.* No. 2,791. [2] *Observatory*, vol. xii. p. 194.

[3] Montigny, *Bull. de l'Acad.* Bruxelles, t. xlvi. p. 613, 2nd series.

[4] *Ibid.* p. 17; *Comptes Rendus*, t. xcvi. p. 573.

[5] *Wochenschrift für Astronomie*, 1880, p. 294.

was extended by Schönfeld to the southern tropic, and is now, under Dr. Gill's direction, being completed by photographic means to the southern pole. When this is done about 650,000 stars will be identified, and (in Professor Holden's phrase) 'indexed.' Each will form a possible object of future investigation, and will have been admitted (so to speak) to the citizenship of astronomy.

Now Mr. Plummer has shown [1] that the lucid stars (or those visible to the naked eye) in Argelander's 'Durchmusterung' give as much light as 7,349, the telescopic stars as 23,337 sixth magnitude stars. So that those singly invisible really illuminate the sky just three times as much as those we can see and distinguish. Summing up the entire light of this grand collection of 324,000 stars, we get for its equivalent $\frac{1}{163}$ of the light of the full moon,[2] and for the total from all the 650,000 stars to $9\frac{1}{2}$ magnitude, $\frac{1}{80}$ full moonlight.

What amount of scattered effulgence reaches us from still fainter stars, there is at present no sure means of estimating; yet in putting the inclusive aggregate at one-tenth full moonlight M. Gustave l'Hermite[3] almost certainly overshot the mark. The number of such objects is entirely unknown. All that can be stated with any confidence is that there are many millions of them. The largest telescopes in the world, it is thought, might show sixty million stars over the entire sky; but the margin of doubt is very wide. M. l'Hermite has, however, computed the population of the stellar universe from his valuation of stellar light-power, and finds it, on the assumption that the scattering of the stars is everywhere just such as it is in our own neighbourhood, to be *sixty-six thousand millions*! This extravagant result indicates, without doubt, that both valuation and assumption are erroneous, although the inquiries based upon them are interesting and ingenious. Especially noteworthy is the fact that each class

[1] *Monthly Notices*, vol. xxxvii. p. 436.

[2] Mr. Plummer's result appears to be vitiated by his adoption of too low a value for the light of Sirius. That given in the text is derived from combining Pickering's photometric determination of Vega with Seidel's measures of the same object, showing it to be equal to $\frac{1}{24000}$ full moonlight.

[3] *L'Astronomie*, t. v. p. 406.

of stars sends us appreciably more light than the class next above it. The light aggregate of second magnitude stars exceeds that of first, of third that of second, and so on. The fainter the stars, in short, the greater is their total luminous power,[1] because their augmented numbers more than counterbalance their diminished *individual* lustre. But this progression, it is evident, cannot go on indefinitely, since otherwise an indefinitely intense radiance would fill the sky. Darkness would be abolished through the shining of invisible stars. It follows that the observed order of the stellar world has assignable limits—that the star-depths, however profound, are not absolutely unfathomable.

The task of exploring them is not then altogether hopeless. It can never indeed be exhausted; but it can fairly be grappled with by finite minds. It does not evade their efforts with the passive scorn of material infinitude. Genuine, if partial, successes have crowned them in the past, and will, it may be hoped, continue to crown them in the future.

We must not, however, in seeking encouragement from the thought that it does not utterly defy our powers, underrate the arduousness of the enterprise we have engaged upon. The nature of our own sun offers a vast and intricate problem, still very far from being solved; but stellar space contains many millions of suns, variously constituted, variously circumstanced, exceeding for the most part perhaps very greatly in size and splendour the body we depend upon for vital necessities. Now each of these millions of suns challenges the closest *personal* attention; no single one of them is exactly like any other, and their differences and resemblances open endless vistas of instruction and interest. Their incredible remoteness in no way derogates from their real dignity. An all but evanescent speck of light in the field of the great Lick refractor may be the life-giving centre of a system of worlds, each abounding as marvellously with proofs of creative wisdom and goodness as the far-away little planet in which our temporal destinies are rooted. Such light-specks are then equally deserving of study with the most effulgent orbs in the sky,

[1] *L'Astronomie*, t. v. p. 409.

although it may never be practicable to bestow it upon them. We can scarcely indeed imagine the amount of telescopic improvement which would be needed in order to bring them within the range of critical examination. For the present, at any rate, physical research must be confined to some thousands of the brighter stars which may serve as specimens of the rest. Nor need we lament the restriction. Generations of workers might expend their energies in gathering facts from the field actually open to them, and yet leave a full harvest for their successors. In all experimental inquiries, it may indeed be said that the reaper, as he garners one crop of knowledge, sows another : so endless are the secrets of nature, so untiring the inquisitiveness of man.

The stars in their combinations demand inquiry no less than the stars in themselves. Stellar systems are to be met with in indescribable profusion and variety, from mutually circling pairs, through groups including thousands of physically related objects, to the stupendous collection of groups which we call the Milky Way. But as yet investigation has barely skirted the edge of this well-nigh infinite region. Before it can be penetrated by so much as a plausible conjecture, statistics are wanted of the distances and movements of thousands, nay millions of stars.

Nor is the amassing of them any longer the desperate enterprise it seemed a short time since. By the unhoped-for development of novel methods, the pace of inquiry has been quickened all along the line. Particulars are accumulated faster than they can be assorted and arranged. Time seems to have multiplied itself for the purpose of gratifying curiosity which becomes keener as its sublime objects loom more distinctly above the horizon of thought. Ten years now count for a century of the old plodding advance. Express trains carry passengers on errands of research, as well as of business or pleasure. Problems ripen as if in a forcing-house, and so numerously as almost to bewilder the attention.

The whole subject of sidereal natural history is indeed wide and intricate beyond what it is easy to convey to those approaching it for the first time. There is scarcely a topic in physical astronomy with which it is unconnected. The progress of discovery has gradually drawn closer the generic relationships of the heavenly bodies. The sun has come to be recognised as the grand exemplar of the stars; meteorites show themselves to be intimately associated with comets; comets incur a 'vehement suspicion' of kinship with nebulæ; while the stellar and nebular realms blend one with the other as indistinguishably as the animal and vegetable kingdoms of organic nature.

The strange cloud-like objects called 'nebulæ' may be considered as wholly of telescopic revelation. Only one of them —the famous object in the girdle of Andromeda—can be at all easily seen with the naked eye; and even that escaped the notice of all the Greek, and most of the mediæval astronomers. The 'nebulosæ' of the ancients were many of them small groups of stars accidentally set close together; but among the seven enumerated by Ptolemy were two real clusters like the Pleiades, only (presumably) much farther away, one in Perseus, and the other in Cancer. Indeed, to extremely short-sighted persons, the Pleiades themselves put on a nebulous appearance, the individual stars running together into one wide blot of light.

Halley was the first to form anything like an adequate conception of the importance of nebular observations. He was acquainted in 1716 with six 'luminous spots or patches, which discover themselves only by the telescope, and appear to the naked eye like small fixed stars; but in reality are nothing else but the light coming from an extraordinary great space in the ether, through which a lucid medium is diffused, that shines with its own proper lustre.'[1] Only two of Halley's half-dozen objects, however—those in Orion and Andromeda— were genuine nebulæ; the rest when viewed with better instruments than his 'six-foot tube,' proved to be magnificent star clusters.

This small beginning of knowledge was followed up by Lacaille in the southern, by Messier in the northern hemisphere. But Herschel's great telescopes opened the modern

[1] *Phil. Trans.* vol. xxix. p. 390.

epoch in the science of nebulæ. They began as the result of his labours to be counted by the thousand instead of by the score. Portions of the sky were found to be crowded with them. Yet the vast majority must always, owing to their extreme faintness, remain imperceptible without powerful optical aid, only sixty-four coming into view with the same telescope which showed Argelander 324,000 stars. Thus access to the nebular heavens is granted conditionally upon making the very most, by the employment of large telescopes, of the little light they send us. No wonder, then, that comparatively little progress has been made with their investigation. Upwards of eight thousand nebulæ, it is true, have been discovered, but for the most part, only to be again neglected. Assiduous and prolonged observation is, nevertheless, indispensable for the detection of the cyclical or progressive changes doubtless proceeding in these inchoate systems. Nor have mere visual records proved sufficient for the satisfactory treatment of so delicate a subject. Other methods, at once more stringent and more expeditious, were needed, and have now happily been made available. With their help, it is already evident that knowledge will advance by rapid strides.

This then is the task of sidereal astronomy—to investigate the nature, origin, and relationships of sixty million stars and upwards of eight thousand nebulæ—to inquire into their movements among themselves, and that of our sun among them—to assign to each its place and rank in the universal order, and, gathering hints of what has been and what will be from what is, distinguish hierarchies of celestial systems, and thus at last rise to the higher synthesis embracing the grand mechanism of the entire—the sublime idea of Omnipotence, to which the stars conform their courses, while 'they shine forth with joy to Him that made them.'

CHAPTER II.

THE METHODS OF SIDEREAL RESEARCH.

SIDEREAL science is, on its geometrical side, of modern development; on its physical side, of modern origin. The places of the stars, as referred to certain lines and points on the surface of an imaginary hollow sphere, are obtained now on essentially the same principles as by Hipparchus, only with incomparably greater refinement. And refinement is everything where the stars are concerned. Significant changes among them can only be brought out by minute accuracy. To a rough discernment their relative situations are immutable; and systematic inquiries into their movements hence became possible only when the grosser errors were banished from observation. Bessel's *discovery* of Bradley's exactitude gave the signal for such inquiries. It became worth while to re-observe stars already so well determined that discrepancies might safely be interpreted to mean real change.

Thus it is only within the last sixty or seventy years that the stars have been extensively catalogued for their own sakes, and no longer in the undivided interests of planetary or cometary astronomy. The scope of such labours now widens continually. For the objects of them are all but innumerable, and the nineteenth century has brought to bear on its large schemes of scientific ambition heretofore undreamt-of facilities for executing them by combination. The project set on foot by the German Astronomical Society in 1867 of fixing the precise places of all stars to the ninth magnitude, has found co-operators in all parts of the world; when it is completed so far as to join on to the southern 'zones,' observed by Gould and Gilliss, not far from 400,000 stars will be not merely recorded, but known in the strict sense of modern astronomy.

Even this is not enough. There is a prospect that, early in the coming century, stars, approximately placed, will be counted by millions, while further unidentified millions still congregate unheeded in the unexplored recesses of the sky.

A star is located in the heavens, just as a city or a mountain is located on the earth, by measurements along two imaginary circles. Its 'declination,' or distance from the celestial equator, corresponds to terrestrial latitude; its 'right ascension' to terrestrial longitude. The astronomical prime meridian passes through the first point of Aries, that is, through the sun's position at the vernal equinox; intervals from it are reckoned eastward from 0° to 360°, or in time from 0^h to 24^h. And since the zero-point retreats slowly westward by the effect of 'precession,' it follows that the right ascensions of nearly all stars increase steadily year by year, apart from any movements 'proper' to themselves.

The diurnal revolution of the sphere furnishes the sole standard of time in sidereal astronomy. Sidereal noon at each spot on the earth is the moment when the first point of Aries crosses the meridian of that spot; the right ascensions of the heavenly bodies indicating the order of their successive culminations. Thus, if the right ascension of a star be two hours and twelve minutes, it will always cross the meridian of any given place two hours and twelve minutes after the first point of Aries has crossed it, *coming up behind it*, to that extent, in the grand diurnal procession. Differences in right ascension signify accordingly differences in times of culmination; and their measures in hours, minutes, and seconds, need only multiplication by fifteen (the number of times that 24 is contained in 360), to appear as measures of arc in degrees, minutes, and seconds.

A transit-circle and a clock are the two essential instruments for determining the places of the stars. The instant, to the tenth of a second, at which a star stands in the meridian, is noted; the vertical circle is read, showing its 'zenith-distance' (giving at once its declination when the latitude of the observatory is known), and the observational part of the work is done. The data thus obtained, after undergoing numerous corrections, suffice to determine the position of the star with reference to some other 'fundamental' star, the *absolute* place of which has been separately and laboriously ascertained.

This business of star-location forms the substratum of the older astronomy. But the precision given to it is altogether new, and alone has fitted it to be the means of eliciting facts so coy as those that relate to stellar movements. For the disclosure of them subtle devices of accuracy are needed which our astronomical progenitors never cast a thought upon. Optical and mechanical skill have, in our days, reached a point of almost ideal perfection; yet when the artist has done his utmost, the instrument is, in a sense, still in the rough. The astronomer then takes it in hand, and his part is often the more arduous and anxious. The investigation of small surviving errors, the contrivance of methods for neutralising their effects, the carrying out of delicate operations of adjustment, the detection of microscopic deformations, tremors of the soil, inequalities of expansion by heat, fall to his share. Even his own rate of sense-transmission has to be determined, and figures, under the title of 'personal equation,' as a correction in the final result. For, between the actual occurrence and the perception of a phenomenon, there is always a gap, more or less wide, according to individual idiosyncrasy, and it is only after this gap—tiny though it be—has been crossed, that electricity can be called upon to play its prompt part as amanuensis to the observer.

This detailed and painful struggle against error has alone made sidereal astronomy possible, by *precipitating* from the mixed solution that held them the minute quantities it deals with. Just because the universe is almost infinitely large, these quantities are almost infinitely small. They are small, not in themselves, but through the incomprehensible remoteness of the bodies they affect.

Geometrical sidereal astronomy deals with the motions of the stars. But these are of different kinds. 'Proper' motions—so-called to distinguish them from 'common' apparent displacements due to the slow shifting of the points of reference on the sphere – advance uniformly along a great circle; orbital revolutions of one star round another are periodical in small ellipses; besides which annual oscillations, varying in extent with the distances from ourselves of the objects performing them, are barely measurable in a few of the nearest stars. For determinations of these orbital and 'parallactic' movements such microscopic accuracy is indispensable as has only recently been attained. The instruments employed in them are the equatoreal with micrometer attached, and the heliometer.

An equatoreal is a telescope so mounted as to follow the diurnal revolution of the heavens. It is connected with an axis directed towards the pole, and revolving by clockwork once in twenty-four hours. So that an object once brought into the field of view remains there immovably for any desired time, provided that the telescope be clamped in position, and the clock set going. The inconvenience of the earth's rotation in producing a continual 'march-past' of the heavenly bodies is thus neutralised.

To the eye-end of an equatoreal is usually attached an arrangement of spider lines constituting a 'filar micrometer.' Two sets of such threads (which in subtlety and evenness of texture far surpass any artificial production), crossing at right angles and some of them movable by fine screws, while the whole can be made to revolve together, afford a most delicate means of ascertaining the distance and direction from each other of any two objects close enough for simultaneous observation. Measures of double stars are executed, and some stellar parallaxes have been determined in this way. But for the latter purpose, the 'heliometer' is the more appropriate instrument.

Its designation is a misnomer, or rather represents the tradition of an original purpose to which it was never effectively applied. The true function of a heliometer is the critical measurement of two adjacent stars, or of a star and planet. Primarily, it is an equatoreal telescope; its micrometrical powers are conferred by the division of the object-glass into two halves sliding along their common diameter, and duplicating by their separation the combined image formed by them when together. The amount of movement given to the segments in bringing about alternate coincidences between opposite members of the pair of stars shown by each, suffices to determine with the utmost nicety the interval between them. That is to say, after endless precautions for accuracy have been taken, and endless care bestowed upon detecting and obviating occasions of infinitesimal error.

The Radcliffe Observatory at Oxford possesses the largest heliometer in existence. The diameter of its object-glass is seven and a half inches. A similar instrument, however, erected at the Royal Observatory, Cape of Good Hope, nearly forty years later, is but slightly inferior in size, and is in other respects considerably its superior. Dr. Elkin, at Yale College, has charge of the only heliometer in the New World, while several are to be found in Germany and Russia. The Repsolds of Hamburg may be said to hold a monopoly in the mechanical part of their production; and Merz of Munich stands almost alone among opticians in his readiness to take the responsibility of sawing a fine object-glass in two. Nor is the aptitude for the use of these instruments by any means universal among observers; hence their comparative scarcity.

The science of the *motions* of the stars is only a part of modern sidereal astronomy. Within the last quarter of a century, a science of their *nature* has sprung up and assumed surprising proportions; a science the reality of which confounds forecast, yet compels belief. Sidereal physics has a great future in store for it. Its expansiveness in all directions is positively bewildering. The 'What next?' is hardly asked, when it is answered often in the least looked-for manner. In following its progress, the mind becomes so inured to novelties, that antecedent improbability ceases to suggest dissent. Some details of what we have thus so far learnt will be contained in the ensuing chapters; the means employed must be briefly indicated in this.

They are of three principal kinds—spectroscopic, photometric, and photographic. The general theory of spectrum analysis has been explained elsewhere;[1] here we need only repeat that it rests upon the constant position in the spectrum of the rays of light given out, when in a state of vaporous incandescence, by each separate chemical substance. These invariable lines serve as an index to the presence of the substance they are associated with, in the sun or in a star, no less than in the laboratory. Whether they be bright or dark, the principle remains the same. They are bright when the ignited vapour originating them is the chief source of illumination; dark, when a stronger light coming from behind is absorbed by its interposition. Their appearance as 'lines' is merely due to the transmission through a narrow slit of the light afterwards prismatically dispersed.

Now a main difficulty in getting starlight to disclose its secrets, is that there is so little of it. It will not bear the necessary amount of spreading-out, but evades analysis by fading into imperceptibility, like a runnel of water that widens only to disappear. Hence the absolute necessity in stellar spectroscopy for large telescopes. The collecting nets have to be widely extended to gather in a commodity so scarce. Could we at all realise, indeed, the portentous expanse of the ever-broadening sphere filled by the stellar beams as they travel towards us, we should be inclined to wonder, not at their faintness, but at their intensity. But the weakening effect of distance is in some degree counteracted by powerful concentration; and this is one of the chief uses of the large telescopic apertures so much in vogue at the present time.

Viewed with the Lick refractor of thirty-six inches, any given star is 32,400 times brighter than it appears to the naked eye, or 324 times brighter than when shown by a two-inch telescope.[2] The large instrument, that is to say, provides 324 times as much material for experimenting upon as the small, or places upon the same level of advantage for purposes of scrutiny objects 324 times as faint.

The interpretation of spectral hieroglyphics, by which we learn the chemical constitution of a star, is a very delicate and

[1] See the author's *Popular History of Astronomy*, 2nd edit. p. 175.

[2] Holden, *English Mechanic*, vol. xlvi. p. 528.

laborious operation. What is called a 'comparison-spectrum' is usually employed as an adjunct to it. Rays from some terrestrial source are reflected into one half of the slit, through the other half of which the stellar rays are admitted. Both sets then traverse the same prisms, and form strictly comparable spectra side by side in the same field of view. Lines common to both can thus easily be identified; and their genuine occurrence leaves no doubt that the element compared —hydrogen, sodium, iron, magnesium, or any other—enters into the composition of the star. But this process of matching can seldom or never be completely carried out. A dozen known lines may be attended by a hundred unknown ones, either too faint to be distinctly seen, or in positions unfamiliar to terrestrial light-chemistry. Nor is it safe to infer the absence of an ingredient from the absence of its representative rays. Many causes may contribute to render the display of lines in stellar spectra *selective*.

Where direct comparisons can be dispensed with, a slit is not essential to stellar light-analysis. For a star, having no sensible dimensions, gives rise to none of the confused overlaying of images produced by grosser light-sources unless superfluous rays be excluded by the use of a fine linear aperture. Hence the possibility of applying a 'slitless spectroscope' to the stars. Their light is then simply passed through a prism, either before it enters, or as it leaves the telescope. The resulting variegated stripe, looked at through a cylindrical lens to give it some tangible breadth, shows the dark gaps or lines significant of the 'type' of the star.

But prismatic analysis is not merely communicative as to the physical and chemical nature of the stars. It can tell something of their movements as well. And, what is especially fortunate, the information that it gives is of a kind otherwise inaccessible. 'End-on' motions, as every one knows, are visually imperceptible; the discovery that the spectroscope has the power to make them sensible is of such far-reaching importance that Dr. Huggins, by bringing the method into effective operation, performed perhaps the greatest of his many services to science. Through the link thus established, geometrical and physical astronomy have been placed in closer mutual relations than could have been thought possible beforehand.

The observations concerned are of great delicacy, and can only be made with a powerful telescope, collecting light sufficient to bear a considerable amount of dispersion. Their object is to measure the minute displacements of known lines due to 'radial' or end-on motion, and proportional in amount to its velocity. These displacements are towards the blue end of the spectrum when the star is approaching, towards the red when it is receding from the earth. The refrangibility of the luminous beams is changed, in the one case, by the crowding together of the ethereal vibrations, rendering them more numerous in a given time, in the other, by their being (as it were) drawn asunder, and so rendered less numerous. The juxtaposition of a standard terrestrial spectrum, such as that of hydrogen, gives the means of measuring deviations thus produced, and so of determining the rate of approach or recession of the star examined. But the process is impeded to a degree hardly imaginable without personal experience by troubles in the ocean of our air. The twinkling of the stars is represented in their spectra by tremors and undulations often permitting only instantaneous estimates of line-positions. But for this inconvenience an unexpected remedy, as we shall presently see, has been found.

Stellar photometry has a two-fold object. It gives the means of investigating, first, the individual nature; secondly, the collective relations of the stars. Stellar lustre is affected by endless gradations of change. It is rarely, perhaps never, really constant. Periodical fluctuations are in many cases obvious; secular variations are suspected. The suspicion can be verified only by precise light-measurements repeated at long intervals.

Their application to the problems of stellar distribution becomes feasible through the dependence of brightness upon distance. The law of the *decrease* of light with the *increase* of the square of the distance, is universally familiar. If all the stars were equal in themselves, their apparent differences would thus at once disclose their relative remoteness. We could locate them in space just as accurately as we could determine their lustre. But in point of fact the stars are enormously unequal, and we can hence reason from distance to brightness only by wide averages. A statistical method alone is available, and its employment involves the establishment of strict principles of light measurement.

The first requisite for this purpose was an unvarying and consistent scale, provided with the least possible disturbance to existing habits of thought, by regularising the antique mode of estimation by 'magnitudes.' Intervals loosely defined and unequal were made precise. A 'light-ratio' was agreed upon. To this proportion of change from one magnitude to the next, the numerical value 2·512[1] has been assigned. That is to say, an average first magnitude star sends us 2·512 times as much light as an average star of the second magnitude, which, in its turn, is 2·512 times brighter than one of the third, and so on. From the first to the third magnitude, the step is evidently measured by the *square of the 'light-ratio'* ($2{\cdot}512 \times 2{\cdot}512 = 6{\cdot}310$); and in general the relative brilliancy of any two stars may be found by raising 2·512 to a power represented by the numerical difference of their magnitudes. One first magnitude star, for instance, is equivalent to one hundred of the sixth rank ($(2{\cdot}512)^5 = 100$); and to no less than a million stars of the sixteenth magnitude.

All this is a matter of pure definition, and definition is a useful leading-string to experiment. It is something to have a clear conception in the abstract, of what a tenth, eleventh, twentieth magnitude star is, even though the conception be not altogether easy to realise. The problem of applying the numerical standard set up was practically solved almost at the same time by Professor Pritchard at Oxford, and by Professor Pickering at Cambridge in the United States. They first systematically and extensively employed instrumental means in stellar photometry, with the result of satisfactorily ascertaining the comparative lustre of all stars visible to the naked eye in these latitudes.

Professor Pritchard adopted for his researches the 'method

[1] Selected as the number of which 0·4 is the logarithm.

of extinctions.' The image of each star was made to vanish by sliding between it and the eye a wedge of neutral-tinted glass, the thickness of which needed to produce extinction was found to give a very exact measure of intensity. In this way the brightness of 2,784 stars from the pole to ten degrees south of the equator was determined and registered in the 'Uranometria Nova Oxoniensis.'

The Harvard 'meridian-photometer' (a modification of Zöllner's) was constructed on the principle of 'equalisation.' The images of the pole-star (adopted as a standard of comparison) and of each star successively experimented upon, were reflected into a fixed telescope, and brought to an exact equality by means of a polarising apparatus. From the amount of rotation given, for this purpose, to the double-refracting prism, the difference of original light-power was easily deduced. The method is of wider applicability than that by extinctions; none the less, the 'wedge photometer,' in the form given to it by Professor Pritchard, has taken its place as an indispensable adjunct to such inquiries. With either instrument the limit of clearly distinguishable difference is about one-tenth of a magnitude.

The Harvard photometry[1] includes all stars to the sixth magnitude as far as 30° of south latitude, to the number of 4,260. But the more southerly among them were evidently observed at a great disadvantage, owing to their low altitude, and the consequent heavy light-tax exacted from them by the atmosphere. A process of 'reduction' was then needed before the stars in different situations could be fairly compared. The recorded brightness of each was 'reduced,' according to a scale of correction laboriously constructed from experiment, to what it would be in the zenith. Even here the loss of light is about twenty per cent., but it increases to thirty-six at 30°, to sixty per cent. at 13½° above the horizon.[2]

In the Harvard photometry only two stars, Aldebaran and Altair, are rated as *strictly* of the first magnitude. Each

[1] *Harvard Annals*, vol. xiv. pt. i. (1884). For a comparison with the Oxford results see *ibid.* vol. xiii. p. 15.

[2] Pritchard, *Memoirs R. Astr. Society*, vol. xlvii. p. 372.

of them gives just three times as much light as the pole-star. They have in the northern hemisphere five superiors—Arcturus, Capella, Vega, Procyon, and Betelgeux—the standing of which has accordingly to be expressed by fractional numbers. Capella and Vega are of magnitude 0·2, signifying that each is eight-tenths of a magnitude brighter than Aldebaran or Altair, while the 0·0 attached to Arcturus means that it excels either of these stars by one whole magnitude. Carrying out the same system of notation, we get negative numbers for the designation of still higher grades of lustre. The fact, for instance, that we receive from Sirius more than nine times as much light as from Aldebaran, corresponding to a superiority of two magnitudes and four-tenths, is compactly expressed by calling its magnitude −1·4. To find a star outshining Sirius, we must go to our own sun, to which a rank can be assigned on the same scale. Its light, as measured by Alvan Clark in 1863, exceeds that of the dog-star 3,600 million times. Bond made the disproportion 5,970, Steinheil 3,840 millions to one. From a mean of the three, Professor Pickering fixed the sun's stellar magnitude at −25·4,[1] its superiority over Sirius amounting to 24, over Arcturus to 25·4, grades of ascent.

A third photometric review of the northern heavens was begun at Potsdam in 1886 by Drs. Müller and Kempf.[2] Since stars down to 7·5 magnitude are embraced by it, it will considerably overlap the two earlier works of the same kind, and will include a re-determination of all their stars, uncertainty as to the light-rank of which will thus be reduced to a minimum. Zöllner's polarising photometer is the instrument employed, perhaps for the last time in such a comprehensive research—not from any intrinsic incapacity for even more extended usefulness, but because better means are at hand.

The invention of the telescope itself does not mark an epoch more distinctly than the admission of the camera into

[1] *Proceedings American Academy*, vol. xvi. p. 2.

[2] *Vierteljahrsschrift Astr. Gesellschaft*, vols. xxii. p. 145, xxiii. p. 124, xxiv. p. 141.

the celestial armoury. All the conditions of sidereal research, in especial, are being rapidly transformed by its co-operation. The versatility of its powers is extraordinary; no task has yet found it unready or incapable. It is the very Ariel of the astronomical Prospero.

This untiring serviceableness was made possible by the substitution in 1871 of gelatine for collodion as the vehicle for the salts of silver, the decomposition of which under the influence of light forms the essential part of the photographic process. The new plates were, however, first used for 'astrographical' purposes by Dr. Huggins in 1876. Since they are five times more sensitive dry than wet, exposures with them can be indefinitely prolonged. They may, besides, be prepared any desirable time before, and developed any desirable time after exposure, thus accommodating themselves in a really wonderful way to the needs of astronomers.

The unique power of the photographic plate as an engine of discovery is derived from its unlimited faculty for amassing faint impressions of light. By *looking long enough* it can see anything there is to be seen. Captain Abney's experiments convince him that no rays are too feeble to overthrow the delicate molecular balance of silver bromide if only their separately evanescent effects get sufficiently piled-up through repetition.[1] By this power of accumulation the camera leaves the eye far behind. With any given telescope much more can be photographed than can be seen, and the opening of a region of research, beyond the range of visual accessibility, appears to be close at hand. Already the depicting of the invisible has, in some isolated cases, been realised; it will perhaps before long be pursued by methods and with results peculiar to itself.

The penetration of space has, indeed, even thus limits. It is unlikely, for instance, that external galaxies (if such there be) will ever reveal themselves on our plates. A *ne plus ultra* is imposed, if no otherwise, by the restricted possibilities of continuous exposure to the sky. Darkness does not last indefinitely; the time during which a given object (unless

[1] *Observatory*, vol. xii. p. 165.

situated very near the pole) is high enough above the horizon for portrait-taking purposes is still shorter. Mr. Roberts has, however, adopted in certain cases and with a measure of success, the expedient of successive exposures given to the same plates on different nights.

The telescope forming the image which imprints itself upon the prepared plate, is always equatoreally mounted, and has a motion given to it exactly concurrent with the revolution of the sphere. Yet the utmost mechanical ingenuity cannot make the concurrence absolutely perfect. Minute inequalities survive and need intelligent correction. Even more sensible are disturbances caused by the changes of atmospheric refraction with the ascent towards, or decline from the meridian of the objects in course of delineation. For these reasons a photographic telescope has, as a rule, a guiding telescope attached to its axis, through which an observer watches to counteract, almost to anticipate, nascent tendencies to displacement. The strain upon the attention is severe; its endurance, upon occasions, for three, even four hours at a stretch, is no small proof of resolution. The effect, however, of long exposures can be obtained in shorter times by giving increased sensitiveness to the plates, which have now been brought to such a pitch of 'rapidity' that their 'fogging' from the general illumination of the sky, or the glare of a great city, is a danger that has to be carefully guarded against.

Exposures can also be curtailed by shortening the focus of the photographic telescope, the image being thus rendered smaller, and—through the closer concentration of the same amount of light—more intense. For simply exploring the skies, sounding their depths, and dredging-up their contents, nothing can be better than the form of an ordinary portrait-lens. With such a one, only two inches in diameter, the picture of the comet of 1882 was taken at the Royal Observatory, Cape of Good Hope, the 'thick-inlaid' background of which was the first palpable revelation of the star-charting powers of the camera; and much of Professor Pickering's admirable work in sidereal photography has been done with a Voigtländer's 'doublet' (two achromatic lenses in combination) of eight inches aperture and about forty-five focus. With this instrument, objects imperceptible through the Harvard fifteen-inch refractor are photographed; a thousand stars within one degree of the pole have been catalogued where some forty were previously located;[1] and to eighteen known nebulæ twelve till then unknown were added within a region about $\frac{1}{250}$ of the area of the heavens. In the 'Bruce telescope,' about to be erected under Professor Pickering's direction on one of the mountains of Southern California, under almost ideal atmospheric conditions, this plan of construction will be carried out on a larger scale. The object-glass will be twenty-four inches in diameter, the focal length eleven feet. Stars fainter than have ever before been reached, visually or photographically, are expected to make their appearance on plates exposed to the powerful concentration of light thus effected. The portrait-form of lens has the additional merit of giving a large field of view. Each photograph taken with the Bruce telescope will cover five degrees square (25 square degrees) on a scale of one minute of arc to a millimetre. The whole heavens could be charted on about two thousand such plates.

Where accurate measurements are aimed at, however, the type of instrument represented by the MM. Henry's photographic telescope is preferable. The object-glass in this is of the ordinary achromatic kind, but corrected with reference to chemical, instead of to visual action. The rays selected to be brought to a focus are those to which, not the human, but the 'photographic retina' is sensitive. The aperture is thirteen inches, the focal length eleven feet; a plate-holder is substituted for an eye-piece, while a guiding-telescope of slightly inferior dimensions is enclosed within the same rectangular tube. The field of view with the Paris photographic telescope within which definition may be considered as virtually perfect, is a circle three degrees in diameter,[2] covering an area of not quite five square degrees. Fully ten thousand of these plates would be needed to picture the sphere.

[1] Mr. Roberts has charted 1,270 stars over the same space from a negative taken at Maghull, August 14, 1888. *Monthly Notices*, vol. xlix. p. 10

[2] *Bulletin Astronomique*, t. vi. p. 303.

A score of such instruments will presently be employed in all parts of the world, in carrying out the international star-charting operations set on foot by the Paris Congress of 1887. Their completion, in four years or less, will place at the disposal of astronomers two distinct series of plates, each in duplicate to obviate errors. The longer-exposed series, descending to the fourteenth magnitude, and including, it is anticipated, twenty-five million stars, may be invoked centuries hence for the decision of cases of suspected change. Were a register of the kind dating from Tycho's or Kepler's time now extant, it is scarcely too much to say that the whole fabric of stellar astronomy would be reared upon comparisons and consultations with it. Realising, then, the value of what it has power to execute, the present generation owes it to posterity to lay up for their use a store of facts that cannot fail to ripen into knowledge.

The second series of international plates will provide the materials for a great catalogue of about a million and a half of stars to the eleventh magnitude. In taking, no less than in measuring them, precision will be safeguarded with the most anxious care. Among the special means for securing it, M. O. Lohse's device of the 'reticle' deserves particular mention. Each plate will carry a latent image of a system of lines at known distances (about five millimetres apart) which, developed with the imprinted stars, will afford a secure index to possible slight distortions of the film. Such distortions are the *bête noire* of the celestial photographer; but once made manifest, they become innocuous.

The situations in the sky of the million and a half stars thus depicted can be ascertained by micrometrical measures of their positions with reference to certain *known* stars suitably distributed on the plates. The accuracy attainable in this way is sufficiently described by saying that it fully meets the requirements of modern research. These million and a half of new recruits will then be definitely entered on the muster roll of the celestial array. Each of them will have a 'local habitation and a name'—or at least a number. They will be individualised and pursued from century to century by curious inquiry. Their changes can no longer lurk unheeded, and the record of their existence, outlasting the destruction or decay of the negatives from which it has been derived, may be said to be for all time.

Should Professor Pickering's plan of work with the Bruce telescope be executed, a third set of plates, showing, it is expected, all stars to the seventeenth magnitude north of thirty degrees south declination, will be added to the astronomer's archives. Their comparison with the international long-exposure plates ought to prove of the highest interest. Stars of seventeenth are, by definition, nearly sixteen times fainter than stars of fourteenth magnitude, and they should be, were there no failure of the star-supplies, sixty-three times as numerous. So that, if the international plates contain twenty-five millions, the Bruce plates *ought* (if they really embrace seventeenth magnitude stars) to contain twelve hundred millions for three-quarters of the sky. There are, indeed, strong indications that no such portentous multitudes will make their appearance; but it will be an immense point gained to have made sure that they are not there.

'Studies of the distribution of the stars,' Professor Pickering remarks, 'can now scarcely be undertaken in any way except by photography.' But photography, to be really instructive on this point, must be combined with photometry. The portrayal of millions of stars projected side by side on a spherical surface tells us little or nothing of their relations to the immensity of space. This can only be found out, for the vast majority of them, by collecting statistics of the amount of light they send us. Hence the importance of the photometry of small stars. Yet no visual means have hitherto proved competent to deal with it. Eye-estimates, however guided and succoured by instruments, break down when pushed too far down the scale. The problem is evidently one of those reserved for successful treatment with the camera.

What is called 'photographic irradiation' affords one means of attack upon it. This arises from the diffusion of light within the substance of the gelatine film. The particles directly meeting the stellar rays reflect them irregularly all round to other particles, thus widening the area of chemical decomposition, and creating circular images which, with the same exposure and on the same plate, are found to vary in size with the magnitudes of the stars they represent.[1] Thus, from a few stars of ascertained brightness, that of the rest imprinted with them may, down to a pretty low grade, readily be inferred. The faintest stars, however, give rise to dots so small that their differences cannot be reliably measured.

The method of 'trails' is a useful adjunct. For this purpose, the stars, instead of being held, so to speak, in a vice, by the following motion of the telescope, are allowed to travel across the plate, recording their passage in tracks of 'reduced' or metallic silver, the width and density of which serve as an index to the lustre of the objects originating them. But the almost instantaneous graphical power needed to produce this effect belongs, as can easily be imagined, only to the brightest stars. Trails can practically be got from the fainter ones only in the immediate vicinity of the pole, where the diurnal motion is excessively slow.[2]

For the lowest ranks of stars, the relative lengths of exposure needed to get impressions of them furnish the most promising means of determining their comparative light. The principles upon which this can be done are however still unsettled. The times, which for moderately bright stars sensibly follow the simple proportion of luminosity, do not appear to do so for fainter objects. A star of the thirteenth magnitude, for example, requires more than twice and a half times as long an exposure (other conditions being identical) to make itself perceptible as one of the twelfth. Exactly how much more must be decided by experience. A triple exposure for each additional magnitude has been provisionally adopted at Harvard College; but no rule in the matter has yet been fully tested.

Photographic is not always the same thing as visual

[1] *Astr. Nach.* No. 2,884. See also, Charlier, *Publication der Astr. Gesellschaft*, xix., Leipzig, 1889; Schaeberle, *Publications Astr. Society of the Pacific*, No 4, p. 51; Holden, *ibid.* No. 5, p. 112.

[2] By changing the rate of the clock so as to let the telescope lag slightly behind the stars, the effect of a diurnal motion as slow as desired can be given; but the trails will be proportionately shortened with the same exposures.

brightness. Colour interposes a distinction; since the quick vibrations at the blue end of the spectrum are those chiefly effective in releasing silver from chemical bonds. Blue stars are consequently far more, and red stars far less conspicuous self-printed than to the eye. Photographs of tinted double stars thus often show curious reversals, a small blue companion coming out superior to its yellow or reddish primary.[1] And stars too faint for chromatic discernment with the telescope can sometimes be picked out on a negative as coloured, simply through discrepancy of relative magnitude. These cases are, however, exceptional. In general, the eye and the sensitive plate agree in estimating gradations of stellar light.

Nebular photometry is still in its infancy. Dr. Huggins ascertained, in 1866,[2] the extreme intrinsic faintness of such objects; and there the matter rested until the universal agency of the camera was made available. At a meeting of the Royal Astronomical Society on March 8, 1889, Captain Abney described a mode of obtaining a scale of photographic density by exposing to a standard light during different intervals of time, a number of small squares on one corner of a prepared plate. These, developed with the subsequently taken picture of a nebula, afford so many terms of comparison for the relative brightness of its parts, 'the effect of increase or decrease of time in determining density of deposit being exactly equivalent to increase or decrease of intensity of light.'[3] A curve of varying lustre can in this way be drawn through the parts of a nebula-photograph, from a reference to which, at any lapse of years, it may be positively determined whether local brightening or fading has occurred. Only internal luminous relations, however, are capable of being thus tested. The contrivance gives the means of comparing different sections of the same nebula, not one nebula with another, nor even the same nebula, as a whole, with itself at different epochs. Its adoption may nevertheless be expected to make a beginning in exactknowledge of the light-changes of nebulæ.

[1] Instances are given by Mr. Espin, *Observatory*, vol. vii. p. 247.

[2] *Phil. Trans.* vol. clvi. p. 392.

[3] *Observatory*, vol. xii. p. 160.

There is scarcely one of the numerous tasks of nebular astronomy that cannot be better performed photographically than visually. In the simple perception of faintly illuminated surfaces, the promptly fatigued living retina is left far behind by the imperturbable gaze of a sensitised plate; in their delineation, the subtlest human hand is at an equal, if not greater disadvantage. Spectroscopic inquiries, both stellar and nebular, are enormously facilitated by the substitution of permanent autographic records for sets of quivering lines, measurable only at critical moments, and constantly liable to be effaced by atmospheric intervention. It is true that the range of observation is not the same in both cases. The plates now in ordinary use ignore the lower end of the spectrum, while reaching up much higher than the eye can follow into the ultra-violet. The use of coal-tar products, however, recently adopted in their manufacture, promotes sensitiveness to yellow and red rays; and 'orthochromatic' plates absolutely free from colour-preferences can be produced by special processes. So that even the part of the spectrum which seemed reserved for visual study will doubtless before long be appropriated by photography.

The wonderful comprehensiveness and adaptability of this method are strikingly apparent in the results obtained of late years at Harvard College. By no other means could the spectroscopic stellar survey partially completed there, have been carried out on so great a scale. Since 1885, 28,000 spectra of 11,000 stars have been catalogued; and similar data for southern stars are now being collected at Peru with the same eight-inch 'doublet' previously in use at Harvard. The mode of work (though suggested by Father Secchi) is a practical novelty. A prism large enough to cover the entire object-glass is placed in front of it. The stellar beams are thus analysed before they are concentrated; every stellar image is transformed into a prismatic riband; and stars by the dozen or by the score print their spectra together on a single plate with a single exposure. Slit and cylindrical lens are alike rejected; the diurnal motion is employed to widen the spectral bands sufficiently to bring out their distinctive features. That is to say, the stars are allowed to 'trail' slightly across the direction in which their light is dispersed. The results are admirable; innumerable lines are clearly recorded; but the difficulty of measuring them in the absence of any system of reference-lines has not yet been fully overcome.

For detailed identifications, the more laborious plan adopted by Dr. Huggins in 1879 is still pursued. The stars are taken one by one; their rays are admitted through the postern-gate of a slit, and record their peculiarities side by side with a comparison-spectrum providing starting-points for measurement. Dr. J. Scheiner is thus, at Potsdam, engaged in collecting data of hitherto unapproached accuracy from the same plates used by Dr. H. C. Vogel for determining stellar motions in line of sight. In this last difficult branch the superiority of the photographic method is perhaps more conspicuous than in any other. The extraordinary precision attainable with it is chiefly due to the virtual elimination of the effects of air-troubles. The lines from which information as to movement has to be gathered depict themselves in their extra-atmospheric places. Their waverings, so baffling to the eye, are left comparatively unnoticed by the sensitive plate.

Nor is it only here that the chemical mode of procedure attains a refinement on a par with its power. The subtlest problem of stellar astronomy is that of annual parallax. Nowhere is there less room for compromise in the matter of accuracy; yet it has been triumphantly solved with the aid of the camera. The issue of Professor Pritchard's careful and persevering experiments has been to render photographic determinations of parallax of equal authority with visual determinations, and so to quicken incalculably the rate of progress in collecting results. It happens moreover that the objects most inviting to the one mode of treatment are precisely those reached with difficulty by the other. Stars too faint for the eye to deal with satisfactorily, come out on the plate in neatly measurable form, much more promising for exactitude than the distended images of brighter stars. Photographic parallax work has not, up to this, been attempted elsewhere than at Oxford.

It has been executed by means of a large reflector (thirteen inches in aperture), the gift of the late Dr. Warren de la Rue, the first example being thus given of the employment in tasks of such delicacy of that class of telescope. For picturing purposes, indeed, mirrors have the special advantage over lenses of being perfectly achromatic; they collect in one focus *all* the rays, visible and invisible, striking them. But even in the very best refractors, this is not the case. However skilful the 'achromatic' combination of different kinds of glass, a large amount of light is necessarily 'thrown away.'[1] Opticians have to choose what sections of the spectrum they will turn to account, and neglect the rest. Photographic refractors are for this reason useless in ordinary observation. The images they give are wholly built up out of blue light, while the light proper for seeing by wanders unserviceably astray. Hence the plates exposed with them must be of an approximately uniform character. No tolerable results could be got with 'orthochromatic' plates in the Henry telescope. These evils, should the new Jena glass prove a workable reality, will be greatly diminished, but they will not be abolished.

They are, however, to a great extent outweighed by countervailing advantages. Refractors are more manageable than reflectors. They are less sensitive to slight strains, less intolerant of unequal pressures; they accommodate themselves better to mechanical exigencies, can be more rigidly mounted, hence made to follow more strictly the circling of the sphere, and so to keep a steadier hold of the objects in the field of view. Where *quantitative* precision is chiefly aimed at, choice is thus naturally directed to them, and they have been stamped as the official instruments of celestial photography by their adoption for the vast star-charting operations decided upon at the Paris Congress. The splendid nebular delineations, on the other hand, achieved with reflectors, by Messrs. Common and Roberts and by M. von Gothard, strongly recommend their application to that and other special purposes. Each telescopic genus thus has its place in the boundless fields of photographic research; each has its own line of superiority; neither excludes the other.

[1] Sir H. Grubb, *Monthly Notices*, vol. xlvii. p. 309.

The foundation of stellar astronomy is, as we have said, in infinitesimal accuracy. It could not otherwise exist, since the quantities concerned are so small as to get buried out of sight amid the errors of rough observations. But for its progress, something more is required. A few scattered items of knowledge do not constitute a science. The word implies the suffusion of a subject with intellectual light derived from large inferences. Large inferences must, however, be based on a wide store of facts; and as yet the facts collected by sidereal study are few indeed compared with its innumerable objects. Reasoning is therefore cramped in cautious minds, or left to run wild in impatient ones, for lack of data. They are needed to wing thought in the one case, to restrain it in the other. And by photography alone it would seem that they can be supplied in the abundance and with the promptitude required. For, to a certain extent, the work has to be done against time. Just as rapid intuitions are necessary for following a long train of mathematical reasoning, because, where the steps are laboured, the wearied faculties at last refuse to continue to take them, so some degree of forward impetus is indispensable for sustaining the universal interest which gives a subject its vitality, but declines to follow too tardy an exchange of one halting-place for another.

Thus, not merely what it can do, but the rate at which it can do it, has to be considered in estimating the value of photography as an ally to astronomy. And there is little fear of its admitting lassitude through sluggishness of pace. It keeps up a very 'Sturm und Drang' of progress. The decuple powers of enumeration desired by Homer for cataloguing the crowd of Greek ships are far outdone by it. Its instrumentality, moreover, came to hand just when the multitudinous character of the problem set by the heavens began to be grasped in all its formidable reality.

The swiftness of the photographic method is due, not alone to the great number of objects it can register together, but to the dispersion and division of labour it makes possible. Records obtained by it have the enormous advantage of being permanent. They fix the flitting incidents of the heavens as the phonograph fixes the transient accents of the human voice. All the scanty hours of unclouded darkness can accordingly be devoted to securing materials for subsequent investigation in daylight or bad weather. Innumerable experts may be employed in this way at remote places and with different ends in view; and a single negative might conceivably serve as the basis for successive inquiries in stellar photometry, chromatics, distribution, parallax, and proper motions. The danger seems rather to be that star-prints may get too little than too much study; nor will zeal in securing them avail aught without industry in discussing them. The possession of pictures of celestial objects does not in itself constitute an increase of knowledge. They contain *latent* information just as the skies themselves do, but the educing process by which it is made *sensible* is as necessary in the one case as in the other. Hence the visible rise into importance of a new branch of practical astronomy which might be designated 'astrometry,' and the obvious likelihood that significant detections will in future more and more be made with the microscope in the study, by investigators who have perhaps never in their lives worked with a telescope.

Not that the telescope is, or ever can be, superseded. On the contrary, the enlargement of its capacities becomes more desirable with every fresh addition to the apparatus used in conjunction with it. Modern sidereal astronomy may be said to *live on light*. Large telescopic apertures are a *sine quâ non* for its growth and activity. Indeed, the objects it has to do with are, in great measure, otherwise invisible.

CHAPTER III.

SIRIAN AND SOLAR STARS.

The stars, speaking broadly, are suns. But what is a sun? We can only reply by taking *function* into consideration. A sun is a great radiating machine, and the obvious criterion for admission to the order is fitness for this office. Qualification to be a centre of light and heat is the dominant characteristic of each of its true members. Now the solar emissive activity is concentrated in a shining shell of clouds known as the 'photosphere,' which the entire energies of the *organism* (so to speak) seem directed to maintain and renew. And with reason, since its efficiency as a radiator depends upon the perpetuation of the condensing process by which this brilliant surface is produced.

The possession of a photosphere must then be regarded as an essential feature of the suns of space. But such a structure can only be formed in an incandescent atmosphere, the action of which modifies, more or less powerfully, the light emitted from it. The spectroscope can then alone decide whether a given sidereal object be, in the proper sense, a sun. For it is not so much the quantity as the quality of its radiations that determines the point. They must be such as can be supposed to emanate from condensed and vividly glowing matter bathed in cooler, though still ignited, vapours. That is to say, they must be *primarily* unbroken from end to end of the rainbow-tinted riband formed by prismatic dispersion, while showing the *secondary* effects of absorptive encroachments. A continuous range of vivid light derived from a photosphere, crossed by dusky lines due to the atmosphere through which the photospheric radiations have to make their way out into space, is hence the distinctive spectrum of a sun.

The enormous light-power, and, so far, the solar nature of the stars, followed as a corollary from the Copernican theory, since at the unimaginable distances implied by their apparent immobility while the earth performed its vast circuit, they should otherwise have been totally invisible. But the analogy could be strictly tested only by spectrum analysis, and it proved practically complete. Complete, that is, for the great majority of the stellar populace. There is a residuum in which it is impaired; there are a few instances in which it is actually overthrown.

This degradation of type shows itself in different ways. Absorption in some cases becomes so immoderate as well-nigh to *smother* the original light of the star; the atmosphere in others outshines the photosphere, giving rise to bright instead of dark lines in the spectrum, while a similar effect may at times be produced rather by a decline of photospheric, than by a heightening of atmospheric radiative intensity. When the failure has gone so far that the light of a seeming star, analysed with the spectroscope, is found to consist chiefly of isolated rays of various colours, then the object approximates more to a nebula than to a star. It certainly has no claim to the designation of a sun.

But as in the other kingdoms of nature, so here; there are no abrupt transitions. Continuity is everywhere maintained. The descent from a perfect sun to an undoubted nebula is effected without interruption. We propose, however, in the present chapter, to consider only bodies of assured status, with radiative machinery in full working order—bodies, as to the essentially sun-like nature of which there can be no difference of opinion. Their continuous spectra are crossed, more or less numerously, by dark lines serving as the 'recognition-marks' of the various chemical substances ignited in their atmospheres. They are inscribed from end to end with hieroglyphics which it is the chief business of the physical student of the stars to decipher.

About eleven-twelfths of all the stars show linear spectra

of absorption. They fall into two great divisions, corresponding to Father Secchi's first and second spectral types, which include respectively 'Sirian' and 'solar' stars.

The Sirian stars are of a brilliantly white colour, sometimes inclining towards a steely blue. Sirius is the exemplar of the class, but at least every alternate star in the sky is of analogous constitution. Among the more conspicuous examples may be mentioned: Vega (α Lyræ), Algol (β Persei), Canopus, α Crucis, β Argûs, Spica (α Virginis), Regulus (α Leonis), Castor (α Geminorum), β, γ, δ, ε, η Ursæ Majoris.

The light of these objects is not materially encroached upon by absorption. There is little or no trace in them of the general *damping-down* through vaporous intervention, by which our own sun is shorn of a large proportion of his more refrangible beams. The Sirian spectra, although not intact, are entire, and are hence especially strong in their ultra-violet sections. To the feebleness in them of absorptive effects there is, indeed, one remarkable exception. The sign-manual of hydrogen is stamped upon them with extraordinary intensity. By their broadened and hazy aspect the lines characterising this substance testify to its existence in these stellar atmospheres at an exalted temperature, and under pressure relatively high, yet certainly inferior to that of the terrestrial atmosphere at sea-level. Besides hydrogen, a number of metals are present, but, it would seem, in insignificant proportions, since they show lines too faint and fine for easy recognition. Fraunhofer's 'D,' the ubiquitous double-line of sodium, is nevertheless unmistakable, as well as the magnesium group *b*, and a line of iron in the green part of the spectrum known as 'E.'

The most striking feature in spectra of this type has, however, been disclosed by photographic means, for the application of which the strength of their ultra-violet radiations peculiarly fits them. These are powerfully absorbed by glass, and Dr. Huggins accordingly rejected this material *in toto* from the apparatus employed by him in the autumn of 1879.[1] All the lenses included in it were made of quartz; a metallic mirror,

[1] *Phil. Trans.* vol. clxxi. p. 669.

eighteen inches in diameter collected the light afterwards sifted by transmission through an Iceland-spar prism. A discovery of singular interest resulted.

A photograph of the spectrum of Vega obtained with one hour's exposure contained twelve strong lines (shown in the accompanying illustration) forming a group in obvious rhythmical connection, and crowding together in an accelerated proportion as their wave-lengths shorten. A common origin for the entire series at once suggests itself: and, on the ground that its two most refrangible members were already known as hydrogen-lines, Dr. Huggins did not hesitate to pronounce hydrogen responsible for all. His inference has since been amply justified. Seven of them proved to have been, a short

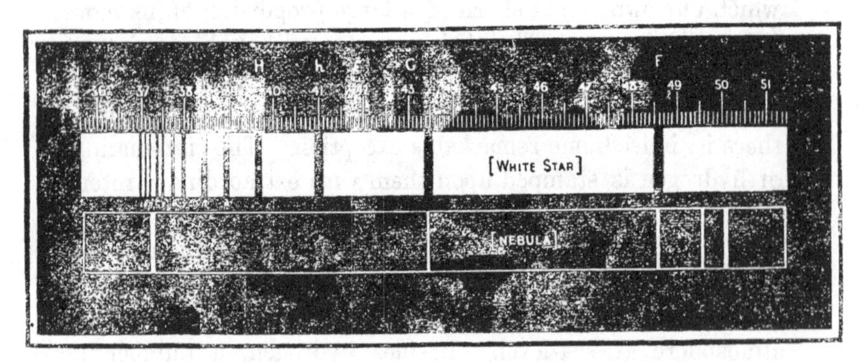

FIG. 1.—Spectrum of Hydrogen in White Stars compared with the Spectrum of the Orion Nebula (Huggins).

time previously, photographed directly from glowing hydrogen, by Professor H. W. Vogel;[1] and the entire series was similarly procured later by M. Cornu[2]. But with extreme difficulty. Although the purified gas filling the capillary tube placed in front of the slit was excited by powerful electrical discharges, its highest radiations took no less than three hours and a half to get satisfactorily printed. Some idea may thus be gained of the intense incandescence reigning in the remote stellar atmospheres from which they were first derived.

The complete spectrum of hydrogen comprises fourteen lines, the designations of which, with the actual lengths of the

[1] *Astr. Nach.*, No. 2301. [2] *Journal de Physique*, t. v. p. 341 (1886).

ethereal waves producing them, measured in ten-millionths of a millimetre, are subjoined as they appear in fig. 1, from which, however, the least refrangible, named C, is omitted as lying too far to the right.

In following the course of sidereal research, it is important to bear in mind at least the order of their succession in the spectrum. The two first, belonging to its visible part, are, it should be remarked, evidently correlated by their wave-lengths with the others, so as to form part of the same progression.

	Hydrogen-lines		Wave-lengths
1	Fraunhofer's	C	6562
2	,,	F	4861
3	near ,,	G(G)	4340
4	,,	h	4101
5		H	3968
6	Huggins's	α	3887·5
7	,,	β	3834
8	,,	γ	3795
9	,,	δ	3767·5
10	,,	ϵ	3745·5
11	,,	ζ	3730
12	,,	η	3717·5
13	,,	θ	3707·5
14	,,	ι	3699

The fact is a memorable one that the true character and full extent of the hydrogen-spectrum became known through 'astro-physical' inquiries. It shows with what curious unexpectedness the obligations of one science to another may be repaid, and exemplifies the advantages to be reaped by terrestrial chemistry from extending its experimental range to the heavenly bodies. The significance, for its purposes, of the 'white star' set of rays has been heightened by Cornu's detection of a corresponding series in the ultra-violet spectra of thallium and aluminium.[1] Not only is the title of hydrogen to rank as a genuine metal thereby ratified, but the existence is hinted at of a general law of spectral emission which will

[1] *Journal de Physique*, t. v. ser. ii. p. 93.

perhaps eventually prove communicative regarding the coveted secrets of molecular structure.

The complete presence of the hydrogen series may be called the badge of the first type of stellar spectra. 'No celestial body,' Mr. Lockyer remarks, 'without all the ultra-violet lines of hydrogen discovered by Dr. Huggins, can claim to belong to it.'[1] Accidental circumstances, it is true, may encroach upon the integrity of the set in any given photograph. Either through atmospheric hindrances, or through inadequacy in the supply of light, partial failures must be expected in attempts to get impressions of such susceptible vibrations. Thus, in the spectrum of Sirius, which comes next to that of *a* Lyræ in fig. 2, the absence of the five uppermost lines is entirely due to the slight elevation above the horizon of the star when photographed. Recalling, indeed, that this series has only by straining the resources of the laboratory been induced to print itself from a light-source close at hand, we can but marvel that its absorptive effects are to any extent perceptible in stellar beams attenuated by distance, and compelled, at the end of their journey, to submit to atmospheric incursions especially directed against their ultra-violet constituents.

All the six 'white-star' spectra then, represented in fig. 2, may be considered to bear, in reality, the full stamp of the emanations of hydrogen. Casual deficiencies are invariably of the most refrangible lines; they occur in the region where want of light first begins to tell, and just when the spectrum becomes characterless through faintness.

There is hence reason to believe the presence of the complete absorption-spectrum of hydrogen to be a searching test of stellar constitution. The stars in which it is apparent are so closely allied as most properly to form a class by themselves. It includes, however, a variety, of which the chief stars in Orion (Betelgeux excepted) are the representatives, distinguished by Vogel for the almost total blankness of their visible spectra. Yet the photographic investigation of their light has not proved barren. All the typical lines of hydrogen

[1] *Proceedings Royal Society*, vol. xlv. p. 382.

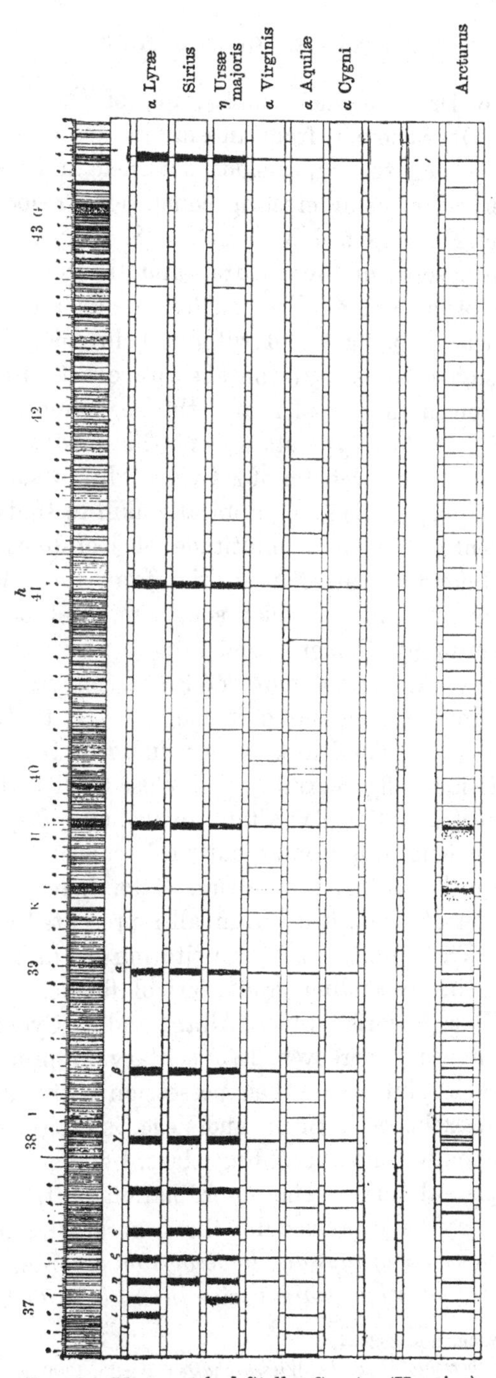

FIG. 2.—Photographed Stellar Spectra (Huggins)

appeared in Dr. Huggins's photograph of the spectrum of Rigel; and Dr. Scheiner, from an examination of the section F to H, to which the impressions obtained at Potsdam are limited, derived some interesting conclusions respecting the stellar species exemplified by Rigel.[1] (See fig. 3.) All the lines in the spectra of these stars—and there are many besides those of hydrogen—are equally broad; and they have the peculiarity of being broad without being hazy. The explanation suggested is that the absorptive action producing them takes place in a cooler and thinner atmosphere than that of ordinary first-type stars; but this is doubtful. Dr. Scheiner found the spectrum of *α* Cygni, while visually of the pure Sirian type, to be photographically akin to that of Rigel; and he identified in it a multitude of iron-lines showing curiously different relative intensities from those belonging to the same lines in the solar spectrum, and originating, it was accordingly thought, under abnormal conditions of temperature.[2] In the spectrum of Sirius, forty-three out of ninety-one fine lines measured in the space F to H are certainly due to iron. The drawings reproduced by Dr. Scheiner's kind permission in fig. 3 exhibit, with close fidelity, the photographed spectra of Capella, Altair, Sirius, and Rigel, over a third of their length, but on an enlarged scale.

Although the hydrogen spectrum is dominant throughout the first order of stars, it is not in all represented with equal emphasis. The diffuse lines constituting it in Sirius and Vega show, in descending gradations of fineness, in Benetnâsch (*η* Ursæ Majoris), Spica, Altair, and *α* Cygni (see fig. 2). They appear, moreover, less solitary in proportion as they become less intense. Their possession of the ultra-violet field, all but exclusive in Sirius and Vega, is progressively encroached upon, in the succeeding stars, by the development of other spectral lines. The co-ordination of the two kinds of change may be expressed by the general statement that *the conspicuousness of rays due to absorption by ordinary metals in the spectra of white stars varies inversely with that of the*

[1] *Astr. Nach.* Nos. 2923–4.

[2] *Sitzungsberichte Akad. der Wissenschaften*, Berlin, 1890, vol. viii. p. 7.

Fig. 3.—Photographed Stellar Spectra (Scheiner).

hydrogen series. When the hydrogen-rays become effaced from the invisible, and cease to be predominant in the visible part of the spectrum, the second, or solar type of stars is reached.

These are about one-sixth less numerous than the first kind. We may take as examples: Capella, *α* Ursæ Majoris, *α* Cassiopeiæ, *α* Arietis, *ε* Argûs, *α* Serpentis, Aldebaran, and Arcturus. The pole-star, Procyon, *α* Aquilæ, and probably the brilliant southern binary *α* Centauri, stand nearly midway between the two groups.

A golden tinge like that of sunlight betokens, in stars of the second order, a spectrum more or less perfectly similar to that of the sun, delicately ruled from end to end through the absorptive effects of a great variety of metallic vapours; non-metallic substances give no sign of being present, unless by a trace of carbon in the sun. The extent to which the lines are crowded together may be judged of from Cornu's photographic solar spectrum depicted in the upper horizon of fig. 2. Dark hydrogen-rays are present, but with no pre-eminence, and the series is apparently not continued above *h* in the violet. We say 'apparently,' because the fifth hydrogen line *may* lurk concealed within the shadow of an obscure diffuse band which covers its place. This band was named by Fraunhofer H, and its companion, a little higher up, is designated K. The pair form the most strongly-marked feature of the spectrum of calcium when raised to the highest pitch of incandescence; and there is much to be said in favour of Mr. Lockyer's view that they emanate, not from calcium in its entirety, but from some of its subtler ingredients. There is at any rate no doubt that they are what is called 'high temperature lines'; the light of ordinarily glowing calcium does not contain them.

In their 'unreversed,' or bright condition, they are peculiarly characteristic of the spectrum of solar prominences. Daylight observations at the edge of the sun show them always brilliant in the chromosphere, and often extending to the very summits of its flame-like extensions.[1] During

[1] Young, *Nature*, vol. xxiii. p. 281.

the total eclipse of May 17, 1882, the violet radiance of H and K flooded the shadowed part of our atmosphere, and, dimly illuminating the purple disc of the moon, was scattered far out among the 'aigrettes' of the corona.

Now, as we have said, the calcium H falls very near indeed to the hydrogen H. The difference of their wave-lengths amounts to no more than one ten-millionth of a millimetre, and it is only under favourable circumstances that they can be seen as distinct. As a rule, either the hydrogen-line is so widened as to mask the calcium line, or the calcium-line as to enwrap the hydrogen-line. Professor Young, however, saw them side by side, *bright*, during his observations of prominences in 1879–80; and they appear *dark* in a photograph of the spectrum of α Cygni taken at Harvard College, November 26, 1886. In this star, the hydrogen rays have thinned down almost to their solar condition, while other metallic lines are fine, yet pronounced. Such a critical balance of conditions alone made possible the separate discernment of the two lines on the Harvard plate.

Thus, while the double origin of H opens the way to misunderstandings of its true significance, the state of the calcium-line at K is a most useful index to the physical condition of a star. Next to the mode of appearance of the hydrogen series, it is perhaps the most significant individual feature in analysed star-light. The substance emitting both H and K is evidently of first-rate importance among the vapours surrounding the sun. But, so far as terrestrial experiments can inform us, it arises as a modification of the metal calcium only under the strongest electrical excitement. This no doubt generates an enormous degree of heat, but is perhaps not identical with it in its method of action upon matter. Professors Liveing and Dewar advert to the probability 'that the energy of the electric discharge, as well as that due to chemical change, may directly impart to the matter affected vibrations which are more intense than the temperature alone would produce.'[1] There is a practical certainty that, under some circumstances, bodies are rendered electrically luminous

[1] *Proceedings Royal Society*, vol. xliv. p. 242.

at comparatively low temperatures; it would be unphilosophical to assume that, under others, of which we know little or nothing, the distinction ceases to be valid. The fact, however we view it, is undoubted, that K is almost effaced from the spectra of what we have reason to believe are the hottest stars, while developing remarkably in cooler bodies. It may be that Dr. Huggins's proposed arrangement of photographed stellar spectra in a continuous series, according to the character in them, or even total absence from them of this line,[1] has a deeper meaning than is at once apparent.

Considerable diversity of detail is to be found among spectra of the second type. The model solar star is Capella; in the invisible, as well as the visible part of its prismatic light, all the characteristic solar groups exist in about their solar strength. Dr. Scheiner identified with extreme precision 255 lines photographed from Capella (between wave-lengths 4124 and 4638), with lines derived from the sun[2] (see fig. 3); and there is every indication of an almost perfect constitutional similarity between the two bodies.

In Aldebaran this standard is pretty widely departed from. The pale rose tint of its light is accounted for by the slightness of absorptive effects in the red end of its spectrum, while numerous lines modify the yellow and green, and the violet rays are so feeble that, with an exposure *fifty times* that required for Sirius, Dr. Huggins obtained only an impression virtually bounded by F and H. This feebleness of chemical action may, however, be due, not to original deficiency of light, but to its stoppage in the vaporous envelope of the star.

Of its spectral lines, seventy were determined by Huggins and Miller in 1863,[3] seventy-two by Vogel ten years later,[4] fifty-four being clearly traced by both observers to the radiations of the following nine substances: hydrogen, sodium, magnesium, calcium, iron, bismuth, tellurium, antimony, and mercury. Beyond a doubtful indication of bismuth, there

[1] *Phil. Trans.* vol. clxxi. p. 679. [2] *Astr. Nach.* No. 2923.
[3] *Phil. Trans.* vol. cliv. p. 424.
[4] *Bothkamp Beobachtungen*, Heft ii. p. 11.

is no sign in the solar spectrum of the arrest of light by any of the last four elements,[1] the high vapour-densities of which it is important to note.

If we were to consider only the visible spectrum of Arcturus, we should certainly suppose it to vary from the sun in the same *direction* as Aldebaran—that is, to be farther removed than Capella from the Sirian stars. Its rays, like those of Aldebaran, have a ruddy tinge, and are seen, when dispersed by the prism, to be powerfully stamped by metallic absorption. But the camera has a different story to tell. In the intensity of its ultra-violet as compared with the rest of its beams, Arcturus falls but little behind Sirius. Their inscription with legible characters extends, indeed, very much higher up (see fig. 2).[2] The characters, however, are unlike what might have been expected. Up to the limits of visibility, the solar analogy is fairly well preserved, H and K (the latter especially) being even more distended than in the sun. But beyond, non-solar groups of strong lines appear; and among them, six out of the nine typical rays of hydrogen.

The star with which Arcturus shows the closest analogy in the ultra-visible part of its spectrum, is α Aquilæ. And it is remarkable that the resemblance concentrates itself in features common to solar appendages and to the two stars. The spectrum of the 'Sohag prominence,' photographed by Dr. Schuster in 1882, contained, besides H and K and the complete hydrogen-set, a number of bright lines of unknown or doubtful origin. In the course of his examination of the plate, Captain Abney was struck with approximate coincidences between ten of them and dark lines photographically recorded by Dr. Huggins from α Aquilæ.[3] Seven of these common lines reappear in Arcturus, besides five others, probably also identifiable with prominence-rays. The preponderance of the violet calcium pair (H and K) is a further point of agreement.

[1] Lockyer, *Chemistry of the Sun*, p. 220.

[2] Dr. Huggins obtained, however, in the spring of 1890, a photograph of the spectrum of Sirius showing a group of six broad lines more refrangible than any of the hydrogen series. Their approximate wave-lengths are 3338, 3311, 3278, 3254, 3226, 3199 ten-millionths of a millimetre.

[3] *Phil. Trans.* vol. clxxv. p. 267.

The peculiarity of the analysed light of Arcturus consists, then, largely in its inclusion of the prominence-spectrum strongly 'reversed,' or turned from bright to dark. This seems to tell two things: first, that this star has very extensive chromospheric surroundings—that it possesses an enhanced equivalent of the ocean of incandescent gases making a rose-red edge to the black moon during total solar eclipses. Next, that it is a hotter star than the sun. We must, however, define what we mean by saying that one star is 'hotter' than another. The expression, otherwise open to misunderstanding, conveys, as used here, simply that the temperature of the emitting surface is higher in one case than in the other. The photosphere is at an intenser glow; the heat is more concentrated at a particular level.

But, it may be asked, how can we judge of the temperatures of the stars? By what criterion can we compare them? Not either by colour, or by the relative extent of their spectra, since both depend upon the kind and amount of absorption produced by their atmospheres, more than upon the original quality of their light. To describe, for instance, the white and red stars as illustrating respectively the white and red stages of heat, would mislead by the use of a totally irrelevant simile. Criteria of other kinds enable us, nevertheless, to discriminate, as to temperature, between star and star. The surest perhaps are those relating to 'reversals.'

The display of dark lines upon a bright background implies that the continuous light emanates from a source hotter than the interposed absorbing vapours. Thus, the photospheres of Sirian stars must (so far as we can see) be at higher temperatures than the hydrogen-strata emitting the series of vibrations thrown, as it were, in shadow upon them. For, if the strata were hotter, the lines would show bright against the photospheric radiance; if they were of equal temperatures, the lines would not show at all, radiation just balancing absorption. Now the ultra-violet members of the series, although they can be derived directly from the solar surroundings during total eclipses, make no mark in the solar spectrum. And why? Either because the solar photosphere is not hot enough to reverse them, or because the solar chromosphere, *in the main*, is not hot enough to emit them. Both conditions are, however, combined in the Sirian stars as well as in Arcturus, the glow of all of which is hence presumably more fervid than that of the sun, Capella, or Aldebaran.

The spectra of several of the Orion stars are believed to be variable.[1] Visual examination, at least, sometimes wholly fails to show the hydrogen-lines in Rigel, while at other times they are distinct, even conspicuous; nor are indications altogether wanting of relative fluctuations affecting C and F.[2] Analogous changes have been recorded by M. von Konkoly in ζ Orionis, the third star in the belt; while its companions ε and δ wear, occasionally, by his observations, the aspect of solar stars.

The two first stellar orders are thus probably connected, not only by *gradation* but by *migration*. The dividing-line, rendered difficult to draw by the occurrence of intermediate examples, is still further effaced by the swinging across it of a few unstable objects. Light, however, is not darkness, because each melts into twilight. Distinctions are none the less real for being established by almost insensible transitions. Let us recall what the distinctions are.

Spectra of the Sirian pattern serve, it might be said, as brilliant screens for the display of the reversed hydrogen rays. But only four out of the fourteen appear, and those comparatively shrunken and insignificant, in Capella and the sun. Mixed absorption, which in Sirian stars is insignificant, has instead become predominant. The light of Vega, for instance, reaches the outskirts of our air substantially as it left the star's photosphere. It retains its native bluish tinge; the proportionate intensities of its variously refrangible portions remain unaltered. But the sun, if its vaporous envelope could be suddenly exchanged for that of Vega, would probably leap up to three or four times its present lustre. Its rays, no longer subdued into benignity, would have a keen edge

[1] Vogel, *Astr. Nach.* No. 2839.

[2] *O Gyalla Beobachtungen*, Bde. viii. p. 5, x. p. 57.

to them, and would dazzle like lightning with their violet gleams.

What kind of change, we may now ask, in the physical condition of the sun might conceivably produce this effect? Under what circumstances can we suppose its investing strata reduced to consist almost exclusively of hydrogen? Certainly not by a simple increase of heat, which, through the resulting more copious metallic vaporisation, should produce, other things being equal, a more complex, and presumably a more strongly absorptive atmosphere. An increase of gravity, were that possible, would come nearer the mark. For it would deprive heavy metallic vapours of power to rise to any considerable distance above the photosphere; most of them, indeed, would probably sink altogether below it; the lighter ones that remained above, but close to it, would, on that account, be too hot for the effectual stoppage of its light. Hydrogen alone, owing to its enormous elasticity, would still float in the cooler regions, though in layers very much more concentrated than at present. Hence the widening of its characteristic lines. On such grounds as these the view was urged some years ago by Mr. Johnstone Stoney,[1] that white stars are actually differentiated from yellow ones by the stronger mastery in them of gravity over temperature. Just the reverse, however, of this appears to be true; the mass of Sirian stars is probably small compared with their radiative energy.

But heat and gravity are not the only forces active in the neighbourhood of the sun. From a careful survey of all the evidence, Dr. Huggins infers the corona to be a product of electrical repulsion similar in nature to the tails of comets.[2] And Professor Young points out that 'no gaseous envelope in any way analogous to the earth's atmosphere could possibly exist there in gravitational equilibrium under the solar conditions of pressure and temperature.'[3] By the effects of their own weight, if unopposed, the vapours surrounding the sun, whatever their temperature, should increase in density downward

[1] *Proceedings Royal Society*, vol. xvii. p. 48. [2] *Ibid.* vol. xxxix. p. 127.

[3] *General Astronomy*, art. 331. See also Prof. Bigelow's *Solar Corona discussed by Spherical Harmonics*, Washington, 1889.

with great rapidity; but rays as fine as if derived from a vacuum-tube are emitted by beds of them hundreds, nay, thousands of miles in thickness. It is certain, then, that gravity meets a counteracting influence, which is inexplicable except as the result of electrical repulsion.

The various gradations of intensity of this counteracting force in the stars may serve in a great degree to explain the generic differences of their spectra. Thus its marked diminution in the sun, equivalent to a virtual increase of gravity, would not improbably be attended by the transformation of the solar into a Sirian spectrum. The heavier metallic vapours would then altogether drop out of sight beneath the photosphere; those that remained would exercise but a feeble absorption; hydrogen would remain concentrated and predominant. The corona, shorn of its rays and streamers, would appear, during eclipses, as a mere rim of vivid light; the rosy prominences at its base would most likely have finally collapsed into a shallow chromosphere. The Sirian stars are, in this view, suns with very scanty appendages. They collect the whole of their luminous energies into the blaze of their photospheres. And it may be surmised that not so much the thermal as the electrical condition of stars is gauged by the fineness or strength of the significant calcium lines in the violet, which develop perhaps solely under the influence of powerful electrical excitement.

CHAPTER IV.

STARS WITH BANDED SPECTRA.

THE shining face of each one of the innumerable suns aggregated into the vast system of the galaxy is, as we have said, veiled by absorbing vapours; but in widely varying degrees. The light of Vega, though indelibly stamped with the characteristic lines of hydrogen, reaches our upper air without sensible general modification. Sunlight is not only charged with significant inscriptions, but throughout toned down and mellowed; while in Aldebaran, the process of stoppage in the blue has been carried so far as to leave the red rays visibly predominant. In Aldebaran, too, the first symptoms begin to appear of a generic change in the manner of absorption through the emergence of dark *bands* in addition to the dark *lines* of the spectrum.

This change is full of meaning. Isolated rays of definite wave-lengths, forming in the spectrum what we call 'lines,' bright or dark, are emitted only at very high temperatures. They represent perhaps the fundamental vibrations of the 'atoms' of each different substance. But these, at lower grades of heat, are not free to thrill separately. Bound together into 'molecules,' they give rise, by their associated vibrations, to complex systems of light-waves, dispersed into sets of prismatic 'flutings,' each with a sharp and a nebulous side. The fluted spectra thus constituted are derived from chemical compounds such as oxides and chlorides, as well as from 'elementary' substances, both metallic and non-metallic, at moderate degrees of heat. But the flutings can always be made to give place to lines by sufficiently raising the temperature of the glowing vapour emitting them.

It appears then that the increase of absorption in passing from the white stars through the yellow to the red, is attended a certain point by the addition of another kind of light-stoppage, and one indicating the action of cooler vapours than those previously manifest. That is to say, a spectrum of bands or flutings is superposed upon a spectrum of dark lines similar to that of the sun or Aldebaran. The alteration obviously implies an augmented extent of absorbing atmosphere. For the bands must originate in a region of less heat, that is, at a greater distance from the stellar photospheres, than the lines. The stars displaying them are hence distinguished by the wide compass and varied heat-levels of their incandescent surroundings, affording the means of arresting radiations by different systems of absorptive effects. And since the blue end of the spectrum invariably suffers most from transmission through vaporous strata, the red colour of these objects is easily seen to be a necessary accompaniment of their spectral peculiarities.

Stars with banded spectra are of two distinct kinds, constituting Father Secchi's third and fourth stellar types. In Type III. numerous shadings terminate abruptly towards the violet, but very gradually towards the red, producing something of a colonnaded effect. They are finely developed in Betelgeux (α Orionis), Antares (α Scorpii), γ Crucis, α Herculis, β Pegasi, α Ceti, Mira Ceti (variable), L_2 Puppis (variable), &c. One star in four hundred may be roughly estimated to belong to this type. About 800 of them are at present known, of which not far from 300 have been discovered by Mr. Espin of Wolsingham with his 17½-inch silver-on-glass reflector. Half a hundred have already been identified by photographic means in the southern hemisphere.[1]

These stars are invariably of a reddish or orange colour, very little of their blue light making its way through the piled-up vapours that surround them. Their spectra can hence be photographed only with great difficulty. With a forty-fold exposure (as compared with that needed for Sirius), Dr. Huggins succeeded in getting from Betelgeux, the brightest star of the entire class, a bare trace of ultra-violet action.[2]

[1] E. C. Pickering, *Henry Draper Memorial Fourth Annual Report*, p. 6, 1890.

[2] *Phil. Trans.* vol. lxxi. p. 886.

Vogel's suspicion of the presence of hydrogen was, however confirmed by the appearance on the negative of a thin '*G*' line. As a rule, no evidence of *absorption* by hydrogen can be found in star-spectra of the third type; *emission* by it, however, leaves its mark in a few. This most universally diffused of substances is not then absent from such stars. It may exist in them as copiously as in the sun, yet by an approximate balance of temperatures neutralise in general its own action upon the light proceeding from their photospheres. Radiation and absorption being thus about on a par, the marks of hydrogen vanish.

In the visible spectrum of Betelgeux no less than ninety-five dark lines were measured by Huggins and Vogel,[1] and these are only the more prominent among a crowd of others.

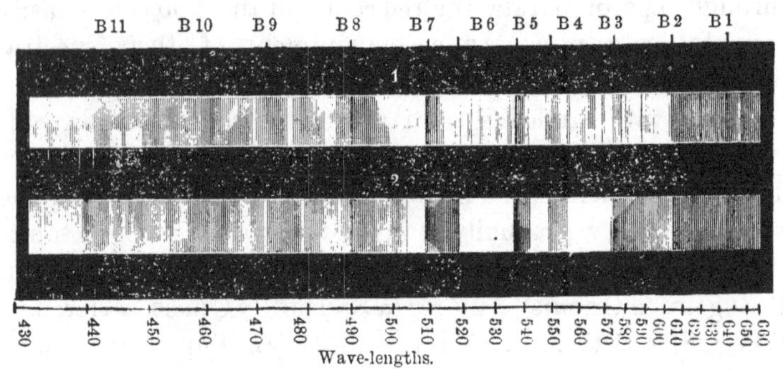

Fig. 4.—Stellar Spectra of the Third Class. 1, α Orionis; 2, α Herculis, (Dunér.)

The chemical substances indicated with certainty as constituents of the star's atmosphere (besides hydrogen) are sodium, magnesium, calcium, iron, and bismuth; silver, manganese, thallium, tin, cadmium, antimony, and mercury were more dubiously recognised. The vapour-densities of several of these metals are significantly high. In *α* Herculis, hydrogen makes no visible sign, but the spectrum is thronged with metallic rays of absorption, some cf them so strongly marked as to show, under Vogel's scrutiny in 1883 with the 27-inch Vienna refractor, across fluted shadings considerably denser in this star than in Betelgeux.[2]

[1] *Phil. Trans.* vol. cliv. p. 426; *Beobachtungen zu Bothkamp*, Heft i. p. 20, i. p. 16.

[2] *Beob. zu Bothkamp*, Heft i. p. 28; *Potsdam Publicationen*, No. 14, p. 21.

The two spectra can be compared in fig. 4, which reproduces drawings made by M. Dunér with the 9½-inch refractor of the Lund Observatory. The red end is on the right, as may be seen by the increase in that direction of the wave-lengths in the scale marked underneath. Of the few lines given in the figure, the majority, the reader will not fail to remark, appear to terminate bands. They seem placed there as if to accentuate the transition from brilliant to dusky tracts of prismatic light. Such coincidences are characteristic of the third type of stellar spectra; they are too frequent, as M. Duner points out,[1] to be supposed casual; yet they are totally unexplained. The origin of the *lines* can indeed generally be traced either to iron, calcium, or magnesium: what is perplexing is that the edges of the *bands* should fall just in the same positions.

The chemical meaning of the latter has not yet been satisfactorily explained. The first point to be noted about them is their essential invariability. All the leading bands are repeated, as if in stereotype, in every star of the class. A few secondary shadings, it is true, are visible in some which cannot (owing perhaps to their inferior brightness) be distinguished in others; but the fundamental *pattern* of the spectra is everywhere the same.[2] So striking is this rigidity of design as to suggest that the seven or eight principal flutings represent a single system of vibrations; that they are due to the absorptive action of one substance. As a combined effect, the strict uniformity observed is extremely difficult to account for. There are, however, great differences, not only in the general intensity of the bands, but in their relative importance. Those numbered 2 and 3 in the figure are usually the most conspicuous, but numbers 7 and 8 predominate in some very red stars, the more refrangible rays of which are besides in large measure cut off by general absorption.[3]

The only systematic interpretation so far attempted of this remarkable class of spectra was offered by Mr. Lockyer, as a part of his 'meteoric theory,' in 1887–8. Of the dark flutings they include, two are ascribed by him to absorption

[1] *Sur les Étoiles à Spectres de la Troisième Classe*, p. 121.
[2] *Ibid.* p. 8. [3] *Ibid.* p. 9.

by manganese, two to magnesium, three more respectively to iron, barium, and lead.[1] But these identifications depend upon coincidences of wave-length which profess to be no more than approximate, and they are hampered by apparently insuperable difficulties. They are coupled with another alleged feature in the spectra of these stars of still greater physical importance. The reader is aware that the distinctive light of comets consists mainly in the bright emanations of carbon, concentrated in a yellow, a green, and a blue band. These Mr. Lockyer asserts to be represented *directly* in the radiations of Betelgeux and its congeners. But the evidence that their spectra include the spaces of enhanced brightness which, if this were so, should just match the cometary flutings, is very far from conclusive. It consists almost wholly in the close agreement in position between the dark edge of a band in the stars (No. 7) graduated towards the red, and the brilliant edge of the green fluting in comets dying away towards the blue. The former, on Mr. Lockyer's hypothesis, is not a genuine band of absorption, but an effect of contrast produced by a bright carbon-fluting *facing the other way*. But the supposition is thereby involved that stars like Mira Ceti, in which this interval is almost black, are totally deficient in continuous light—a supposition contradicted by the photographic, as well as by the visual brilliancy of their spectra. The invisibility of the yellow and blue carbon-flutings,[2] always seen in comets, makes the presence of the green fluting appear still more improbable.

The point is of crucial importance as regards the constitution of these bodies. In Mr. Lockyer's view, they are 'not masses of vapour like our sun, but swarms of meteorites.' They are composed of millions of separate rocky or metallic fragments each circulating independently round the centre of gravity of the entire. Circulating independently, but with numerous jostlings and collisions, in the course of which heat is developed, and

[1] *Proceedings Royal Society*, vol. xliv. p. 54.

[2] There are grave objections to admitting the reality of the 'masking' effects of manganese and lead absorption, by which Mr. Lockyer seeks to explain the non-appearance of the fluting at wave-length 5634.

glowing vapours evolved. Each meteorite is encompassed with its own distinct little atmosphere, by the action of which certain characteristic sections of its light are absorbed, while the 'interspaces,' throughout the whole vast extent of the system, are filled with glowing hydrocarbons. Hence the supposed compound nature of a spectrum integrating the *radiation* of carbon and the *absorption* of manganese and lead,[1] and suggesting that the objects it characterises should be looked upon as comets on a prodigious scale, rather than as suns.

FIG. 5.—Spectrum of Betelgeux (α Orionis) photographed at Harvard College.

But let us consider how the matter really stands. Beneath the series of shadings which differentiate stars of the third from stars of the second class, lies a spectrum of lines of precisely the same general character with that of the sun, and bearing witness to a sun-like incandescence. Photographic testimony is strongly to the same effect. Fig. 5 shows the blue and violet parts of the spectrum of Betelgeux as imprinted upon a highly sensitive plate exposed at Harvard College. Like the solar spectrum, it is closely and delicately ruled throughout with fine lines of metallic absorption, and bears no trace of radiation by carbon or any other vapour. The underlying continuous light is obviously, as in the sun, photospheric—it

[1] *Proc. R. Soc.* vol. xliii. p. 130.

is derived from the brilliant condensation-surface of the star. Indeed, spectra of the second and third types are, in their photographic portions, so nearly identical, that difficulty has been experienced, in the course of the great cataloguing work carried out at Harvard College, in distinguishing them. Dr. Scheiner, too, remarks that all the chief lines visible between F and H in solar stars reappear in fluted spectra, with the addition, however, of numerous very narrow dark streaks diffuse on one side, the strength of which seems to increase with the more pronounced development of the type.[1]

In the manner of visibility of hydrogen in their spectra, however, these stars exhibit a marked deviation from the solar model. Most, as we have said, give no sign of containing that element at all; in one (α Orionis), it shows feebly obscure, while a few, all of them subject to great fluctuations of light, bear the stamp of its direct incandescence. Photographs of 'Mira' Ceti and of a similarly variable star in Orion (near χ^1 Orionis), taken at Harvard College in the autumn of 1886, gave the earliest unquestionable proof of the occurrence of bright lines of any kind in banded stellar spectra of a normal type. We are enabled, by the kindness of Professor Pickering, to reproduce in fig. 6 a beautiful enlargement of the photographed spectrum of Mira. It includes six vivid hydrogen-rays, namely, G, h, and four of the ultra-violet set. But the intermediate line at H is missing, its place being occupied by the closely adjacent calcium-line, which, with its comrade K, is just as dark, as broad, and as hazy as in the solar spectrum. There can be no question, however, but that the hydrogen-series is *originally* complete; the H-line is radiated with the others, and would show equally bright were it not stopped-out by the overlying metallic vapour vibrating in approximately the same period.[2]

The fact thus recorded makes several points perfectly clear. We learn from it, to begin with, that the blaze of hydrogen registering itself in the spectrum occurs quite close to the photosphere of Mira, since the stratum of less intensely glowing calcium, which absorbs the small quantity of its light

[1] *Astr. Nach.* No. 2924. [2] *Observatory*, vol. xi. p. 84.

it is capable of touching, lies above it. The relatively cool calcium, however (as the character of its absorption shows), is at just such a pitch of thermal or electrical excitement as prevails in the hottest part of the sun's surroundings, while the hydrogen beneath proves itself, through its emission of the typical ultra-violet rays, to be at the enormous temperature of the Sirian atmosphere. Conclusive evidence seems thus to be provided that stellar spectra of the third type originate at various heat-levels in powerfully ignited vaporous envelopes.

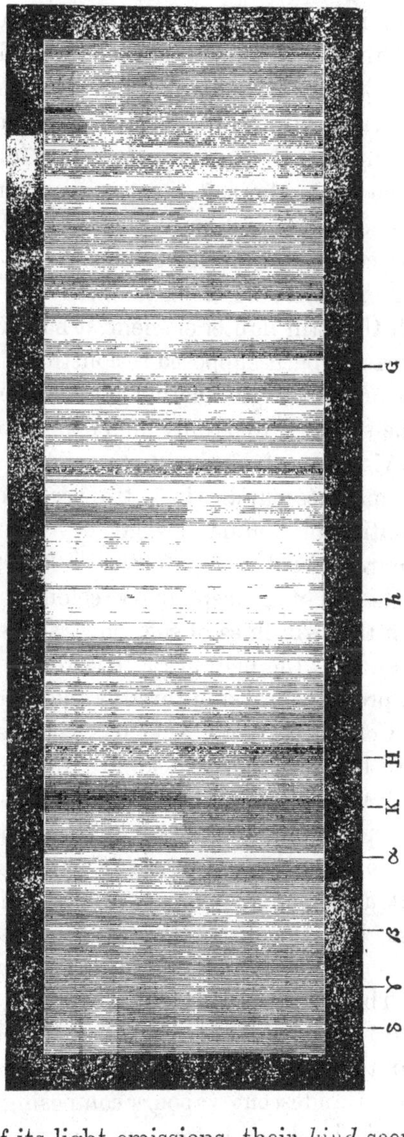

Fig. 6.—Spectrum of Mira showing bright lines of Hydrogen and dark lines of Calcium (Pickering).

The dark blue hydrogen line (near G) was *seen* bright in Mira by Mr. Maunder at Greenwich, October 5, and by Mr. Espin October 23 and 30, 1888.[1] The corresponding green and red rays (F and C) have not hitherto been detected. Notwithstanding the amazing variability of this star in the *amount* of its light-emissions, their *kind* seems

[1] *Monthly Notices*, vol. xlix. pp. 18, 300.

to remain tolerably constant,[1] but further inquiries on this head are desirable. The bright lines lately discovered by Mr. Espin in several objects of the same class come into view, it would seem, about the time of maximum general brilliancy, and die out as the stars fade. In producing these remarkable emanations one other substance is frequently associated with hydrogen. This is the enigmatical 'helium,' known only through spectroscopic observations at the edge of the sun, where it occurs in enormous profusion in the chromosphere and prominences. Except 'coronium' (the chief component of the solar corona), it is the most ethereal form of matter we are acquainted with, and in chemical union with that still lighter element, it has, with some plausible show of reason, been supposed to constitute hydrogen.[2] The feebleness of its absorptive action allows its presence in sidereal bodies to be signified to us only under the rare conditions occasioning the visibility of its *direct* radiations. One of these—perhaps the most essential—is a transcendental temperature. The evolution of helium lies far beyond the range of terrestrial resources for heat-production. Its spectroscopic badge is a single line of a deep yellow colour, a little more refrangible than the twin-lines of sodium. And since these are known as D_1 and D_2, the helium ray has received the appellation of D_3, the progression being towards the blue end of the spectrum.

Collecting, then, what has been learnt about stars of the third type, we find that all are more or less red, that the light of all is powerfully absorbed, and that all exhibit highly complex spectra comprising superposed orders of effects undeniably due to very different stages of temperature. It may be added that a large proportion are of conspicuously variable brightness, but the discussion of this peculiarity will be more conveniently entered upon in a subsequent chapter.

These facts suggest the all but certain inference that the absorbing atmospheres of these stars are much more extensive than that of the sun, and include a wider range of temperatures. The incandescent vapours composing them are, at the photospheric level, as hot as—in some cases much hotter than—

[1] *Henry Draper Memorial, Third Annual Report*, p. 5.

[2] Grünwald, *Astr. Nach.* No. 2797.

those in a corresponding position in the sun; but they ascend to heights far beyond the limit of the solar mixed metallic strata, and are hence competent to produce a cooler kind of absorption than any affecting the sun's prismatic light.

Of this difference we can offer only a speculative explanation. If it be admitted as probable that the spectroscopic transition from Sirian to solar stars represents an enhancement of electrical repulsive action producing a partial neutralisation of the power of gravity over their gaseous surroundings, we may be permitted to conjecture that the same cause works still more energetically in red stars. Their atmospheres are inordinately distended because the materials composing them are *virtually* very light; they are to a great extent emancipated from the thraldom of their own weight; even the massive atoms of bismuth and mercury, buoyed up, as it were, by a levitative influence, float at comparatively high levels. The flaming appurtenances of these bodies must (should this view prove correct) be on a prodigious scale; it is not impossible that visual traces of their abnormal development have been perceived. For unless the intermittently hazy aspect of certain red stars [1] be due to instrumental causes, it indicates a vast diffusion of glowing matter in their neighbourhood. Coronas, in fact, hundreds of millions of miles in extent, are on such occasions made manifest. An approach towards a nebulous condition is thus plainly hinted at.

In the extreme members of the class, accordingly, the solar analogy is weakened to the verge of abolition. The function of a 'sun' can hardly be regarded as in any sense discharged by such a body as Mira Ceti. But the stage of Mira is only reached by insensible gradations. At the other end of the scale we meet radiating centres, the efficiency of which is not perceptibly compromised by the shaded bands modifying their light. As dependents upon Antares (α Scorpii), for instance, planets like ours might be warmed and clothed with vegetation, and lavishly stocked with life. At the same distances they would indeed receive probably several hundred times as

[1] See Mr. Peek's observations *Knowledge*, vol. xii. p. 126; also Pogson, *Monthly Notices*, vol. xvi. p. 187.

much heat and light as from our sun, and they would receive it with approximate constancy. Daylight upon them would be of a warmer glow than here; there would be no cold grey dawns; every landscape would be suffused with a flush as of autumn; an universal mellowed illumination would soften and harmonise all the features of the scenery. If such phenomena as total eclipses occurred at all, inconceivably brilliant displays of red flames and coronal streamers might, during their progress, astonish intelligent spectators, whose observations at other times would perhaps disclose the presence of spots on their sun, compared with which the sun-spots we often wonder at are but puny objects; but unless their science had put them in possession of some method of light-analysis, they could never know that the body sovereign in their sphere was a 'star with a banded spectrum.'

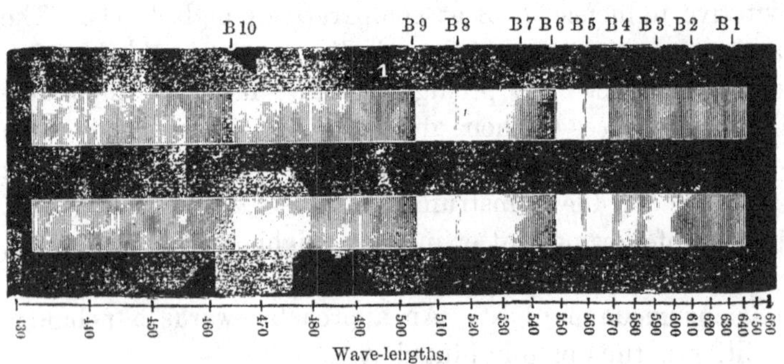

FIG. 7.– Stellar Spectra of the Fourth Class; 1, 152 Schjellerup; 2, 19 Piscium (Dunér).

A comparison of figs. 7 and 4 shows at once the chief points of difference between stellar spectra of the third and of the fourth types. The intenser bands of absorption in the latter, it will be noticed, are only three in number, and they face the opposite way from the more frequent shadings in the allied genus. Their sharp sides are towards the red, their shaded sides towards the violet end of the spectrum. No doubt exists as to their origin; all three agree accurately in position with the bright bands given out by burning alcohol and other hydrocarbons. They are the identical cometary bands vainly looked for in spectra of the third type, reversed;

instead of being prismatically coloured, they are intensely dark. The stars showing them might be called 'carbon stars'; they bear, as their discoverer Father Secchi perceived, the unmistakable signature of that protean substance which more than any other deserves to be called the material basis of life. The best authorities agree that the signature is that of pure carbon, though there is no practical way of obtaining it except through the medium of a hydrogen compound. And since this is also very likely to be the case in the stars, it may be assumed that carburetted hydrogen, or some closely analogous gas, forms an important constituent of their atmospheres. The absence from their spectra of all traces of free hydrogen is thus only what might be expected. Stars of Class IV. are rendered the more interesting by their scarcity. About one hundred and twenty of them are now known, recent extensions of the list being largely due to Mr. Espin's diligent explorations; and he calculates that if as many proportionately are found in the southern hemisphere, there may exist in all 168 brighter than 8·8 magnitude.[1] For the most part they are telescopic objects. No more than seven or eight can be made out with the naked eye. Of these the most brilliant is a star in Canes Venatici of 5·5 magnitude, numbered 152 in Schjellerup's Catalogue of Red Stars[2] (fig. 7, No. 1). From the extraordinary vivacity of its prismatic rays, it was named by Father Secchi 'La Superba,' and in many such stars a magnificent effect is produced by the sudden alternation of dazzling 'zones' of red, yellow, and green light, with spaces of profound obscurity. Another fine specimen of the class is 19 Piscium (fig. 7, No. 2), averaging about sixth, but variable to the extent of one magnitude. To it also belongs Hind's 'Crimson Star' (R Leporis) and many other noted variables.

The faintness of fourth type stars can be accounted for by the extraordinarily powerful atmospheric absorption exercised upon their light. The vapours closing in upon them leave only an aperture here and there for their rays to escape through; and those more refrangible, as usual, bear the brunt of their attacks. The violet end of the spectrum is either cut

[1] *Monthly Notices*, vol. xlix. p. 364. [2] *Astr. Nach.* No. 1591.

off, it might be said, wholly, or greatly enfeebled. Hence a special depth of colour in these stars. Some appear like 'a drop of blood' freshly fallen upon the field of the telescope; others gleam like clear flame, or glow like carbuncles; none are indifferent or undistinguished in tint. They have accordingly only of late been rendered accessible to photographic investigation, four new specimens having been detected at Harvard College through the peculiarities of their self-imprinted spectra.

To the question whether this deficiency of blue rays is to any degree innate, or must be altogether attributed to atmospheric quenching, no categorical reply can at present be made. Absorption carried far enough *might* produce, even in Sirius, the observed result; but it is more likely that imperfect condensation accompanies strong absorption, and that the photospheres of 152 Schjellerup and 19 Piscium really fall short, in intrinsic brilliancy, of the solar, still more of the Sirian, standard. At the same time their powerful incandescence is undoubted. Through the obscurity of the carbon bands can be distinctly seen a 'Fraunhofer spectrum,'—a spectrum, that is to say, composed, like that of the sun, of dark metallic lines thrown out upon a continuous background. Among the substances originating them, sodium is certainly, iron probably, recognisable;[1] but particular identifications are of extreme difficulty in objects with so little, and such unequally distributed light. Enough, however, is known to assure us that near the photospheres of fourth type stars metallic vapours are suspended in a state of ignition, while the reversal of their characteristic rays proves the temperature of the photospheres themselves to be still higher. The great depth of their atmospheres is certified by the origin within them of two strongly marked systems of absorptive effects due to widely different degrees of heat; bright lines of any kind, however, have so far not been certainly recorded in these spectra.

They are, not only in general outline but in particular details, of an unvarying pattern; at the most, some difference in the relative intensity of the flutings distinguishes one from

[1] Dunér, *Recherches* &c. p. 123; Vogel, *Potsdam Publicationen*, No. 14, p. 27.

another. Notwithstanding their obvious affinity to banded spectra of the third class, they stand emphatically apart from them. Intermediate specimens are completely wanting. Between the first and second, and the second and third types, they abound; but the third and fourth are sharply separated. No stragglers are picked up between the camps. The objects collected in them have nevertheless many qualities in common, and among the most striking, that of luminous instability. About twelve per cent. of stars with banded spectra are well-known variables, the proportion being about equal for each kind. Thus, the differences of their constitutions are not such as materially to affect their inherent tendency to luminous fluctuation.

Stars with banded spectra are not found indifferently in all parts of the sky; but incline to collect into groups, separate for each order. Eight regions of concentration, the principal in the constellation Lyra, are indicated for objects of the third type; those of the fourth have their chief gathering-places near γ Cygni and about the tips of the Horns of Taurus.[1] Both these spots lie in the Milky Way, the zone of which has an overwhelming attraction for the 'carbonaceous' stars. All small stars, indeed, are strongly condensed towards it, and M. Dunér is of opinion[2] that the objects in question merely conform to the universal law governing sidereal structure. But their obedience to it at least shows them to be situated in the region of small, that is, of excessively distant stars, hence to be copious radiators, and by no means on the verge of extinction.

[1] Espin, *Observatory*, vol. x. p. 259.
[2] *Spectres de la IIIe Classe*, p. 125.

CHAPTER V.

GASEOUS STARS AND NEBULÆ.

THE fifth spectral class is at present the most restricted of all, but its numbers are being rapidly augmented both by photographic and by improved visual means. The objects belonging to it are distinguished by the display in their spectra of isolated *bright* lines on a more or less perfectly continuous background, sometimes, however, also interrupted by dark lines or bands. They present us then with a triple combination—a *direct* gaseous spectrum, a *reversed* gaseous spectrum, and a spectrum due to glowing solid or liquid matter, all simultaneously made manifest by the unrolling, as it were, of a single scroll, yet each originating under very different conditions. The investigation of what those conditions are constitutes one of the most important tasks of physical sidereal astronomy.

The state of bright emission is in some stars normal, in others it only supervenes as part of a great general increase of light. This is the case with many 'temporary' and periodical stars, their blazing atmospheric constituents being almost invariably hydrogen and helium. Objects, on the other hand, showing bright lines with approximate constancy, can be discriminated into two varieties, according as they give or withhold evidence of the presence in them of hydrogen.

The first specimen of a 'gaseous star' was made known by Father Secchi's discovery, August 19, 1866, of the green line (F) of hydrogen conspicuously bright in γ Cassiopeiæ,[1] the middle star of five of the second and third magnitudes grouped into the shape of a W on the opposite side of the pole from

[1] *Sugli Spettri Prismatici delle Stelle Fisse*, Mem. i. p. 10, Mem. ii. p. 62.

the Great Bear. Soon afterwards the same peculiarity revealed itself in β Lyræ, the emission-spectrum of both stars consisting of three rays of hydrogen and one of helium. But these cannot always be seen, even under the best optical circumstances. As early as 1872 Vogel was struck with their apparently unaccountable caprices of visibility;[1] and M. von Gothard watched vainly for them during two years before he caught sight in γ Cassiopeiæ of the crimson twinkling of C *in particularly unfavourable weather*, August 13, 1883.[2] His subsequent observations, and those of M. von Konkoly,[3] fully established the occurrence, in both stars, of rapid spectral fluctuations, which have recently been studied, with attentive curiosity, in widely separated parts of the world.

The variability disclosed is of a most singular kind. It might have appeared a safe forecast that the brightness of the lines—of those, at any rate, emitted by the same substance—would change in concert, if it changed at all; experience shows, however, that each flares out and dies away, to a great extent independently of the others. They take it in turns to be brilliant. The *stress* of variability, however, is in γ Cassiopeiæ laid upon C,[4] in β Lyræ upon D_3, usually the leading feature of this star's gaseous spectrum.[5] Dazzling at times, it often fades to extinction at intervals of a few days; but no definite law has yet been found for its changes. The disappearance of the hydrogen-lines is occasionally emphasised by reversal. They were seen *dark* by Von Gothard at Herény, September 5, 1882,[6] and probably again by Mr. Maunder at Greenwich, October 19, 1888. In γ Cassiopeiæ, the turning of the balance appears never to go beyond effacement. Its rays do not, however, escape the effects of mixed absorption. Mr. Keeler perceived, with the Lick refractor, the green part of its spectrum to be 'full of very fine, delicate, dark lines, seen only under good atmospheric conditions, the *b* group' (due to magnesium) 'being somewhat more prominent than

[1] *Bothkamp Beobachtungen*, Heft ii. p. 29.

[2] *Astr. Nach.* Nos. 2531, 2589.

[3] *Observatory*, vol. vi. p. 332; *O Gyalla Beobachtungen* vol. viii. p. 5.

[4] Copeland, *Monthly Notices*, vol. xlvii. p. 92.

[5] Maunder, *ibid.* vol. xlix. p. 300.

[6] *Astr. Nach.* No. 2581.

the others.'[1] The sodium-lines at D have been seen dark in both stars, but their appearances are temporary. Of a general arresting action there is little or no trace. The white light of these objects remains practically unmodified.

Stellar bright lines have been explained in several different ways. One is by the extent of surface radiating them. Stars possessing enormous self-luminous atmospheres would, it has been supposed, send us a sum-total of light in which the gaseous beams predominate simply through quantitative excess over the continuous radiance of comparatively small nuclei. And Dr. Scheiner alleges, in support of this view, an effect of shading about the sharp, bright hydrogen-rays in γ Cassiopeiæ, due, in his opinion, to absorption by the denser strata near the photospheric level.[2] But it is difficult to believe in the reality of the state of things thus presented to us. In the first place, the fineness of other dark lines in the same spectrum implies an atmosphere tenuous throughout. Again, the phenomena of spectral variability appear entirely inconsistent with this rationale. Fluctuations in atmospheric extent, even if we could admit their occurrence on the incredible scale and with the incredible swiftness required, would not account for the *relative* variability of the bright lines. Finally, the extreme complexity lately observed by Dr. Becker in the emissive spectrum of β Lyræ[3] affords a reasonable certainty that it originates in moderately shallow vaporous layers. An atmosphere composed solely of hydrogen and helium might conceivably rise to vast altitudes, but not one of the heterogeneous constitution implied by authentically recorded facts.

It seems, then, as if the display of bright lines must depend upon a real excess of atmospheric brilliancy, involving, it would seem, the prevalence of a higher temperature in the gaseous layers than in the photosphere covered by them. Unless, indeed, the same result be brought about by a peculiarity of electrical condition which we are at present unable

[1] *Publications Astr. Soc. of the Pacific*, vol. i. p. 80.
[2] *Sitzungsberichte*, Berlin, 1890, viii. p. 8.
[3] *Monthly Notices*, vol. l. p. 191.

either to define or to reason about. The spectral variability of γ Cassiopeiæ and β Lyræ can only arise from thermal or electrical instability. Messrs. Frankland and Lockyer showed in 1869 that the luminosity of very rare hydrogen can be reduced to consist of the green line (F) alone.[1] The appearance of C is a symptom of enhanced excitement, so that the epochs of its brilliancy in the stars we are considering correspond presumably to temperature maxima. The substance, on the other hand, of which D_3 is the spectroscopic 'device,' is, as the reader is aware, unknown to terrestrial chemistry, but thrives in the very hottest part of the solar furnace. It is pretty clear, however, that changes of temperature will not alone suffice to account for its appearances and disappearances. They show no disposition in β Lyræ, a well-known 'short period' variable, to follow the light-changes of the star, tending, indeed, to obey, if any, an inverse law connected possibly with the increased effect of contrast as the general spectrum becomes enfeebled.

As to the local origin of such emissions, one item of positive evidence is afforded by the Harvard photograph of Mira (fig. 6). The stopping-out shown in it by calcium-absorption of one line of the otherwise unbroken hydrogen-series, gives a perfect assurance that the calcium-vapour lies between us and the hydrogen.

The same state of things doubtless exists in all similar stars. That is to say, the bright-line emission associated with the light-maxima of variable stars has its seat in a comparatively shallow layer of vividly incandescent gases at the lowest atmospheric level. But it would be unsafe to generalise this conclusion too widely. Miss A. C. Maury's examination of some Harvard photographs of the Pleiades showed the spectrum of one member of the group (Pleione) to contain narrow, bright hydrogen-lines superposed upon the dusky stripes characterising the Sirian type of stars.[2] Here, then, an absorbing stratum is placed beneath a more vividly incandescent bed of the same substance, and the condition partially indicated in γ Cassiopeiæ is fully attained.

[1] *Proc. R. Society*, vol. xvii. p. 454. [2] Pickering, *Ast. Nach.* No. 2934.

Bright hydrogen, frequently associated with helium and other substances (Dr. Becker has laid down seventy bright lines in the spectrum of χ Cygni), may now be regarded as a common feature of the Mira class of variables. It has also been detected through photographic researches executed in the southern hemisphere by delegates from Harvard College, in η Argûs, distinguished for extraordinary light-changes, in δ and μ Centauri, as to which there is no suspicion of inconstancy, and, among northern stars, in P Cygni (Flamsteed's No. 34 in that constellation). This last star has a curious history. First seen by Janson as a 'new star' in 1600, the *revenante* of the Swan (as Huygens called it) remained steadily of the third magnitude for nineteen years, after which it gradually lost light and disappeared. It rose, however, to a second maximum in 1659,[1] and, after some irregular fluctuations, settled down, towards the close of the seventeenth century, at about its present fifth magnitude status. No present suspicion of variability attaches to it; but Bessel observed it as of 6·7 magnitude, September 14, 1825.[2] The biographies of other gaseous stars, were they made known, might possibly be found to include similar vicissitudes. The hydrogen lines in this object, as in Pleione, are doubly reversed.

The finest specimen of the fifth spectral class is invisible in these latitudes. First detected by Respighi at Madras, December 24, 1871,[3] the peculiarities in the light of γ Argûs were studied with some care and much delight at the 'extraordinary beauty' of the spectacle they present, by Dr. Copeland at Puno in the Andes, April 24, 1883.[4] An 'intensely bright line in the blue,' he remarks, 'and the gorgeous group of three bright lines in the yellow and orange, render the spectrum' of this star 'incomparably the most brilliant and striking in the whole heavens.' There is no sign that it is in any degree variable. Its appearance to the present writer, at

[1] *Journal des Sçavans*, Dec. 1666, p. 288.
[2] Gore's *Catalogue of Variable Stars*, *Proc. R. Irish Acad.* vol. iv. ser. ii. p. 199.
[3] *Comptes Rendus*, t. lxxiv. p. 516.
[4] *Copernicus*, vol. iii. p. 206.

the Cape, in October 1888, tallied precisely with Dr. Copeland's description, only that the additional feature of a deep band of absorption below the cobalt line seemed unmistakable.[1] A vivid continuous spectrum extends into the violet as far as the eye has power to follow it, and accounts for the brilliant whiteness of the star.

Its chemistry, however, remains mysterious. Only negative assertions can be safely made about it. Hydrogen has certainly nothing to do with producing any of the bright lines recorded; and there is great difficulty in accepting Mr. Lockyer's identification of the blue ray with a modified carbon-fluting. He may, however, turn out to be right in ascribing to sodium-vibrations the faintest and least refrangible of the γ Argûs quartette.[2] Its position, at any rate, lies suspiciously near to that of D.

In the course of his brief explorations from Puno, Dr. Copeland came across five small stars of the γ Argûs type; which is also conformed to by two objects discovered by Professor Pickering in 1880–1,[3] by ten southern stars 'spectrographically' investigated at Harvard College from plates exposed at Chosica in Peru, and by a whole group in Cygnus (see Appendix, Table I.). The first three of these were noted for the singular quality of their light at Paris in 1867 by MM. Wolf and Rayet,[4] whose names are hence often used to designate this stellar variety; others were distinguished by Dr. Copeland and Professor Pickering. Accurate measurements of the three original Wolf-Rayet stars, of 'Argelander-Oeltzen 17681,' and 'Lalande 13412' (so called from their numbers in different catalogues), were made with the great Vienna refractor by Professor Vogel in 1883, and we have his kind permission to reproduce the drawings in which he depicted his results (see fig. 8, in which the blue end is to the right). In spectrum No. 2, belonging to the eighth magnitude star Argelander-Oeltzen 17681, we see the prismatic pattern of γ Argûs reduced to its simplest form.

[1] *Observatory*, vol. xi. p. 430. [2] *Proc. R. Society*, vol. xliv. p. 33.
[3] *Nature*, vols. xxii. p. 483, xxiii. pp. 338, 604.
[4] *Comptes Rendus*, t. lxv. p. 292.

There is the fundamental yellow line at 5810, with, far away towards the other end of the spectrum, a space of profound absorption (between wave-lengths 4880 and 4700) followed by a vivid effusion of bright light. This appears in γ Argûs as a well-defined, though somewhat wide line; but its diffuseness in other similar stars occasions considerable uncertainty as to its real position on the scale of wave-lengths. In Lalande 13412 Vogel places its maximum brightness at 4690, in one of the Wolf-Rayet stars, at 4640; yet the same substance is certainly glowing in both.

The close relationship, indeed, of all the objects belonging to this spectral variety is apparent from a glance at the figure.[1] None fail to show the two persistent rays visible alone in No. 2; but additional ones drop in and out very curiously. The orange 'D' is confined to γ Argûs; its 'citron' line, however, at 5680 comes out in the Cygnus stars, accompanied, in two of them, by the hydrogen F, as well as by an unidentified radiation at about 5400, common to most members of the group. The Wolf-Rayet stars are noticeable telescopically for the yellow tinge of their light; they are of 8 and 8·5 magnitudes, and have remained without change of any kind during twenty-three years. The bright lines in their spectra, instead of being half-drowned in photospheric emissions, as in γ Cassiopeiæ and β Lyræ, stand out prominently from a background[2] which, shown as continuous by moderate instruments, was seen by Mr. Keeler with the Lick telescope 'as an extremely complicated range of absorption-bands and faint bright lines.' The preponderance has, in such objects, passed over to the gaseous elements of their light. Although shining like distant stars in the telescope, they are in reality of a nebular rather than of a stellar nature.

Thus, by insensible degrees, we have descended into the borderland between the two great sidereal kingdoms—a region of nondescript objects, where distinctive qualities are almost effaced, and definitions cease to be valid. The brilliant stars

[1] Vogel, *Potsdam Publicationen*, No. 14, p. 20.

[2] Vogel, *Berichte der Sächsischen Gesellschaft der Wissenschaften*, 1873, p. 557.

FIG. 8.—Spectra of Gaseous Stars compared with the Fourth_Type Spectrum of 152 Schjellerup (Vogel).

1, 152 Schjellerup; 2, Argelander-Oeltzen 17681; 3, Lalande 13412; 4, Wolf-Rayet, No. 1; 5, Wolf-Rayet, No. 2; 6, Wolf-Rayet, No. 3.

of Orion may be said to mark the first stage on the road towards nebulosity. For their spectra appear at times unbroken by the traces of hydrogen-absorption more or less strongly impressed upon them at others; and the transition is an easy one from this state of things to that existing in β Lyræ, where the same sort of fluctuating balance of temperature inclines preferentially the other way. That is to say, the hydrogen atmosphere of this star tends to rise above the thermal level of its photosphere. Emissive superiority is substituted for neutrality, and gains more and more the upper hand as gaseous stars merge into undoubted nebulæ.

In Table I. of the Appendix will be found a list of all the stars in which bright lines have, up to the present, been detected. They are fifty-one in number, and nearly all are situated in the very thick of the Milky Way. The only gaseous star, indeed, besides a few variables, which can be said to lie even slightly apart from that great collection is θ Muscæ. The tendency of these objects to collect in groups is very marked, especially among the Wolf-Rayet variety, eight specimens of which are congregated in Cygnus, five within a restricted area in Scorpio, while two occur about half a degree apart, in the Keel of the Ship. Father Secchi's conjecture as to the localisation of spectral peculiarities is thus proved to have a solid foundation of fact.

There is only one recognised exception to the rule that the lines flashing out in long-period variables are those of hydrogen and helium. R Geminorum is a star changing in 371 days from above the seventh to below the twelfth magnitude. It might be called the 'comet-variable,' since the three bright bands of carbon usually seen in comets were most probably recorded in its spectrum by Vogel as it rose towards a maximum on April 7, 1874.[1] Their display, moreover, was combined with that of the typical yellow ray of the Wolf-Rayet stars. The significance of the emission by even one star in the heavens of distinctively cometary light was accentuated by Mr. Lockyer's striking reproduction of its range of bright lines in a 'meteoritic glow' at South Ken-

[1] *Astr. Nach.* No. 2000.

sington.[1] Unfortunately, however, no thoroughly satisfactory observations of this unique spectrum are extant, for Vogel's were made under unfavourable conditions, and have not been repeated. They leave, nevertheless, little or no doubt that R Geminorum makes no exception to the rule that hydrogen and carbon are, in a spectroscopic sense, mutually exclusive—that where one appears the other remains imperceptible.

The distinction between gaseous stars and stellar nebulæ is slight. Both kinds of object present the telescopic appearance of small stars, and have, in several instances, been registered as such. Both, when a prism is applied, disclose analogous peculiarities. Analogous, not identical. The light of a gaseous star so examined is ordinarily concentrated in two points, or where a cylindrical lens is employed in two short lines, yellow and blue respectively; that of a stellar nebula gathers into one green knot. Its rays are, in a sense, incapable of analysis; they are so nearly monochromatic that they can be refracted without being dispersed. Objects of this nature can be picked out at a glance from ordinary stars by Professor Pickering's method of 'sweeping' with what we may call a prismatic eye-piece. Above a score of them have in this manner been found since 1881 ; [2] and it is remarkable that the exclusive preference for the Milky Way of gaseous stars is shared by stellar nebulæ.

The green ray of the latter is the characteristic token of gaseous nebulæ. From the great Orion 'portent' to the faintest 'planetary,' all without exception show it ; and in many it is so predominant as virtually to stand alone. But its origin remains an enigma. In position it is almost coincident with an important line of nitrogen. A trifling divergence, however, shows them to be certainly distinct. A different view has been urged by Mr. Lockyer. He found that fragments of meteorites gently heated in vacuum-tubes lit up by the passage of an electric current gave out, pro-

[1] *Proc. R. Society*, vol. xliii. p. 133.

[2] Pickering, *Observatory*, vols. iv. p. 81, v. p. 294 ; *Astr. Nach.* No. 2517 ; Copeland, *Monthly Notices*, vol. xlv. p. 91

minently and persistently, a fluting due to magnesium burning at about the temperature of the bunsen-flame.[1] Now the sharp edge of this fluting appeared, with the slight amount of dispersion used in the experiments, to agree sensibly with the chief line of nebulæ; and it was inferred (some other approximate coincidences pointing, it was thought, towards the same conclusion) 'that the nebulæ are composed of sparse meteorites, the collisions of which bring about a rise of temperature sufficient to render luminous one of their chief constituents, magnesium.'[2]

Precise inquiries have nevertheless failed to ratify the suggested identification of the fundamental meteoric and nebular lines, found by careful measurements to stand very slightly apart in the spectrum. But spectroscopic agreements must be absolute if they are to be reckoned significant. And the disagreement, in the present case, was rendered palpable by Dr. and Mrs. Huggins's application to it of the laborious method of direct comparison. The *injection* of the light of burning magnesium into the apparatus simultaneously displaying the spectrum of the Orion nebula had the effect of at once producing a well separated doublet. The fluting-edge lay markedly lower down towards the red than the nebular line.[3] Nor is it likely that a fluting, however attenuated the vapour emitting it, could be reduced to so thin a 'remnant' as the sharp, fine line seen in nebulæ.

Nebular radiance cannot then, it seems, be imitated in the laboratory. Possibly it signalises a modification of matter arising only under extra-terrestrial conditions. The line at 5005 has, however, been seen in two comets and in one star. On January 9, 1866, Dr. Huggins observed the spectrum of Comet I., 1866 (of the November meteors), then within two days of perihelion, but at a distance from the sun not greatly inferior to that of the earth. No trace was seen of the usual carbon-bands; the continuous line of prismatic light being interrupted only by a vivid point matching the chief nebular line in position, and suspected to be accompanied by the

[1] *Proc. R. Society*, vol. xliii. pp. 124, 127.
[2] *Ibid.* vol. xliv. p. 2. [3] *Ibid.* vol. xlvi. p. 48.

second nebular line (at 4957).[1] Comet II., 1867, apparently shared, under closely similar circumstances, the same uncommon quality.[2] Again, the 'new star' of 1876 shone as it faded with a green ray identical with that of stellar nebulæ.

These bodies, judging by the two illustrative instances just mentioned, are not greatly heated. The condensed or nuclear portions probably included in them give but slight signs of incandescence; spectroscopically (whatever they may be physically) they are mere spheres of glimmering gas. Now planetary nebulæ (so called by Sir William Herschel because they exhibit a planet-like disc) are not intrinsically different from the stellar sort, but they either are, or, owing to their greater vicinity, appear larger and brighter, and hence offer better facilities for the analysis of their light.

It is found to consist of three rays composing the fundamental spectrum of all gaseous nebulæ. Although of varied relative intensity in different individuals, the predominance uniformly remains to the line at 5005, which accordingly survives alone in such a dearth of light as that created by the combined distance and faintness of the monochromatic objects detected by Professor Pickering's method.

The nebular trio of lines (the three to the right in the lower part of fig. 1), since they lie adjacent to each other in the middle or green part of the spectrum, inevitably give a resultant green or bluish colour to the objects they characterise. Two of them are unidentified; the third, or most refrangible, is the familiar F line, the never-failing, and frequently solitary emanation of the ubiquitous gaseous metal. The undoubted and universal presence of hydrogen in nebulæ constitutes the one link as yet recognised between nebular and terrestrial chemistry. But cometary and nebular chemistry are not (as Mr. Lockyer first pointed out) wholly disconnected; and some nebular spectra include traces of stellar and solar affinities which must prove of the highest importance for determining the place in creation of cosmical cloudlets.

[1] *Proc. R. Society*, vol. xv. p. 5.

[2] *Monthly Notices*, vol. xxvii. p. 288.

Those of the planetary sort display in a few cases,[1] besides the usual three lines, an additional cobalt-blue one most likely identical with the blue ray of γ Argûs. It seems also to have been very faintly visible, in the spectrum of the Orion nebula, to Mr. Albert Taylor, at Sir Henry Thompson's observatory at Hurst Side, October 13, 1888.[2] But the chemical meaning of the line has not been determined. We only know that it appears in the majority of gaseous stars, and in a very few nebulæ and comets. Solar spectroscopy ignores it.

Although the spectrum, both visual and photographic, of the last-named marvellous object has been diligently studied, our knowledge of it is still far from complete. It includes, over and above the normal three, a considerable number of lines, the relationships and conditions of appearance of which offer a tempting subject of investigation. In other nebulæ hydrogen attests its presence solely by its green ray. But F is here attended by G, h, H, and the two first of the ultra-violet series,[3] while the crimson C line remains imperceptible. The hydrogen, then, entering into the composition of the Orion nebula is at a different, perhaps a higher, stage of excitement than in the sun itself. It is associated with luminous helium. The D_3 line detected by Dr. Copeland, December 28, 1886, and subsequently measured by him thirty times,[4] had its identity confirmed by Mr. Taylor's observations.

The diffused light about the central group of rays in this object shows, Dr. Copeland remarks, 'some indications of resolvability into lines or bands;' and Dr. and Mrs. Huggins express the opinion that the faint continuous spectrum visible in most gaseous nebulæ might, were more light available, 'be found to consist, in great part at least, of closely adjacent bright lines.'[5]

[1] Copeland, *Copernicus*, vol. i. p. 2. The 'four-line' planetaries are numbered in Dreyer's New General Catalogue, 7662, 7026, and 7027. The two last belong to the singular group of allied gaseous objects in Cygnus.

[2] *Monthly Notices*, vol. xlix. p. 124.

[3] The lines h and H were photographed by Mr. Lockyer early in 1890, the entire series a little later by Dr. and Mrs. Huggins.

[4] *Monthly Notices*, vol. xlviii. p. 360.

[5] *Proc. R. Society*, vol. xlvi. p. 60.

Nearly all that is known as to the photographic spectrum of the Orion nebula has been elicited by the same eminent inquirers, and many features in their results are of extreme interest. A plate exposed in 1882 contained a strong ultra-violet ray of about wave-length 3725 (see fig. 1) which reappeared February 5, 1888, with a number of fainter companions as shown in the upper section of fig. 9. Now the light admitted on this occasion to the camera was obtained from the vivid central mass of the nebula, the slit in fact lying across two of the four bright stars grouped into the 'trapezium,' which forms apparently the core of the entire formation. The pair of stellar spectra thus included in the photograph are seen in the figure as two unbroken strips of light, crossed by the

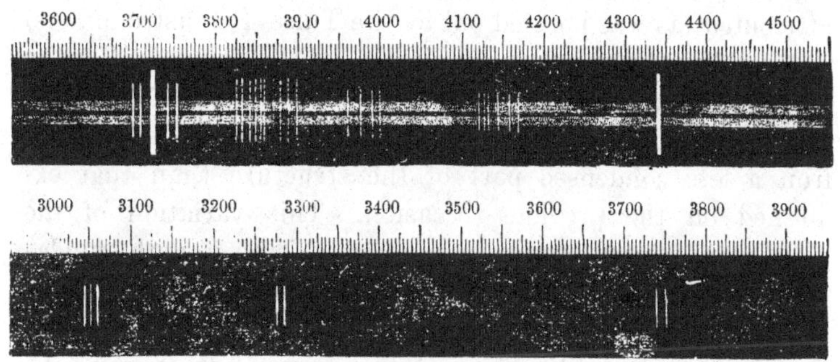

FIG. 9.—Photographic Spectrum of the Great Nebula in Orion (Huggins).

gaseous rays of the nebula. But not of the nebula alone. Three groups of thin and feeble lines belong, so far as it is possible to judge, primarily to the stars, though they are also emitted by the nebulous stuff in their immediate neighbourhood. The stars then, since an indefinite extent of circumjacent vaporous material partakes of the nature of their glowing atmospheres, are *in* the nebula, and not merely projected *upon* it.[1] They form an integral part of its structure.

Their gaseous nature was unsuspected until disclosed by this negative. No bright lines had been *seen* in their spectra. It remains to be decided whether the trapezium stars are to be regarded as *sui generis*, or whether they will prove to be the first known members of a new class. Several coincidences,

[1] *Proc. R. Society*, vol. xlvi. p. 41.

some approximate, some exact, between the ultra-violet rays emitted by them and those absorbed in the atmospheres of Arcturus and *a* Aquilæ (as photographed by Dr. Huggins in 1879) are apparent, and may serve as an index to their physical condition.[1] They, at any rate, suggest a practicable path for further investigation.

The next successful photograph of the nebular spectrum was taken a year later than its predecessor, February 28, 1889. Its curious unlikeness to it can be seen in fig. 9 (lower part), where the spectrum of 1888 survives only in the subordinate pair of lines on the extreme right (identical with those on the extreme left above); even the predominant ultra-violet ray at 3725 has vanished. The change can only be attributed (as was pointed out by the Tulse Hill astronomers) to local differences in the constitution of the nebula. For the slit lay in the second experiment at some distance from the trapezium, and the light it transmitted originated consequently from a less condensed part of the structure than that examined on the previous occasion. This variation of the photographic spectrum is the more surprising from the fundamental sameness of the visible spectrum of the nebula.

We learn from it the important lesson that the spectra of the great nebulæ, like those of the 'zoned' stars, must be considered as integrating the results of emanations taking their rise under notably diverse circumstances. Innumerable strata of nebulous matter are piled one upon the other in the same line of sight. The eye is impotent to discriminate between them; even the spectroscope can do so only indirectly. For, at the centre of the nebula, the lines coming from all its depths are seen or photographed together; their respective local origins are unnoticed. Light, on the other hand, taken from near the edges of the same object, emanates exclusively from its higher regions, and its characteristic peculiarities may safely be referred to that circumstance. The possibility, then, seems at hand of dividing in this way the Orion nebula and others of the same class into various spectroscopic levels, distinguished by minor emissive differences. The D_3 line, for

[1] *Observatory*, vol. xii. p. 367.

instance, may prove separable from the γ Argûs blue line, and this again from Dr. Copeland's line at 4476. As to the fundamental trio, they are doubtless common to all the heights and depths, nodes and convolutions, of gaseous nebulæ. The chemical composition of the much scrutinised one in Orion, so far as we yet know it, agrees, we cannot fail to remark, with that of the solar prominences; a result partial indeed, yet tolerably secure, even if somewhat difficult to reconcile with any hitherto admitted theory.

Only the species designated as stellar, planetary, annular, and irregular nebulæ, give unmistakable signs of gaseity. The rest, and they are the great majority, shine with continuous light; yet the distinction is perhaps less profound than it seems. The gap is at any rate partially bridged. One 'frontier instance' seems to be supplied by the great 'looped nebula' in the southern constellation of Dorado, observed by Mr. C. E. Burton[1] in 1874 to yield a strongly continuous spectrum crossed by the unfailing green nebular ray at 5005. Its gaseous nature is thus shown to be modified by the presence of an unusually large proportion of dense material, and where this predominates, as in the Andromeda nebula, the spectrum, though nominally 'continuous,' is still markedly different from the continuous spectrum of a star.

The light of this 'queen of the nebulæ,' prismatically dispersed for the first time by Dr. Huggins in 1864, was found to end abruptly in the orange; it included no red rays, nor was it anywhere uniform, but seemed *mottled* throughout, whether by the effects of absorption or of irregular emission, it was impossible to decide.[2] Recent observations leave the point still unsettled, the existence of various alleged bright lines or spaces needing further confirmation. In the investigation, nevertheless, of the distinctive peculiarities of 'continuous' nebular light, a field lies open which can hardly fail to be worked with profit to the rapidly advancing science of cosmical physics.

[1] *Monthly Notices*, vol. xxxvi. p. 69. [2] *Phil. Trans.* vol. cliv. p. 441.

CHAPTER VI.

SIDEREAL EVOLUTION.

ARE the stars subject to growth and decay? We might almost as well ask, Are they subject to the laws of nature? There can be in either case no doubt about the reply. We are perfectly assured, both from reason and revelation, that a time was when they were not, and that at some future date they will have ceased to be. And we may further confidently affirm, guided by the analogy of all other creative processes with which we are acquainted, that their present condition has been gradually attained and will gradually become modified.

Each has then a life-history. It is what it is, because it has been what it was. Nor is it conceivable that all should have arrived simultaneously at the same stage of development. A contemporaneous universal origin can by no means be assumed as a postulate; and even if it could, the rate of progress of individual stars must have been indefinitely varied. There is hence a strong probability that the present state of some represents the past of others, the future of many more. Among the hosts of heaven we may expect to find stars in embryo, stars half formed yet chaotic, full-grown stars in orderly and equable working order, stars still effective as radiators though of declining powers, and stars on the verge of decrepitude. Their comparative study ought then, under certain conditions, to enable us to compile, as it were, the typical biography of an average star.

This is a grand idea, but one not easy of legitimate realisation. The criteria of stellar 'age' or 'youth' are far from obvious, and hasty conclusions about them are more likely than not to prove illusory. There are certain principles, however, which we can hardly be mistaken in adopting as our guides. In all cosmical masses, a perennial contest is in progress between heat and gravity. Heat strives for expansion, gravity for contraction. And gravity must in all cases win in the end, for this simple reason that, whereas heat is continually wasted by radiation, gravity remains unalterably the same. Its action is (for what reason we cannot tell) without expenditure. Condensation then attends upon the efflux of time, and its degree, other things being equal, measures antiquity. One of its effects is to renew, up to a certain point, the thermal stores upon the gradual dissipation of which it ensues. For condensation implies a fall of each particle of the condensing body towards its centre, and all their arrested motions turn into heat as inevitably as percussion evokes sparks from a flint. Moreover, the production of heat by contraction exceeds the loss by radiation so long as the body in question remains purely gaseous, but begins to fall short of it with the approach of liquefaction. A gaseous mass condensing in space under the influence of its own internal forces will thus rise progressively in temperature until a change of its state is imminent, radiation going on for ages without *cooling*, because the loss of heat is more than compensated by the attendant transformation of gravitational energy.

The course of development of a star leads it then slowly but surely from a rare to a denser condition, its volume perpetually diminishing as it advances in life. Its changes of temperature obey a more complex law. They at first, and during countless millions of years, take an upward direction; at last a maximum is reached, the tide turns, and a decline sets in, unintermitted until the former sun, reduced to an arid rock, has sunk to the level of the unimaginable cold of space. It must not be forgotten, however, that all this, though true, may not be the whole truth. There are forces in nature powerfully influential, doubtless, in determining phases of development, as to the working of which we are almost wholly ignorant. What do we know, for instance, of the electrical state of gaseous nebulæ? That their shining depends upon it is forcibly suggested by the quality of their light; but as to the conditions prescribing electrical changes in cosmical bodies, we are without a clue. Only their effects can here and there be hesitatingly traced.

We must accordingly be prepared, in examining the heavens, to meet with apparent anomalies. All our specimens will not suit the labels we have prepared for them. Nor can it be supposed that all the stars are made on one pattern, and follow each other along the same rigid groove of change. There are to be found not only many different species of the heavenly bodies, but almost endless varieties of the same species. The universe is no mechanical workshop, turning out objects by the score in blind pursuance of one original intelligent arrangement, but a scene of continuous and exquisite adaptation of means to the most subtly various ends.

The task of settling the mutual relationships of the sidereal classes, ranging them in orderly sequence, and interpreting historically the physical links that unite them, is an arduous one. But at least on one point there is unanimity of opinion. The nebulous state preceded the stellar. The generation of stars from nebulæ was inferred by Sir William Herschel in 1811 and 1814 as the upshot of 'a critical examination of the nebulous system,' showing extreme instances from each of the sidereal kingdoms to be 'connected by such nearly allied intermediate steps as will make it highly probable that every succeeding state of the nebulous matter is the result of the action of gravitation upon it while in a foregoing one.'[1] This conclusion, based upon purely telescopic evidence, has been fully ratified by the spectroscope. In quality of light, as well as in general aspect, distinctions between stars and nebulæ are shaded off by numerous minute gradations. There is no real breach of continuity anywhere. The line-spectra of one division of nebulæ include continuous radiance; the continuous spectra of the other division possibly include bright lines. Gaseous stars take their rise almost insensibly from planetary nebulæ, and themselves merge into unmistakable suns. That the great nebulæ are the parent forms of stellar clusters is rendered highly probable by their common possession of

[1] *Phil. Trans.* vol. ci. p. 330.

luminous as well as structural peculiarities; nor can any definitive separation between them be established. The trapezium-stars in Orion, like crystals embedded in their rocky matrix, are still thickly folded in the generating cosmical stuff. By Dr. Huggins's photograph, they may be said to be 'caught in the act' of completing their transformation, a partial survival of the original community of gaseous nature being made apparent through their self-recorded bright lines. In the stars of the Pleiades a further stage of advance is exemplified. Nebulous appendages are, in their case, reduced to a subordinate position; the group is essentially a collection of most effulgent suns.

But when we come to the various classes of stars, the order of their succession is less easily determined. The earliest and most obvious idea on the subject was based on a false analogy between the colours of the stars and the colours of glowing terrestrial solids. Red stars, it was thought, should be regarded, because they had cooled from a condition of white heat, as *older* than white stars.[1] The colours of stars, however, depend primarily upon the quality and extent of their absorbing atmospheres, and quite secondarily upon their stage of incandescence. A more fruitful suggestion was made by Ångström in 1868.[2] It was that of seeking a test both of age and temperature in the chemical composition of stellar atmospheres. Mr. Lockyer acted upon it in his Bakerian Lecture in 1873,[3] and it formed the basis of Vogel's scheme of classification published in 1874.[4] Here the ancestral type of all other stars was found in those of the Sirian pattern of spectrum. Our sun was assumed to be a somewhat decayed and declining Vega; Betelgeux to represent a still further deteriorated luminary. The two varieties of banded spectra, however, marked two alternative routes to extinction. Between them a choice had, at some certain epoch, to be made, since no transition from one to the other was admitted. *Hic locus est, partes ubi se via findit in ambas.*

[1] Zöllner, *Photometrische Untersuchungen*, p. 243.
[2] *Recherches sur le Spectre Solaire*, p. 38.
[3] *Phil. Trans.* vol. clxiv. p. 492.
[4] *Astr. Nach.* No. 2000.

The objection, however, could not fail to present itself that, in this arrangement, growing suns had no place.[1] Ignoring the rising branch of the temperature-curve traversed by all condensing and radiating bodies, it regarded only the latter half of stellar careers. From the acme of splendour it traced them downward to the dimness of impending incrustation, but gave no account of the stages by which the acme of splendour was reached. Yet the nebulous affinities, of late photographically ascertained, of certain white stars, give so strong, if only a partial support to Vogel's scheme, as to preclude its unqualified rejection. What had seemed its essential deficiency was supplied by a totally different arrangement proposed by Mr. Lockyer in the Bakerian Lecture of 1888.[2]

This remarkable work might be described as a project of unification of all the 'celestial species'—suns, planets, and comets, stars 'of sorts,' and nebulæ. All are exhibited in it as the outcome of the aggregation in space of stony and metallic fragments, some specimens of which the earth yearly intercepts and captures in its orbital course, and delivers over to the inquisitive scrutiny of *savants*. Hence the possibility of producing such evidence as that with which Mr. Lockyer supported his views. It consisted in a number of apparent spectroscopic coincidences between characteristic stellar and nebular rays, brought out by causing vaporised meteoritic particles to glow by electricity in the laboratory. But the coincidences professed to be no more than approximate, and were accordingly admissible as suggestions of great interest, but not as demonstrations.

The general conclusions arrived at may be thus summarised.[3] All self-luminous bodies in space are composed of meteorites variously aggregated and at various stages of temperature depending upon the frequency and violence of their mutual collisions. Comets, nebulæ, gaseous stars, and stars showing banded spectra of the third type, are veritable meteor-swarms—they are made up, that is to say, of an indefinite multitude of separate, and, in a sense, independent solid

[1] *Nature*, vol. xxxiii. p. 585. [2] *Proc. R. Society*, vol. xliv. p. 1.

[3] *Observatory*, vol. xi. p. 84.

bodies, bathed in evolved gases, and glowing with the heat due to their arrested motions. Their component meteorites, however, eventually become completely vaporised, when stars of the solar and Sirian types are produced. These—the only true 'suns'—owe their high temperatures to the surrendered velocities of the original myriads of jostling particles drawn together to constitute them by the victorious power of gravity.

To the meteor-swarms thus serving as the basis of stellar transformations, a collective origin, and a collective history, are necessarily attributed. The theory before us postulates a 'curdling process,' by which from some unimaginably subtle kind of matter, an 'infinitely fine' metallic dust is formed, and this infinitely fine dust 'becomes at last, in the celestial spaces, agglomerated into meteoric irons and stones.'[1] The 'eddying' of these round self-constituted centres gives the meteor-swarms supposed to represent the protoplasm of stars.

But their members, so far as the specimens ranged on the shelves of our museums enable us to judge, by no means display the primitive character which, on this hypothesis, they ought to possess. They are, on the contrary, minerals of highly complex constitution, the products, apparently, of lengthy and intricate processes such as have been going on for ages in the bowels of the earth.[2] Among them are found (with certain minor, yet characteristic differences) genuine breccias, serpentines, and lavas; a diamond-bearing meteorite recently fell in Siberia;[3] while in the Deesa meteorite we have a splinter from a vein of iron injected, it would appear, into a previously existing rock on some unknown planetary globe.[4] The geology illustrated by these samples from distant rock-factories, so strangely dropped out of the clouds at our feet, belongs indeed to a pre-aqueous epoch;[5] they were probably formed in the presence of free hydrogen. But we need not here stop to discuss the obscure question of their origin. What immediately concerns us is the strong presumption

[1] *Nineteenth Century*, Nov. 1889, p. 787.

[2] Stanislas Meunier, *Comptes Rendus*, t. cv. pp. 1038, 1095; t. cvii. p. 834.

[3] *Ibid.* t. cvi. p. 1678.

[4] Meunier, *Géologie Comparée*, p. 83.

[5] Félix Hément, *Les Étoiles Filantes et les Météorites*, p. 96, 1888.

afforded by their structure that they are either débris or ejecta, and were *not* separately formed in space. Nor can processes of vaporisation through mutual impacts followed by re-crystallisations, be brought in to explain the complexities of their constitution. For this reason, if for no other, that since heat is only produced at the expense of motion, vapours expelled as the result of collisions could not condense into *circulating* members of the system, but would slowly fall to its centre. Individual meteorites, which by the hypothesis subsist only by virtue of their circulatory movements, could not then be formed by them.

Bodies of this class doubtless form a link in the cosmical chain, and are united by relations of great consequence with larger cosmical masses, but data are still wanting for the satisfactory determination of what those precise relations are. Spectroscopic agreements are not decisive; to a certain extent they are, indeed, inevitable, through the fundamental unity underlying (there is reason to suppose) the chemistry of all the celestial species. About thirty terrestrial elements, for instance, occur in meteorites; and many, if not all of these, are, we may be sure, in one way or another universally diffused in nature.

Mr. Lockyer's classification of the heavenly bodies may, however, be considered independently of his meteoric hypothesis. To a great extent it stands on its own footing; and it well deserves thinking about. Essentially an evolutionary scheme, it is the first of the kind with any pretension to completeness. Stellar destinies are traced in it, so to speak, from the cradle to the grave. The whole history is placed before us, of how, by the ceaseless advance of condensation, nebulæ are transformed, first into gaseous stars, then into stars with banded spectra of the type of Betelgeux, from which solar stars, and from these again Sirian stars, gradually emerge. Here the ascent ends; the maximum of temperature is reached, and a descent begins, the initial stage of which is marked by a second group of objects like our sun and Capella, distinguished from the first by the circumstance that they are losing, instead of gaining heat; while lower still the condition immediately antecedent to solidification and obscurity is represented by Father Secchi's 'carbon-stars.'

The notion that red stars are more aged—nearer to their end as suns—than white, has little to recommend it. Mr. Lockyer's principle that condensation measures age is really fatal to it. For stars with banded spectra, of both kinds, are almost demonstrably less condensed bodies than stars of which the light is interrupted by linear absorption. Those of the third and fourth types alike possess complex spectra originating at various heat-levels in glowing atmospheres, which permit the escape of but little of their photospheric radiance, and that chiefly of the less refrangible qualities. They are red, not necessarily in themselves, but because we see them as if through a shade of tinted glass. Each type of red star moreover furnishes examples of the intermittently hazy appearance earlier alluded to; and each shows an equal tendency to wide fluctuations of light, connected by a reasonable surmise with the vast compass of their gaseous surroundings. They cannot then, with any probability, be set far apart in the developmental series. The place near its close assigned by Mr. Lockyer to fourth-type stars, is indeed avowedly only provisional.

There are many indications that these objects are not only in an elementary phase of growth, but of exceptional character. The peculiarities of their distribution separate them from ordinary stars; their spectroscopic isolation leaves us without the means of tracing their relationships. With comets, and with at least one gaseous star, they have indeed a close affinity. Thus the chief distinction between 19 Piscium and R Geminorum would be abolished by heightening the atmospheric incandescence of the one, or the photospheric incandescence of the other body, so as to turn dark bands to bright in the first case, bright bands to dark in the second.

But if the fourth stellar type must for the present be looked upon as a sort of *cul de sac*, a main road obviously puts the third in communication with the second type, and the second with the first. The transition from each to the next is by such easy steps that it is impossible to say precisely where the boundary is crossed. They are moreover invariably of the same tenour. As the fluted shadings exemplified by Betelgeux die out, they uniformly leave behind them a line-spectrum of the solar, never one of the Sirian pattern. Solar stars then exhibit the transition from such stars as Betelgeux to such stars as Sirius and Vega. Their intermediate position is unmistakable, and has long been recognised. The only difference of opinion is as to the direction of progress.

On this point the consideration of relative atmospheric extent may serve as a guide. The vaporous envelopes of glowing globes almost necessarily become less voluminous with time. As cooling progresses they slowly subside, and eventually all but disappear. Our own earth, when its temperature was above 212°, kept its oceans in the form of dry water-gas suspended above its surface; and that the steadily advancing process of atmospheric attenuation has not reached its term, we have only to look at the moon to feel assured. Star spectra slightly impressed by absorption ought then (so far as can be determined from this single point of view) to rank as posterior to those strongly so impressed.

This involves the question of the sun's standing. How far has its development proceeded? What will be its next stage? What place should be assigned to it in the grand temporal procession of the skies? Categorical answers to these questions cannot, of course, be given; but we can surmise that at some indefinitely remote epoch the sun was a red star with a banded spectrum like that of Betelgeux; that after many ages it came to resemble Aldebaran; and that as the veil thrown over its violet light grew thinner, its present mellowed splendour was at last attained. But there is a future to be considered as well as the past. The solar transformations are certainly not terminated. Will they include a 'white star' phase? It seems possible that they may, since the intense electrical excitement by which the sun's eclipse-appendages are now maintained will presumably relax; with results perhaps in the subsidence of metallic vapours, and the concentration of hydrogen-strata, issuing in the production of the characteristic spectrum of Vega or of Castor.

We know of no reason why the sun should not be growing hotter instead of cooler. There is not a shadow of proof that the earth received from it, in past geological ages, more heat and light than it does now; and whatever changes may be in progress are beyond doubt immeasurably too slow for detection within the historical period. Mr. Lockyer, it is true, feels the difficulty to be insuperable of attributing to the sun a prospective condition analogous to that now prevailing in Sirius; and he accordingly places our luminary with Capella, Arcturus, and some other selected solar stars, on the descending branch of his curve, while a corresponding post on the ascending branch is occupied by the rest of their compeers. It is, however, scarcely conceivable that a state abolished as an effect of condensation should be restored by its further progress.

Under what spectral category then are stars past their prime to be disposed? The phenomena of double stars may help us to a reply. The comparative development of members of binary systems was first systematically considered by Mr. Lockyer in a suggestive inquiry as to their origin.[1] Since the capture of one star by another is virtually impossible, it may be assumed that coupled objects are products of a single nebula, that they are related by especially close affinities, physical and chemical, and that they have grown up side by side under identical conditions. Inequality of mass constitutes the one certain difference between them. But inequality of mass involves an unequal rate of development, since a small star must, it would seem, run through its changes more rapidly than a large one. We have only then to compare the spectrum of a satellite-star with that of its primary in order to learn the chronological succession of the types they respectively present. But this presupposes a knowledge of their relative masses far from easy to be obtained. Light measurements are indecisive; the less massive may, during æons of time, be the brighter object; and other methods are usually

[1] *Proc. R. Society*, vol. xlv. p. 250.

arduous, if not impracticable. In only one revolving system, indeed, has mass been as yet satisfactorily apportioned. It is that of Sirius, the faint companion of which contains just half its amount of matter. Here then we have a star which ought to be running down hill, and which, judging from its extremely feeble luminosity, actually has advanced far towards extinction.

The spectrum of the Sirian satellite may hence with confidence be taken as typical of the spectra of waning stars. But here again, observations are wanting, nor can they be procured during the present close conjunction of the pair. There is a strong probability, however, that the light of the dim component will prove, on analysis, to be of an undistinguished character, interrupted neither by bands nor conspicuous dark lines, and feeble, not through effects of absorption, but intrinsically. The same dull uniformity may be expected to belong to the spectra of all stars of impaired splendour; but on this subject we shall know more when some progress has been made in determining the relative masses and spectra of binary stars.

To resume. The line of stellar evolution indicated by recent inquiries is from red stars with banded spectra through yellow stars with metallic line-spectra to white stars distinguished by almost exclusive hydrogen-absorption. In these the accentuation of the photosphere as a radiating surface culminates, and its temperature reaches a maximum. Eventually, however, the heat of contraction begins to fall short of the heat sent abroad through space, supply ceases to cover expenditure, and definitive cooling sets in. Stars in this condition *probably* emit light especially feeble in the upper prismatic reaches, and give spectra of scarcely determinable type. They may plausibly be regarded as doomed to complete extinction.

Returning to the opposite end of the scale of stellar being, we find its elements universally in nebulæ of one kind or another. Yet the steps by which stars with banded spectra have taken form out of a primitive diffuse substance, remain largely hypothetical. On the other hand, certain groups of white stars appear to have arisen directly from parent nebulæ, traversing probably no preliminary phases save that marked by a bright line spectrum. Such are the stars in Orion and the Pleiades. The latter, though evidently more fully developed than many of the former, have scarcely yet, if we may say so, been turned out of the workshop, where scraps and shavings of the material used in their construction cling to them and strew surrounding space. These singular relations are not, it must be acknowledged, easily reconciled with the ideas we had, on other grounds, been led to form as to the status of red stars. They warn us to suspend our conclusions. In the mean time, it is permissible to remark that the whole stellar army are not bound to advance along the same road. their individual constitution includes much that is inscrutable to us, nor can it be identical for all.

We should expect to find the prevalence of the several stellar types a rough measure of their durability, and their durability determined by the emissive intensity characterising them. The longest stage ought, on this view, to be the most numerously represented, and that stage ought to be the longest in which cooling and the changes consequent upon cooling make the slowest progress. But this anticipation is not realised. The red stars, although their state should be relatively permanent owing to the feebleness of their radiation, are few compared with Sirian stars, the reckless profusion of which in the dissipation of energy seems premonitory of an imminent decline; while solar stars occupy in both respects an intermediate position. A rapidly increasing series is in fact substituted for the looked for rapidly diminishing one. The inference suggested is that the numbers of objects simultaneously belonging to the various spectral classes is a question of epoch. The world is not now as it always was. There has been a beginning, there will be an end. Why should we not admit that between the beginning and the end progress may be measured by definable landmarks? We have reached a time when the majority of the stars are at their highest lustre; there may have been a time when they were mostly red and variable, and there will perhaps be a time when only the red stars of to-day still shine in the sky, and the Sirian stars of to-day have sunk into obscurity.

For the best part of a century the nebular hypothesis, stellar and planetary, satisfied scientific curiosity as to cosmical origins. It has, however, ceased to do so. The consciousness has become more and more insistent that it is not enough to refer stars to nebulæ, while nebulæ themselves remain unaccounted for. The need presses 'to explain our explanation.' 'Pre-nebular' theories have accordingly come to be of the order of the day. One such is afforded by Mr. Lockyer's meteoric views; another, due to Dr. Croll,[1] supposes space primitively occupied by cold dark masses moving with great but random velocities, and producing very hot nebulæ by their collisions. But it may be remarked, in the first place, that the odds against the occurrence of any such collisions are literally 'beyond arithmetic'; in the next, that the requisite stock of heat might just as well be created directly in the form of molecular motion, as indirectly in the form of molar motion. Such efforts to get nearer to an absolute beginning illustrate the incapacity of the human mind to rest finally in any purely material conception. It wanders vainly from theory to theory, piling Ossa upon Pelion in the shape of hypotheses, vainly hoping that, with the help of the last and latest, the empyrean may at length be scaled. Harmony can only be established between its aspirations and the outer show of the world, and science can only become truly rational, when the fount of all things is reached in an Intelligence akin to, yet infinitely transcending its own.

[1] *Stellar Evolution and its Relations to Geological Time.* London: 1889.

CHAPTER VII.

TEMPORARY STARS.

The facts connected with the light-changes of stars are in the highest degree strange and surprising; and wonder is not lessened by our daily-growing familiarity with them. They are of everyday occurrence, they can be predicted beforehand, in many cases with nearly as close accuracy as an eclipse of the sun or moon, and they affect in manifold ways a great number of objects. Stellar variability is of every kind and degree. With the regularity of clockwork some stars lose and regain a fixed proportion of their light; others show fitful accessions of luminosity succeeded by equally fitful relapses into obscurity; many waver, in appearance lawlessly, about a datum-level of lustre itself perhaps slowly rising or sinking. The rule of change of a great number is that of an evident, though strongly disturbed periodicity; a few seem to spend all their powers of shining in one amazing outburst, after which they return to their pristine invisibility or insignificance.

The amount is as much diversified as the manner of fluctuation. Changes of brightness so minute as almost to defy detection are linked on by a succession of graduated examples to conflagrations in which emissive intensity is multiplied a thousand times or more in a few hours. The range of variation is in some stars sensibly uniform; they subside during each crisis of change to the same precise point of dimness, and recover, without diminution or excess, just so much light as they had before. In others it is widely irregular. The limits of fluctuation in one period furnish no precedent to be conformed to in the next. Nothing is predetermined; the intensity of each phase seems to depend upon a complex set of conditions unlikely to recur twice in the same precise combination.

The first effort to rationalise the phenomena of variable stars was made by Professor E. C. Pickering in 1880.[1] His five classes, though often enough (as might be expected) confused at the borders, are still sufficiently distinct to form a useful framework for the facts. They are as follows: Class I. includes temporary or 'new' stars; Class II., stars like 'Mira' Ceti, strikingly variable in periods of several months; Class III., stars showing slight and irregular fluctuations; Class IV., variables with periods of a few days exemplified by δ Cephei and β Lyræ; Class V., 'Algol-variables,' or stars resembling Algol in Perseus in rapidly losing and regaining a determinate amount of light at intervals measured by hours. We will take each in turn, beginning with the first.

A temporary star may be defined as a variable attaining one brief, vivid maximum. An extraordinarily swift rise to such an extent as to constitute a virtually 'new' object, followed by a slower yet prompt decline, characterise these outbursts, above a score of which have been more or less credibly recorded within historical times. Those connected with the following years[2] are some of them probably, most of them certainly, genuine.

134 B.C. in Scorpio; the star of Hipparchus.

123 A.D. in Ophiuchus.

Dec. 10, 173, between α and β Centauri. Conspicuous; scintillated strongly; visible eight months.

386 (April to July), between λ and ϕ Sagittarii.

389, near α Aquilæ, said by Cuspinianus to have equalled Venus; vanished after three weeks.

March 393, in the Tail of the Scorpion.

827 (?), in Scorpio. Observed during four months at Babylon. There is some uncertainty about the date, none about the fact.

May 1012, in Aries. Described by Epidamnus, the monk of St. Gall, as 'oculos verberans.'

[1] *Proceedings Amer. Acad.* vol. xvi. p. 17.

[2] See Humboldt's *Cosmos*. vol. iii. p. 209 (Otté's translation).

July 1203, in the Tail of the Scorpion, said to have been in aspect like Saturn.

1230, in Ophiuchus.

1572, Tycho's star in Cassiopeia.

1604, Kepler's star in Ophiuchus.

1670, in Vulpecula.

1848, in Ophiuchus.

1860, in Scorpio.

1866, in Corona Borealis.

1876, in Cygnus.

1885, in Andromeda. Making eighteen in all, besides four or five questionable instances mentioned in Chinese annals.

The most noteworthy feature of this list is the curiously partial distribution of the objects enumerated in it. Nearly all of them lie in the thoroughfare of the Milky Way, and half are clustered together in the section of it marked by the stars of the Scorpion and the Serpent-tamer. In time also the grouping of the apparitions is strikingly unequal. The occurrence of three within the seven years 386 to 393 A.D. was succeeded by a blank of four and a half centuries. Kepler's came pretty close upon Tycho's star; and no less than five 'Novæ' blazed out during thirty-seven recent years, while the latest previous record of the kind was of Anthelm's star in 1670.

The brightest sidereal object known to us by authentic description was the 'stranger-star' in Cassiopeia, observed by Tycho Brahe. He first saw it November 11, 1572, but it had already been noticed by Lindauer at Winterthür, November 7, and Maurolycus entered upon its systematic study at Messina November 8. From equality with Jupiter it rose in a few days to be the rival of Venus, showing to keen eyes at midday, and at night through clouds thick enough to obscure every other star. After about three weeks, however, it began to fade, and in March 1574 disappeared finally. Its colour was at first dazzlingly white, then for a while ruddy, and from May 1573 onward, pale with a livid cast. Rapid scintillation distin-

[1] Wolf, *Geschichte der Astronomie*, p. 414; Kaiser, *De Sterrenhemel*, Part i. p. 582.

guished it throughout.[1] There is no reason to suppose its outburst other than solitary. The appearances in the years 945 and 1264 connected with it by a Bohemian astrologer named Cyprian Leowitz,[2] were almost certainly apocryphal.[3]

The 'new' star (designated 'B Cassiopeiæ') can still be perceived smouldering in the spot where it once blazed. Tycho's measurements, reduced and discussed by Argelander, located it within one minute of arc of a reddish, eleventh-magnitude star, first noticed by d'Arrest in 1865, the character of which, as disclosed by the observations of Hind and Plummer in 1870–4,[4] fully warrants the inference of its identity with the famous 'temporary.' Not only is it variable to the extent of nearly a magnitude, but it frequently seems nebulous, with occasional lurid flashes of momentarily increased brightness. Its non-appearance in a photograph taken by Mr. Roberts January 12, 1890, showing above 400 stars where d'Arrest charted 212,[5] may be due to the actinic feebleness of its light.

The star of 1604 ran a parallel course to that of 1572. Discovered by John Brunowski October 10, it quickly overtopped Jupiter, but by the end of March 1605 had sunk to the third magnitude, and a year later vanished. Kepler describes it as sparkling like a diamond with prismatic tints,'[6] but says nothing of progressive changes of colour. 'Nova Serpentarii' has left behind no clearly identifiable representative.

The next 'new star' was discovered near β Cygni on June 20, 1670, by Anthelmus, a Carthusian monk at Dijon. It was then of the third magnitude, but its decline, unlike that of others of its class, was interrupted by two reappearances separated by intervals of invisibility. Between March and May 1671, it rose from the fourth to the third rank, then died out, only flickering up to the sixth magnitude in March 1672.[7] Almost exactly in its assigned position, Mr. Hind picked up, April 24, 1852, a star between the tenth and eleventh magni-

[1] Tycho, *De Novâ Stellâ anni* 1572, p. 302.
[2] *Judicium de Novâ Stellâ.*
[3] Lynn, *Observatory*, vol. vi. pp. 126, 151; Sadler, *English Mechanic*, vol. xxx. p. 402; Tycho Brahe, *Progymnasmata*, p. 331.
[4] *Monthly Notices*, vol. xxxiv. p. 168; Gore's *Catalogue of known Variables*, p. 164.
[5] *Monthly Notices*, vol. l. p. 359.
[6] Kepleri *Opera*, t. ii. p. 620.
[7] J. Cassini, *Eléments d'Astronomie*, p. 69.

tude, which, when reobserved in 1861, had lost more than half its light, and gave the blurred image characteristic of many superannuated 'Novæ.'[1] The triple maximum of Anthelm's star assimilates it to Janson's variable P Cygni,[2] which has itself often been classed as a Nova.

Undoubtedly such was an object detected by Mr. Hind in Ophiuchus, April 28, 1848, when it was of 6·7 magnitude, and intensely reddish yellow.[3] Four days later it had mounted above the fifth magnitude, from which eminence it slowly descended, making no lasting halt until, in 1874–5, it had got down to the thirteenth magnitude.[4]

With the spectroscopic study of temporary stars, a fresh chapter in our knowledge of them opened. Through the magic of the prism, more was ascertained as to their essential nature in five minutes than could have been learned in as many centuries with the telescope alone. On May 12, 1866, Mr. John Birmingham, of Millbrook, near Tuam in Ireland, was amazed to perceive an unfamiliar star of the second magnitude shining in the constellation of the Northern Crown. On May 16, the application of Dr. Huggins's spectroscope showed the object to be wrapt in a mantle of blazing hydrogen. Five bright lines (three of them due to hydrogen) stood out from a range of continuous light broken up into zones by flutings of strong absorption.[5] The incandescence of the star was hence largely atmospheric, and for the rest, from the rapid rate at which it fell away, could have been only 'skin-deep.' That the compound nature of its spectrum testified truly to an immense diffusion of vaporous material in its neighbourhood was certified by Dr. Huggins's visual observation of a singular glow round the star on May 16 and 17. Although its light decreased by a daily half magnitude, and its colour changed from white to orange, no alteration took place in the character of the spectrum. The bright rays, however, faded somewhat less promptly than the continuous light.

[1] *Monthly Notices*, vol. xxi. p. 231; *Nature*, vol. xxxii. p. 355. The star is No. 1814 in the Greenwich Catalogue for 1872.

[2] See *ante*, p. 70.

[3] *Astr. Nach.* Nos. 636, 638, 672.

[4] *Monthly Notices*, vol. xxi. p. 232.

[5] *Proceedings Royal Society*, vol. xv. p. 146.

The visibility of the object to the naked eye lasted only eight days, and already, in the beginning of June, it had sunk to the ninth magnitude. Its slow subsequent decline was interrupted by fluctuations, thought by Schmidt to be periodical in about ninety-four days.[1] When observed by Vogel, March 28, 1878, it was of the tenth magnitude, and gave an ordinary stellar spectrum.[2] Virtually, it had resumed the conditions of its existence when Schönfeld entered it as of 9·5 magnitude in the 'Bonn Durchmusterung.' Its leap upward to the second magnitude, involving a *thousand-fold* gain of light, was accomplished with extraordinary suddenness. Two hours and a half previously to Birmingham's discovery, Schmidt surveyed at Athens the constellation in which the blaze was about to occur, and noticed nothing unusual. He was certain that the star could not then have been as bright as the fifth magnitude.

The name of 'T Coronæ' was bestowed upon it in conformity with Argelander's system of nomenclature, by which the variables in each constellation are designated, in the order of their discovery, by the Roman capital letters from R onward. Only stars otherwise anonymous, however, are included in the distinctive series thus created, so that many variables are still entitled in the ordinary way by Greek letters.

The stellar apparition that ensued after ten years was, in some of its features, the most remarkable of all. Dr. Schmidt noticed at Athens, November 24, 1876, a star of the third magnitude near ρ Cygni, in a spot till then vacant, so far as recorded observations went. The weather having been cloudy during the previous four days, there was no possibility of tracing the steps of its ascent, but it ran down very rapidly, and ceased to be visible to the naked eye on December 15. Its changes of colour pursued an inverse order to those of its predecessor. From golden yellow it turned white, and eventually bluish.

The earliest spectroscopic examination of 'Nova Cygni' was made by M. Cornu, at Paris, December 2 and 4.[3] Just

[1] *Astr. Nach.* No. 2118. [2] *Monatsberichte*, Berlin, 1878, p. 304.
[3] *Comptes Rendus*, t. lxxxiii. p. 1172.

the same range of bright lines was measured by him which would start into view in the solar spectrum upon a considerable

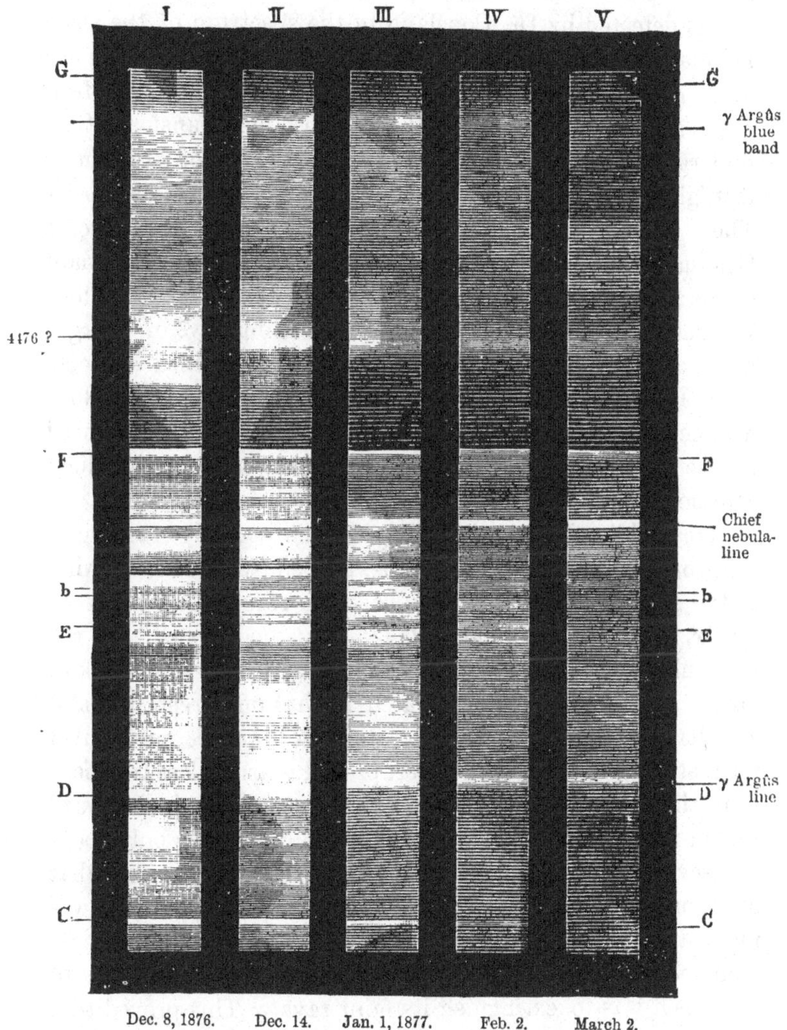

FIG. 10.—Changes in the Spectrum of Nova Cygni (Vogel).

augmentation of incandescence in the sun's gaseous surroundings. Besides three, if not four hydrogen lines, there were the green coronal ray (wave length 5316), the yellow helium ray (wave length 5875), and the magnesium group *b*. With these were very curiously associated the fundamental nebula-line at 5005, and possibly a line at 4476, long afterwards detected by Dr. Copeland in the spectrum of the Orion nebula. The changes affecting the light emissions of the new star between December and March, shown in fig. 10, from Vogel's drawings, indicate the remarkable transformations undergone by the fading object.[1] At the height of its outburst it might be called of the type of β Lyræ and γ Cassiopeiæ. The C of hydrogen was vivid; the continuous spectrum strong. Gradually C yielded its supremacy to F; only a faded remnant of the general prismatic light survived in the yellow and blue; D_3 (helium) vanished;[2] and the two leading rays of γ Argûs made an unexpected appearance. First determined by Copeland, January 2,[3] they became more prominent as the Nova declined, the likeness to the Wolf-Rayet group being enhanced by the presence of the intense absorption-band in the blue,[4] common to them all.

Meanwhile the nebula-line had been steadily creeping to the front; and when observations, suspended in March owing to the encroachments of daylight, were resumed at Dunecht by Dr. Copeland, September 2, 1877, it stood alone.[5] All the remaining light of the object (which had by that time sunk to 10·5 magnitude) was concentrated in that one green ray, and a stellar or planetary nebula was, to all appearance, substituted for a star. This phase, however, proved scarcely less transient than the rest. So far as could be made out with the Harvard fifteen-inch refractor in 1880 (when the Nova gave only about one-seventh as much light as in 1877), the spectrum was that of an ordinary star.[6] A confirmatory *negative* observation was made at Dunecht, February 1, 1881. Nothing could then be seen spectroscopically of the nondescript object; the effect of the prism was to extinguish its faint rays. This proved that it dispersed them—in other words, that they were of various

[1] See Lockyer, *Proc. R. Society*, vol. xliii. p. 139.
[2] It was last seen by Copeland, Jan. 9, *Copernicus*, vol. ii. p. 111.
[3] *Ibid.* pp. 102, 112.
[4] *Monatsberichte*, Berlin, 1877, p. 255 (Vogel).
[5] *Copernicus*, vol. ii. p. 106.
[6] *Annual Report*, 1879–80, p. 7.

refrangibilities. For had they been monochromatic, the image that they gave would have remained intact. The latest observations of Nova Cygni were made by Mr. J. G. Lohse in 1885–6, with the 15½-inch refractor of Mr. Wigglesworth's observatory near Scarborough. They showed it as of fourteenth or fifteenth magnitude, bluish in colour and nebulous in aspect. No spectrum of any kind could be obtained from it.[1]

The steps by which the famous 'temporary' of 1876 descended toward approximate extinction are the same by which, in Mr. Lockyer's scheme of development, cosmical masses ascend from a nebulous towards a truly sun-like condition. It is indeed unproved, and perhaps unprovable, that these remarkable migrations are repeated, in an inverse order, as a normal result of condensation; but it is scarcely doubtful that their relations of temperature have at any rate been rightly assigned. The Nova was certainly at its hottest when the emissions of helium were conspicuous in its spectrum; the 'Swan-star' lines emerged upon a considerable advance in cooling, a further stage of which was marked by the disappearance of hydrogen, and the solitary survival of the nebula-line. This, be it remarked, was visible from the first, so it seems to stand all temperatures. Nor was its display in the midst of a cosmic conflagration wholly without precedent. The spectrum of T Coronæ, there is reason to believe, included it,[2] as well as the ill-defined blue band distinctive of stars on the spectral pattern of γ Argûs.

The two outbreaks were then of essentially the same character, notwithstanding some variety in the phenomena ensuing upon them. The difference may perhaps be explained by the precipitous nature of the descent of T Coronæ. The orderly sequence of changes undergone by Nova Cygni did not get time to develop in the earlier object. All qualities of light radiated by it faded with nearly equal rapidity.

The Novæ of 1866 and 1876 were not more clearly set apart together by the identical peculiarities of their light,

[1] *Monthly Notices*, vol. xlvii. p. 494.

[2] See Vogel's reductions (*Monatsberichte*, Berlin, 1877, p. 242) of Stone's and Carpenter's measures (*Monthly Notices*, vol. xxvi. p. 295).

than the pair we are now about to describe by the extraordinary circumstances of their situation. On May 18, 1860, a nebula in Scorpio, numbered 80 on Messier's list (6093 in Dreyer's New General Catalogue), was observed by Dr. Auwers at Berlin.[1] It presented its usual appearance of a somewhat hazy ball of light, brightening gradually inward, and resolvable with difficulty into separate stellar points, together constituting a closely-packed, and most likely excessively remote globular cluster. Three nights later he looked again, and saw that these minnows had a triton in their midst. A seventh magnitude star shone close to the centre of the stellar group. The existence of the new-comer lasted visibly just three weeks. Before May 25 a decline set in; it had made considerable progress when, on May 28, Mr. Pogson (uninformed of Auwers's discovery) was 'startled' by the apparent *substitution* of a star for the nebula,[2] the dim luminosity of which seemed actually obliterated by the keen stellar radiance emanating from within it. It recovered, however, very speedily from this merely optical effacement. On June 10 its normal aspect was almost restored, and has never since been disturbed.

But after the lapse of a quarter of a century, the significance of this event was accentuated by the occurrence of a similar one elsewhere. This time the great nebula in the girdle of Andromeda was the scene of the outbreak. The unlooked-for addition to it of a 'star-like nucleus' was announced by Dr. Hartwig at Dorpat, August 31, 1885; but it turned out that the change had already been perceived by Mr. Isaac W. Ward of Belfast, August 19, and two nights earlier at Rouen by M. Ludovic Gully, who, however, set it down as an effect of bad definition.[3] Concordant observations by Tempel at Florence, Max Wolf at Heidelberg, and Engelmann at Leipzig showed decisively that the strange object made no show down to 10 P.M. on August 16;[4] and a photograph taken by Mr. Common in August 1884 gave positive assurance that its place had then no stellar occupant as bright

[1] *Astr. Nach.* No. 1267.
[2] *Monthly Notices*, vol. xxi. p. 32.
[3] *Ciel et Terre*, Oct. 1, 1885.
[4] *Astr. Nach.* Nos. 2682, 2683, 2691.

as the fifteenth magnitude.[1] What were virtually the first rays of the Nova reached the earth August 16, 1885.

Between that date and August 31, it mounted from the ninth to the seventh magnitude; then without delay entered upon nearly as swift a downward course, checked, however, by one decided pause. Even the largest telescopes failed to keep it in view after March 1886. The full yellow colour, by which the star at first contrasted effectively with the greenish nebular background it was projected upon, faded with its light. No haze or glow blurred its image, which remained sharply stellar with a power of 1100 on the great Princeton refractor, when the adjacent nucleus of the nebula melted into a confused luminous blot.[2] Attempts, incomplete from the nature of the case, made by Dr. Franz at Königsberg, and by Professor Hall at Washington, to determine the parallax of Nova Andromedæ, gave only negative results.[3] So far as they were significant at all, they indicated its immeasurable remoteness from the earth; nor should it be overlooked that Sir Robert Ball's similar experiment upon Nova Cygni had suggested a similar conclusion.[4]

The spectrum of Nova Andromedæ was of a dubious character. It bore witness to a completely different order of incandescence from that of the 'blaze stars' in the Northern Crown and the Swan. The bright rays which it probably included were inconspicuous. Dr. Huggins, on September 9, was nearly sure of the presence of several in the green and yellow;[5] and Dr. Copeland succeeded with difficulty, on September 30, in getting rough measures of three accessions of light,[6] one of them near the place both of the chief carbon-fluting (at 5164), and of a 'maximum' measured by Mr. Taylor, long after the disappearance of the star, in the spectrum of the Andromeda nebula. Resemblance in quality of light was indeed one of many arguments proving the physical relationship of the two objects. This was, however, superfluously

[1] *Nature*, vol. xxxii. p. 522. [2] Young, *Sidereal Messenger*, vol. iv. p. 282.
[3] *Astr. Nach.* 2816. [4] *Dunsink Observations*, Part V. p. 24.
[5] *Report Brit. Association*, 1885, p. 935.
[6] *Monthly Notices*, vol. xlvii. p. 54.

certain. It is barely conceivable that *one* stellar conflagration should by chance be projected almost accurately upon the core of a nebula in reality quite disconnected from it; but that *two* such highly improbable events should occur within twenty-five years of each other may fairly be called impossible. The Novæ of 1860 and 1885 were then each situated within the substance of the nebulæ they temporarily illuminated.

This collocation obviously falls into line with the galactic affinities of other temporary stars. All of them, except the apparitions of 1012 in Aries and of 1866 in Corona, were Milky Way objects. Now the Milky Way is a plane of condensation for all small stars, but more especially, and in a marked degree, for stars as well as nebulæ of a gaseous nature. Temporary stars are closely cognate with these, not merely through the brief gaseous incandescence bringing them to our notice, but through the symptoms of nebulosity which survive it. They are not ordinary, full-grown suns overwhelmed by some sudden catastrophe. The catastrophe, if determined by some external event, is prepared and rendered possible by their own peculiarities of constitution. It is true that a sun like ours might be led, in the course of its onward sweeping through space, to traverse one of the vast diffused nebulosities with which the heavens abound, when a conversion of a large amount of its motion into heat might, with formidable results, be expected to ensue. But the occurrences hitherto actually observed are of a different kind; and although a plausible way of accounting for them is at hand, it is one not to be admitted without great reserve.

Collisions in some form seem, at first sight, the only explanatory resource available. But we must distinguish. An encounter between two condensed masses (apart from its extreme improbability) would produce quite other, and more permanent, effects than those connected with the outbreak of new stars. Such transient flashes cannot be due to *bodily* collisions; only *atmospheric* collisions, minutely exemplified in the gleam of every shooting star that gets entangled in our upper air, can be concerned in them. Substantially this inference was arrived at by Mr. Lockyer in 1877, from a study of the appearances presented by Nova Cygni.

'We seem,' he then remarked, 'driven from the idea that these phenomena are produced by the incandescence of large masses of matter, because, if they were so produced, the running down of brilliancy would be exceedingly slow. Let us consider the case then on the supposition of small masses of matter. Where are we to find them? The answer is easy:—in those small meteoric masses which an ever-increasing mass of evidence tends to show occupy all the realms of space.' [1]

In his recent cosmical scheme, accordingly, the far-away cataclysm represented to our senses by the appearance of a 'new star,' takes shape as a 'collision between two meteor-swarms.' If we call the smaller of the two a comet, the larger a nebulous star, we shall get rid of much that is hypothetical, and may succeed in realising the situation more distinctly. It is evident that nebulous stars must be more subject than others to encounters of the kind. Our sun, if distended so as to fill the orbit of Mercury, would engulf innumerable comets that now slip clear round it at perihelion. In past times, it probably has absorbed hundreds, nay thousands of these bodies. Enormous comets, moving with high velocities towards bodies (consequently) of great attractive power, should indeed be called into action to produce the conflagrations of 'new stars;' but there is no reason known to us why these conditions should not be fulfilled. We must recall, however, that such conflagrations only exaggerate the changes of other variable stars, and should accordingly be referred to an intensification of their cause. And since that cause can scarcely (as we shall see in the next chapter) be found in actual collisions, the doubt arises whether we are entitled to assume their occurrence in the case of temporary outbursts. If not, then we should substitute for them grazing encounters with nebulous masses revolving in hyperbolic orbits, and overthrowing by their proximity to the attractive body a thermal equilibrium already eminently unstable.

[1] *Nature*, vol. xvi. p. 413.

CHAPTER VIII.

VARIABLE STARS OF LONG PERIOD.

About two hundred and fifty stars have been formally registered as variable, and many more are open to the like suspicion. Gore's 'Revised Catalogue'[1] includes 243 entries, besides 39 provisional additions; Chandler's nearly contemporaneous list[2] enumerates 225 objects. Of these 160 are reckoned as 'periodical,' the rest as 'irregular' or 'temporary.' Periodical stars are further divided into those with 'long,' and those with 'short' periods. Nor is the distinction by any means arbitrary. The stars seem to separate of themselves into two principal groups, undergoing fluctuations in cycles of respectively less than fifty, and between two and four hundred days. The paucity of stars with periods of intermediate lengths is shown graphically in fig. 11, where the height of the curve represents the numbers of stars subject to changes proportionate in duration to the horizontal distance from left to right.

Variations requiring several months for their completion differ both in degree and kind from those run through in a few days. They are of much greater amplitude, ranging over five to eight instead of, at the most, two magnitudes; they are accomplished with less punctuality; and they are frequently attended by symptoms of atmospheric ignition entirely foreign to quicker vicissitudes. Most important of all, they affect bodies of peculiar constitution. Nearly all long-period variables are red stars with banded spectra; those of short period are white or yellowish in colour, and display Sirian or solar spectra. Quality of light is thus the pre-

[1] *Proc. R. Irish Acad.* vol. i. ser. iii. p. 97. [2] *Astr. Journal*, Nos. 179-180.

dominant factor in determining the law of stellar light-changes, and is itself dependent, as we have seen, upon atmospheric conditions. Hence we reach the generalisation that slight absorption accompanies short, strong absorption long periods, and that extensive gaseous surroundings not only favour variability in general, but almost absolutely prescribe its type.

Periods of between one and two hundred days may be called 'long'; but as fig. 11 shows, they do not commonly

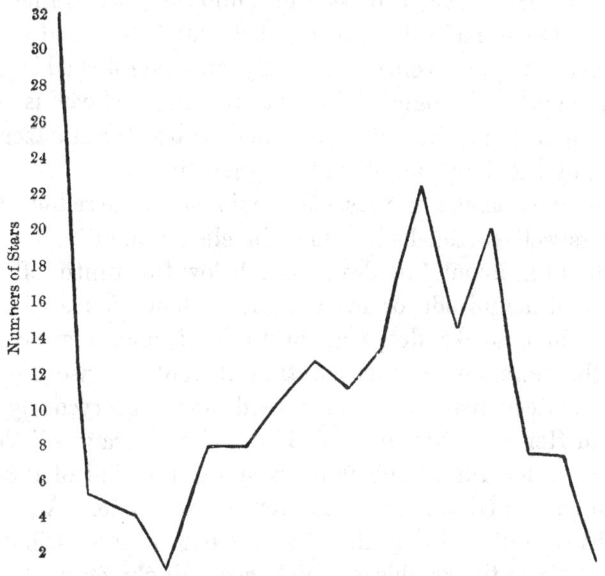

Fig. 11.—Distribution of 171 Periods of Variable Stars from Gore's Revised Catalogue (1888).

occur. Such fluctuations as now engage our attention usually demand more than two hundred days for their accomplishment, and are seldom prolonged beyond 410. Mr. Chandler considers 320 days as the average duration of change for long-period variables;[1] the prevalence, however, among them of periods of about one year is remarkable, and cannot be accounted for by mere accidents of observation. The first and

[1] *Astr. Journal*, No. 193

best known specimen of the class anticipates by about a month the rule of annual recurrence.

When Bayer, in 1603, affixed in his charts the Greek letter *o* to a small star in the neck of the Whale, he had no suspicion of its identity with a supposed 'Nova' which had disappeared seven years previously, after blazing up to the second magnitude. Its discoverer, on August 13, 1596, was David Fabricius, of Osteel, in East Friesland; but though he saw the object again February 15, 1609, he left it to John Phocylides Holwarda, Professor of Philosophy at Franeker in Holland, to ascertain its true character in 1639; and the repetition of the phases once in 333 days was established in 1667 by Boulliau.[1] The name 'Mira' bestowed by Hevelius upon the changing star in Cetus, commemorates the amazement excited by the detection of stellar periodicity.

The phenomena it presents would seem incredible were they less well established. Once in eleven months the star mounts up in about 110 days from below the ninth often to the second magnitude or even higher; then after a pause of two or three weeks, drops again to its former low level in twice the time, on an average, that it took to rise from it. The brightest maximum on record was observed by Sir William Herschel, November 6, 1779, when Mira was little inferior to Aldebaran;[2] the faintest minimum, that of 1783, is said to have carried it below the tenth magnitude. An extent of eight magnitudes may then be assigned to the oscillations of this extraordinary object, which accordingly emits, at certain times, fully fifteen hundred times as much light as at others. That each maximum is a genuine conflagration has been proved by spectroscopic observation; and the conflagrations recur yearly, with approximate regularity, and, after three centuries of notified activity, give no signs of relaxation!

The height of the maxima, however, varies greatly. The two adjacent ones of 1885 and 1886 (represented in fig. 12) showed a nearly fivefold difference of intensity; but Heis's remark that high and low maxima tend to alternate, has not in the long run proved consonant with facts. There is no

[1] *Monitum ad Astronomos*, p. 7. [2] *Phil. Trans.* vol. lxx. p. 338.

rule by which the brilliancy of impending phases can be predicted. That of November 1868, in which the star just failed to reach the fifth magnitude, was, it is true, preceded by a high maximum, but several average or low maxima followed it. All that can be said is that exceptionally bright apparitions are isolated; they do not come in sets, but one by one, at considerable intervals.

The minimum brightness of Mira is about 9·5 magnitude, and tolerably uniform; though Schönfeld stated in 1875 that he had never seen the variable inferior to its ninth magnitude companion.[1] Its redness during low phases, however, embarrasses estimation. Their nature, too, is to a great extent undetermined.

During the four months of each period that the star is 'out of sight,' it is apt to slip 'out of mind' as well. Most pro-

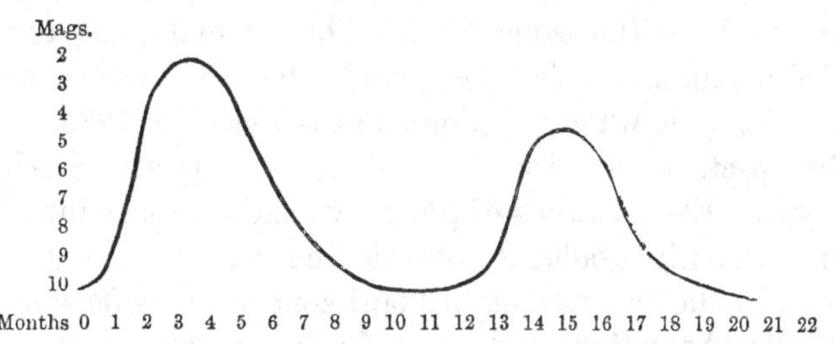

FIG. 12.—Maxima of Mira in February 1885 and January 1886.

bably, perceptible change is suspended for at least a couple of weeks before and after minimum; but details as to this interval of inertness are wanting. Yet it would be of especial interest to ascertain its length, and to note the symptoms indicating its termination. How does recovery set in? is a question that has yet to be answered. Is it with a sudden start, or by a gradual revival? Is it accompanied by changes of colour, or of spectrum? Does telescopic definition remain the same after as before the critical epoch? All these points have a theoretical importance likely to grow with time.

The periodicity of Mira obeys a highly complex law. Deviations to the extent of a fortnight from the mean period of 331 days are common, and the maximum of September 29,

[1] *Jahresbericht Mannheimer Verein für Naturkunde*, Bd. xl. p. 49.

1840, was a full month late.[1] Its perturbations are, indeed, probably themselves periodical, but so many exist which have so far not been formulated, that prediction is often at fault. Argelander detected the influence of a wave of disturbance with an amplitude of twenty-five days, and embracing eighty-eight periods;[2] Schwab's observations indicated subordinate oscillations in six and a half days;[3] and there are half-effaced traces of several besides.[4] The *shape* of the light-curve, too, varies notably. Its peaks are sometimes much blunter than at others; and the star, which usually retains its full lustre during a fortnight, has been known to remain twice that time stationary. Still more singularly, the otherwise invariable rule of an increase more rapid than the ensuing decrease was reversed in 1840. Sixty-two days were occupied in ascending from the sixth to the third magnitudes, forty-nine only in sinking back to the same level. The anomaly, due to the retarded maximum of that year, recalls the abnormal course of the sunspot cycle which culminated at the close of 1883.

The spectrum of Mira is a splendid example of Secchi's third type. Eleven bands of profound shadow, sharp towards the violet, gently gradated towards the red, throw out into strong relief the intervening brilliant zones; while dark lines of metallic absorption, and vivid hydrogen-rays, vary the effect, and add to the intricacy of the characters to be deciphered. Only the more refrangible members of the hydrogen series appear to be brightened in this star; no trace of C or F betrays itself to the most attentive scrutiny. Yet they can hardly be in reality absent. We can only suppose that these slower vibrations are partially absorbed by overlying strata of cooler hydrogen, which, by reason of their lower temperature, are inactive as regards the quicker vibrations. The detailed study of the bright lines that do appear can best be carried on by photographic means. It may be assumed that they become extinct with the approach of each minimum, and are re-kindled during the ascent towards the

[1] Argelander, *Astr. Nach.* No. 416.

[2] *Bonner Beobachtungen*, Bd. vii. p. 332.

[3] *Astr. Nach.* No. 2731.

[4] Argelander in Humboldt's *Cosmos*, vol. iii. p. 234.

ensuing maximum; but positive evidence to this effect is still a desideratum. Absorption certainly increases with the fading of the star. The spectral bands, though very intense at maximum, deepen and widen remarkably after it is passed.[1] The diminution of brightness, however, is not entirely due to this cause. Light intrinsically fails, besides being additionally intercepted.

A reddish-orange star of the sixth magnitude was discovered by Mr. J. E. Gore, of Ballysodare, in Ireland, on December 13, 1885. Its situation, just where the Milky Way streams across the club of Orion, might have seemed confirmatory of the temporary character imputed to it mainly on the ground of the improbability of so bright an object having so long escaped notice. But the decline which carried it below the twelfth magnitude in July 1886 was succeeded by a renewal of light, and a second maximum occurred within a day of the anniversary of the first. The star has since conformed pretty regularly to a period of about a year, not always, however, filling its full measure of change. This, indeed, is the case with all variables of the Mira-type, to which 'U Orionis' (a name substituted for its original designation of 'Nova Orionis') unmistakably belongs. Its spectrum might be called a replica of that of Mira, and both include the same photographic bright lines. The helium ray in addition was probably seen by M. von Konkoly in the Orion variable.

The history of the two stars has also points in common. Each, taken at first for a 'Nova,' was only on further acquaintance recognised as periodical. Each, too, unaccountably escaped astronomical notice for centuries; yet it is perhaps easier to accept the usual explanation of this difficulty by a series of coincidences between epochs of minima and epochs of observation, than to suppose each star to have newly entered upon its vicissitudes at the time of its discovery.

The second long-period variable recognised was a star in the neck of the Swan, which Bayer, ignorant of its changing character, set down in his maps as of the fifth magnitude. It still retains the name he gave it of 'χ Cygni.' Missed by

[1] Maunder, *Monthly Notices*, vol. xlix. p. 303.

Gottfried Kirch in July 1686,[1] it reappeared October 19, and subsequently disclosed to his vigilant watch fluctuations even wider than those of the 'wonderful' star in Cetus. It descends nearly to the thirteenth and rises to the fourth magnitude, sometimes indeed stopping short when barely visible to the naked eye, but more commonly remaining lucid for a couple of months. Nor is its course much better regulated as regards time. Errors up to forty days often attach to its phases, and the attempt to correct them by the introduction of cyclical disturbances has proved only partially successful.[2] The period, estimated at 402 days by Kirch, now averages 410. Olbers noticed that it had been steadily lengthening down to 1818,[3] and it is lengthening still; the compensatory process anticipated by him has not set in. As usual in such cases, the ascent to maximum is much more rapid than the descent from it, occupying at present about 185 days.[4]

The spectrum of χ Cygni is colonnaded like that of Mira. But on May 19, 1889, when the star was near a maximum, Mr. Espin perceived evidence in it of direct radiation by hydrogen and helium, confirmed by Mr. Taylor's subsequent observations at Ealing with Mr. Common's giant reflector. The absorption bands appeared at the same time to have lost their determinate character, and as the general light of the variable waned in the autumn, an immense number of bright lines, carefully observed by Dr. Becker at Dunecht, came visibly into view.

In R Hydræ, too, the advent of maximum brightness is attended by a blaze of hydrogen.[5] The cycle of change by which this star oscillates from above the fourth to the tenth magnitude is rapidly shortening. In 1708 it extended to 500 days, it is now comprised in 434.[6] The period (about 229 days) of S Ursæ Majoris, on the other hand, has lengthened notably since 1855;[7] while a sudden, very considerable abridgment of that of S Aquilæ, towards the close of 1883, was

[1] *Miscellanea Berolinensia*, t. i. p. 208.

[2] *Bonner Beob.* Bd. vii. p. 336; *Mannheimer Jahresbericht*, Bd. xl. p. 110.

[3] Schumacher's *Jahrbuch*, 1841, p. 93.

[4] Gore's *Catalogue*, p. 197.

[5] Espin, *Astr. Nach.* No. 2889.

[6] Chandler, *ib.* No. 2463.

[7] J. Baxendell, jun. *Journal Liverpool Astr. Soc.* vol. iii. p. 52.

followed by great irregularities, with a tendency, on the whole, towards restoration of the *status quo ante*.[1] Nor is it likely that any such disturbances will prove indefinitely progressive.

It is not easy to decide off-hand whether U Geminorum should rank as a periodical or as an irregular variable. Habitually of extreme faintness (below fourteenth magnitude), it gains light at uncertain intervals with marvellous rapidity, approaching, or even slightly overpassing, the ninth magnitude.

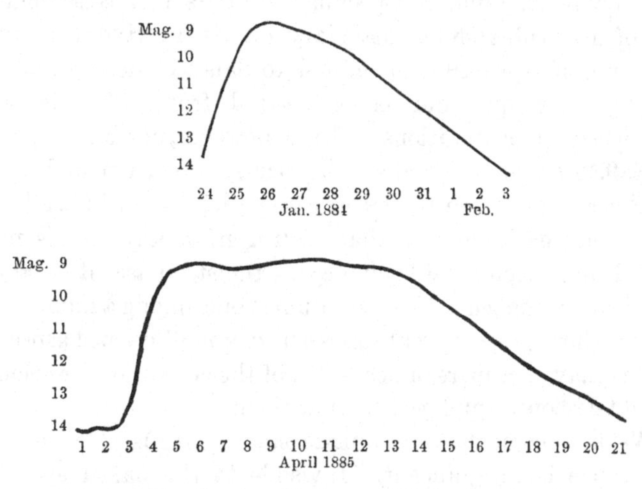

FIG. 13.—Two Types of Maximum of U Geminorum (Knott).

Thus its brightness in February 1869 increased sixteenfold within twenty-four hours;[2] and Mr. Knott estimated its magnitude February 20, 1877, at 13·2; but twenty-six hours later at 9·8![3] Two types of maximum are, according to his observations, exhibited by this star.[4] In one the entire swing to and fro between the fourteenth and ninth magnitudes is completed in ten days, or even less; in the other it occupies fifteen to twenty. A pause at the summit, in the latter case,

[1] Baxendell, *Observatory*, vols. ix. p. 122; x. p. 261.
[2] Schönfeld, *Sirius*, Bd. x. p. 62; *Mannheimer Jahresbericht*, Bd. xl. p. 85
[3] *Monthly Notices*, vol. xxxvii. p. 279.
[4] *Journal Liverpool Astr. Society*, vol. iii p. 11.

with a tendency to a secondary minimum, accounts for the difference. The light-curve takes more or less the form of a double peak with a saddle between (see fig. 13).

Three of these extraordinary crises are, as a rule, undergone yearly; but the star remained quiescent during 617 days in 1860–1, while some of its maxima are separated by no more than 65 to 75 days. Usually undistinguished in colour, it was described by Hind on the night of its discovery by him, December 15, 1855, as shining 'with a very blue, planetary light,'[1] and the peculiarity of its tint (resembling that of a cœrulean-hued planetary nebula in Hydra) at the maximum of April 1881 suggested to Safarik the probability of a gaseous spectrum being derived from it.[2] Neither Dr. Copeland's observations in 1880, however, nor Mr. Espin's in 1889, afforded more than the barest suspicion of bright lines,[3] accompanied in the latter case by persistent indications of a strong dark 'F'; so that hydrogen, at any rate, is not kindled into exceptional brilliancy as the steep ascent towards maximum is climbed. U Geminorum is one among a number of variable stars, presenting at times a hazy and ill-defined aspect; but this may be a mere optical effect of the correction for colour of the telescopes employed to view them.

We now come to that unique star, η Argûs. Its actual appearance is insignificant. Invisible to the naked eye, its reddish colour and slightly superior brightness alone distinguish it in the telescope from the crowd of small stars embroidering one of the finest of the southern nebulæ, sometimes named the 'Key-hole Nebula,' from the aperture of that shape with which it is centrally perforated. Close to an edge of the aperture, in the densest part of the nebula, η Argûs is placed. Nor can we suppose its position fortuitous. Although probably constituted for variability, its situation, plunged (as appears certain) in nebulous substance, combines, we are led to suppose, with its essential nature to produce the exceptional character of its changes.

The first observation of η Argûs was made by Halley at St.

[1] *Monthly Notices*, vol. xvi. p. 56. [2] *Astr. Nach.* No. 2391.

[3] *Observatory*, vol. v. p. 110; *Astr. Nach.* No. 2919.

Helena in 1677, when it was of the fourth magnitude; the next by Père Noël, a Jesuit missionary, in China, about ten years later.[1] The second rank was assigned to it both by him and by Lacaille in 1751; yet the discrepancy with Halley's appraisement remained unnoticed. The higher estimate was besides confirmed by those of Fallows, Brisbane, and Johnson in 1822, 1826, and 1832 respectively, and only the traveller Burchell, familiar with the star as of the fourth magnitude in 1811–15, was surprised one night, at San Paolo in Brazil, to see it temporarily raised to a level with the finest brilliants of the sky. Another, and a still more vigorous outburst, was witnessed by Sir John Herschel at the Cape, December 16, 1837. Without previous note of warning, the star all at once nearly tripled its light, and before the end of the year fully matched α Centauri. Since then it has been kept under strict surveillance as a notorious character, and not without reason. After a partial decline and several preliminary 'flutterings,' it reached a final maximum in April 1843, when Sirius alone among the fixed stars slightly outshone it. This high position was moreover fairly well maintained for nine or ten years. Gilliss, at Santiago in 1850, found it very little inferior to Canopus in light, and in colour more deeply tinged with red than Mars.[2] Still of the first magnitude in 1856,[3] it fell to the second in 1858, to the third in 1859, and ceased to be visible to the naked eye early in 1868.[4]

For sixteen further years the slow ebb of light continued, and the magnitude of the once effulgent η Argûs, carefully determined by Mr. Finlay at the Cape, was in March 1886 only 7·6.[5] This proved to be the lowest point touched. Slight symptoms of recovery became apparent in the two following years. 'A glow and lightening of colour' first attracted the attention of Mr. Thome at Cordoba, March 20, 1887.[6] From 'dull scarlet' the rays of the variable had vivified to 'bright orange.' And in May 1889 Mr. Tebbutt perceived and

[1] Winnecke, *Astr. Nach.* No. 1224.

[2] Abbott, *Monthly Notices*, vol. xxi. p. 230. [3] Moesta, *Astr. Nach.* No. 1054.

[4] Tebbutt, *Monthly Notices*, vol. xxxi. p. 210.

[5] *Monthly Notices*, vol. xlvi. p. 340.

[6] *Astr. Nach.* No. 292

announced, from New South Wales, a decided increase of light,[1] which has since been steadily maintained. Eta Argûs stands now at about the seventh magnitude, and may before long resume its place among lucid stars.

It is, nevertheless, far too soon to decide the question of periodicity. Quite possibly the history may be one that 'does not repeat itself.' Our continuous knowledge of it is embodied in the accompanying diagram (fig. 14), in which a single vast oscillation is indicated, occupying about a century for its completion, and diversified by secondary fluctuations of a very conspicuous character (innumerable minor ones are ignored in the figure). The data at present available, however, afford no grounds for concluding this oscillation to occur regularly.

FIG. 14.—Light-curve of η Argûs, 1810–1890.

Attempts to assign a period to the variations of η Argûs have, so far, signally failed. Wolf's of forty-six,[2] and Loomis's of seventy years,[3] are both palpably too short; an allowance of at least ninety is demanded by the tardy advent of the recent minimum. This would imply the occurrence of minima in 1796 and 1706. But we have complete certainty from Halley's and Père Noël's observations, that the star was on the rise towards the close of the seventeenth century; and if its behaviour then resembled that observed in the nine-

[1] *Astr. Nach.* No. 2849. [2] *Monthly Notices*, vol. xxiii. p. 208.
[3] *Ibid.* vol. xxix. p. 298.

teenth century, it must have continued to rise during at least twenty years after 1685, thus reaching a *maximum* at a calculated epoch of minimum. A centennial or any longer period would encounter difficulties no less insurmountable, besides being in itself improbable, since no genuine stellar cycle has hitherto been found to embrace two complete years. Much more extended periods have sometimes been suggested, but never ascertained. The stars they were ascribed to, when the time came for a repetition of their presumed cyclical changes, showed a total want of conformity with what was expected of them. As examples may be mentioned 63 Cygni, to which Mr. Espin attributed a period of five, and R Cephei, thought by Mr. Pogson to obey one of seventy-three years.

Thus a considerable future increase in the light of the great southern variable, but scarcely a revival of its past splendours, may be looked for. Two kinds of unsystematic stellar fluctuation can be distinguished. In one, a quickly compensated change, either upward or downward, takes place at uncertain intervals; in the other, the shifting from one to a higher or lower order of brightness is more or less permanent. A star may, under this regimen, not only by a sudden start double its light, but continue for years to shine with twofold intensity. We cannot even say that an eventual return to its former status is inevitable. Now in η Argûs, features of both these methods seem to be combined. It has made quick springs, and held its ground, but it has also often kindled into evanescent brilliancy. Its stationary epochs have been followed by epochs of instability; at some times it has shown a tendency to establish itself at halting-places, at others to slip along an inclined plane of change. In all this it differs materially from temporary stars, which leap up, as if by a single impulse, to their solitary maximum, after which they lose in a few months the whole of the light they had acquired.

The range of variation of η Argûs is, indeed, without precedent except among new stars. It amounts to fully eight and a half magnitudes. When Sir Thomas Maclear observed this marvellous object in 1843, it was emitting 2,500 times as much light as when Mr. Finlay observed it in 1886! If asked to pronounce whether rising so high or sinking so low should more properly be regarded as an 'accident' of its strange career, we should be inclined to say the latter, since it figures as of second magnitude in most early observations.[1] The *visible* spectrum of η Argûs is not strongly characterised; but photographs taken in Peru by Mr. J. E. Bailey, of Harvard College, show bright hydrogen lines, which were most likely *seen* with the great Melbourne reflector by Le Sueur in 1870.[2] Continued spectroscopic observations, as the star gains light, may prove of great value.

Pickering's third class of variable stars may conveniently be made to include those subject to irregular fluctuations of every degree, and among them a variety described by Mr. Espin in 1887,[3] of which the range is one and a half magnitudes, and the mode of progression by quick and seemingly casual bounds. As examples, 19 Piscium and several other stars belonging to the fourth spectral type are mentioned. Closer attention has, however, been bestowed upon the slighter changes of some third-type brilliants. Sir William Herschel added in 1795 α Herculis to the list of seven variable stars then known.[4] But the period of two months which he assigned to its oscillations between 3·1 and 3·9 magnitudes has not been ratified. During some years they appear indeed almost to cease, then are hurriedly resumed, but with no settled order. The analogous variations of Betelgeux and β Pegasi are equally unmethodical.

The extraordinary character of a star long known as 'Variabilis Coronæ,' now called 'R Coronæ,' was discovered by Pigott in 1795; a near neighbour of the 'blaze star' of 1866, its changes are of the nature of extinctions rather than of outbursts. Ordinarily of the sixth magnitude, it occasionally drops out of sight with small telescopes, and after lingering near the thirteenth magnitude for many months, slowly

[1] It is marked so in Bayer's charts (1703) probably on the authority of Petrus Theodorus, who voyaged southward in 1595. Winnecke, *Astr. Nach.* No. 1224.

[2] *Trans. R. Society of Victoria*, vol. x. pp. 11, 23.

[3] *Observatory*, vol. x. p. 430.

[4] *Phil. Trans.* vol. xxxvi. p. 452.

regains its lost light. But its phases at times cease wholly, as during the seven years 1817–24,[1] at others are ill-marked. Thus, at the minimum observed by Sawyer, October 13, 1885, the star was still of 7·4 magnitude.[2] It shows no decided peculiarity either of colour or spectrum.

R Cephei is a star which, since the beginning of the present century, has lost on an average $\frac{39}{40}$ of its radiance, and at present in no way tends towards recovery. In the time of Hevelius it was of the fifth magnitude, and Groombridge's observation of it in 1807 showed it to be then still unchanged. By 1840, however, it had sunk to the tenth, and has never since risen above the eighth magnitude.[3] Its identity with the '24 Cephei' of Hevelius, tracked out by Pogson in 1856,[4] is universally admitted. Although its light was considered by Schönfeld to be tinged with red, it appeared bluish to Farley in 1838, and its analysis has disclosed no features of interest.

Among genuinely red stars such shiftings of photometric standing are, perhaps, of ordinary occurrence. Of twenty-two such, kept in view by M. Safarik, at Prague, from 1883 to 1888, for the express purpose of testing their constancy, only nine remained without noticeable change, two were found periodically, six irregularly variable, and five either vanished or lost great part of their light. Earlier observations of several of these objects certified the progress of their decline during twenty to twenty-five years.[5] An example of a sudden acquisition of lustre is afforded by a small red star in the same field of view with γ Cygni. Between December 1885 and June 1886, Mr. Espin perceived it to have risen in rank by a whole magnitude,[6] that is, to be giving out two and a half times as much light as six months previously. And, so far as is known, the gain has been kept.

We must now give some brief attention to the various

[1] Argelander, *Bonner Beob.* Bd. vii. p. 374; Olbers, *Berliner Jahrbuch*, 1841, p. 100.

[2] Gore's *Revised Catalogue*, p. 132.

[3] Schönfeld, *Mannheimer Jahresbericht*, Bd. xl. p. 113; Gore's *Catalogue*, (1884), p. 200.

[4] *Monthly Notices*, vol. xvii. p. 23.

[5] *Astr. Nach.* No. 2874.

[6] *Journal Liv. Astr. Soc.* vol. v. p. 2.

explanations offered of the extraordinary phenomena described in this and the preceding chapters. Our task in this respect has been simplified by recent discoveries. The systematic occurrence of bright lines in the spectra of periodical stars near their maxima brings them, in the first place, into such close physical relationship with temporary stars as absolutely to prohibit the speculative separation of the two kinds of change they respectively exhibit. A theory stands self-condemned which deals with them on different principles. In the next place, the association of variability with processes of luminous change in stellar atmospheres is now placed beyond doubt; and this at once disposes of hypotheses of slag-formation, nebulous eclipses, and axial rotation showing alternately bright and dark sides. Nor can we regard as admissible Dr. Brester's opinion that the sole cause of stellar fluctuations is to be found in intermittent chemical associations and dissociations taking place at the atmospheric outskirts of cooling bodies.[1] For the increase of light at maximum certainly ensues upon a real access of incandescence, and is not a mere appearance due to the dissipation of absorbing vapours. Apart altogether from apparitions of bright lines, M. Dunér's observations discountenanced the supposition of a balanced inverse relation between the growth and decay of brilliancy and absorptive action in variable stars.[2] They often proceed together, but not *pari passu*.

Attempts have several times been made to explain the periodicity of stars through the influence of satellites revolving round them in highly eccentric orbits. Klinkerfues suggested great atmospheric tides, raised at successive perihelion-passages,[3] as a means of bringing about periodic obscurations. Plassmann's[4] view of tidal effects is wider, and perhaps embraces a partial truth. For, just as in the earth the unequal attractions of sun and moon on its centre and surface sometimes

[1] *Essai d'une Théorie du Soleil et des Étoiles Variables*, Delft, 1889.

[2] *Sur les Étoiles à Spectres de la Troisième Classe*, p. 137.

[3] *Göttingische Nachrichten*, 1865, p. 3; see also Dr. Wilsing's comments in *Astr. Nach.* No. 2960.

[4] *Die Veränderlichen Sterne*, Köln, 1888.

provoke, though they could not produce, earthquakes, so the tide-raising power of bodies making very close approaches to stars in a critical state of heat-equilibrium may serve as the occasions of luminous outbursts of a temporary or recurrent nature.

Mr. Lockyer's meteoric hypothesis includes a novel rationale of long-period variability [1] which deserves more serious attention than any previously suggested. Light-fluctuations are regarded in it as closely dependent upon stellar constitution, and the bodies affected by them as exceptionally diffuse—

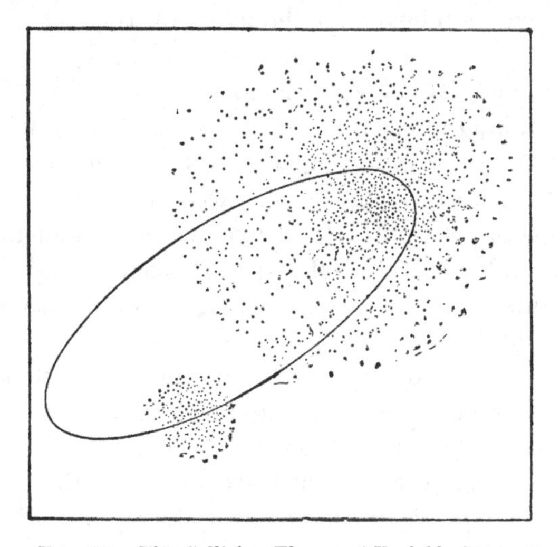

FIG. 15.—The Collision Theory of Variable Stars.

postulates, both of them, fully warranted by observation. It has the further merit of treating temporary and periodical displays from the same general point of view, explains without constraint their quick development of light and more gradual decline, and fits in naturally enough with the gaseous blaze by which they are commonly attended.

Its principle is briefly this. Variable stars are to be regarded as 'incipient double stars'; they are actually meteoric swarms with double nuclei, one of which moves swiftly round the other (see fig. 15), and at its nearest approach

[1] *Proc. R. Society*, vol. xliv. p. 80.

dashes right through its outlying portions, with the result of innumerable collisions between individual meteorites, accompanied by a vast increase in the combined light of the swarms. But the objection inevitably arises that this state of things could not long subsist. Even if set on foot, it should prove transient. By mechanical necessity, the satellite-swarm should speedily become extended into a ring, with, of course, complete effacement of variability. Thus each maximum of a star like Mira, if produced in the way supposed, would be feebler and more prolonged than its predecessor, until maxima and minima were brought to the same uniform level.

The periodicity of variable stars is, besides, of far too disturbed a kind to be thus accounted for. Systemic stability would assuredly prove incompatible with the enormous irregularities it discloses. The abrupt acceleration or retardation, for example, by a month of the hypothetical attendant-swarm of Mira, would be impossible without such a total change in the elements of its circulation as would unmistakably break the continuity of its returns.

But there are other objects far more recalcitrant than Mira to this mode of explanation. Take the outbursts of U Geminorum. They are not wholly capricious. There is a certain disorderly order about them by which they are manifestly akin to the changes of more strictly periodical stars. We cannot then relegate them into a class apart, and invent a fresh hypothesis to suit them; the collision theory, to be acceptable in the one case, must be capable of meeting the other. But we can scarcely conceive any combination of assumptions by which such an extension of its powers could be effected.

Periodical cannot be sharply divided off from irregular variables. Every degree of disturbance, up to the total subversion of laws of change, is met with among them. Many stars seem at times disposed to conform to a period which they later ignore. In others, method is indicated, though too vaguely to be defined; while the majority oscillate, with wide allowance of amplitude, about a period itself often subject to periodical or secular change. It is evident that the immediate and unmodified interaction of revolving masses cannot explain breaches of regularity widening out to its total destruction.

The time has not come to formulate a theory of stellar variability, but we may, at any rate, try to render our ideas on the subject coherent, and thus realise with some distinctness the conditions under which alone any such theory could be regarded as adequate.

As long ago as 1852, M. Rudolf Wolf adverted to the analogous character of the curves representing sunspot frequency and stellar light-change.[1] They are not only of the same general form, but they are marked by precisely the same kind of irregularities. Both are steeper in ascent than in

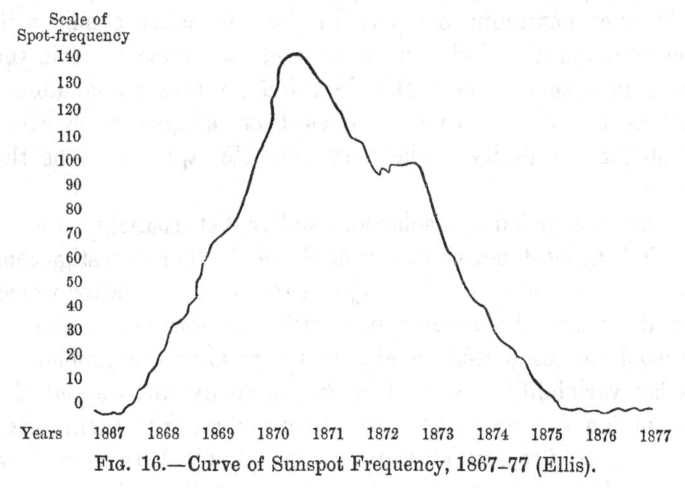

FIG. 16.—Curve of Sunspot Frequency, 1867–77 (Ellis).

descent; both rise into peaks of unequal heights at unequal distances apart. Mira, χ Cygni, R Hydræ, and the rest, have, like the sun, retarded and accelerated, high and low, or abortive maxima. The representation in fig. 16, from the Greenwich observations, of the changes in sunspot frequency during the decennial period 1867–1877, is the very counterpart of the light-curve of a variable star. Especially characteristic is the break in the descending branch reflecting a partial

[1] *Mittheilungen Naturforsch. Gesellschaft,* Bern, 1852, p. 261.

recovery after maximum to which variables of all classes are prone. Moreover, the flow of the vicissitudes of stars and sun alike is broken and disturbed by the superposition upon the normal period of subordinate and superior cycles ranging from a few days, perhaps, to centuries.

The presumption then of their similar origin is very strong; nor are we wholly without evidence of a physical nature to the same effect. The development of bright lines in the spectra of variable stars near their maxima is paralleled in the sun by the increase of emissive intensity in the corona as sunspots increase. Atmospheric incandescence is thus in both cases heightened, although in immensely different degrees; and confirmation is afforded to what was already certified by the congruous shapes of the two curves, namely, that the maximum of spots in the sun corresponds with the maximum of light in stars, and *vice versâ*. It is the more necessary to bear this in mind because actual obscurations by spots have sometimes been alleged as a cause of stellar variability. The very opposite appears to be the truth.

The conclusion that solar and stellar disturbances are alike in kind, at least clears the ground for further investigation. For, besides obliging us to reject causes for the latter which are demonstrably unconcerned with the former, it renders sunspot studies directly available for solving the problem of stellar variability. Now what do we really know about the production of sunspots? Mr. Lockyer regards them, with much show of reason, as rents in the photosphere caused by downrushes upon it of cooled materials, these downrushes being themselves part and parcel of a great system of solar atmospheric circulation.[1] They are, in fact, only local intensifications of the gentle *rain* falling always and everywhere over the surface of the sun, and indispensable doubtless to the maintenance of its luminosity. But of the alternate waxings and wanings of this mode of action, no complete explanation has so far been offered. Endeavours to connect them with meteoric influences have met with no success;

[1] *Chemistry of the Sun*, chap. xxviii.

the periodicity of the sun seems, however, to some extent dependent upon the situation of the planets.

The likelihood, meanwhile, grows continually stronger, as the strange possibilities of close stellar combination are unfolded by new methods, that variable stars owe some of their peculiarities to complex modes of action upon them of satellite bodies. Mr. Lockyer may thus be right in supposing them of an essentially multiple nature, although the manner of regulation of their changes must be far less direct than that assumed by him. Tidal influences, largely appealed to by M. Plassmann, cannot be wholly inoperative, but had better be left to the future for detailed interpretation. The native tendency of red stars to instability of light is apparently connected with the great extent of their atmospheres, and we may further conjecture that the *immediate* cause of their luminous accessions is an enormous addition to the intensity of atmospheric reaction upon their photospheres. That is to say, prodigious falls of cooled matter take place, as in the sun, with the approach of a spot-maximum, but on an immensely enlarged scale. Accompanying electrical phenomena are doubtless proportionately developed, and must exercise a powerful influence on the quality of luminous emission in such stars as Mira.

It cannot be too strongly insisted upon, that while the recurrences of stellar light-change may be prescribed from without, its nature depends almost wholly upon atmospheric conditions. Extensively incandescent vaporous envelopes appear well-nigh inconsistent with real stability in shining. Where they are present, a trifling impulse may suffice to start an important disturbance. Thus a very large proportion of red stars are variable, and nearly all variables of long period are red. The length of the period, too, is very distinctly connected with the intensity of the colour. This was first noticed in 1873 by Dr. Schmidt, of Athens; [1] it has been emphatically confirmed by Mr. S. C. Chandler, who concludes, after an elaborate study of all the facts, that 'the redness of variable stars is, in general, a function of the lengths of their

[1] *Astr. Nach.* No. 1897.

periods of light variation. The redder the tint, the longer the period.'[1] The redder the tint, also, the more profound (we are led to infer) the atmosphere, and the greater the distance, consequently, at which an attendant body should be situated in order to revolve free from it. This seems at least a possible reason for the correlation of duration of change with colour in variable stars of long period.

[1] *Astr. Journal*, Nos. 186, 193.

CHAPTER IX.

VARIABLE STARS OF SHORT PERIOD.

We have seen, in the last chapter, that stars varying their light in periods of less than fifty days stand apart in several important respects from those undergoing slower changes. The distinction is accentuated by the tendency apparent in each class to group its members as far as possible from the frontier-line of separation from the other. Thus, long periods for the most part exceed three hundred days, while a large majority of short periods fall below ten. Thirty-eight stars in all are reckoned as variable within fifty days; of these thirty-two complete an oscillation in less than twenty, twenty-seven (including two with imperfectly ascertained periods) in less than ten days. A comparison of figures 17 and 18 shows that, among short periods taken *en masse*, those of three to four days predominate; those of five to eight days when Algol variables are excluded.

Variables of short period are, as we have said, nearly all white or yellow stars. A very few are reddish; and one—W Virginis—is *suspected* to possess a banded spectrum. R Lyræ, a star of about 4·5 magnitude, with a superb spectrum of the third type, is nominally variable in forty-six days, but its changes are so trifling and so imperfectly rhythmical as to suggest that its proper place is with β Pegasi and α Herculis among stars affected by *abortive* periodicity. Those characterised by the description of variability we are now studying, seem to be bodies of more finished organisation than long-period variables, which (perhaps for that reason) they largely surpass in light-giving power. It has been remarked [1]

[1] Espin, *Observatory*, vol. v. p. 79.

that fluctuations are, on the whole, quick and slight in proportion to the brightness of the objects they affect; and quick fluctuations are executed with much greater precision than slow ones. In many stars the light ebbs and flows like clockwork as to time, and as to measure, with deviations scarcely of the tenth of a magnitude from a settled standard. These remarkable changes progress gradually and continuously in Pickering's fourth class of variables; in his fifth class they

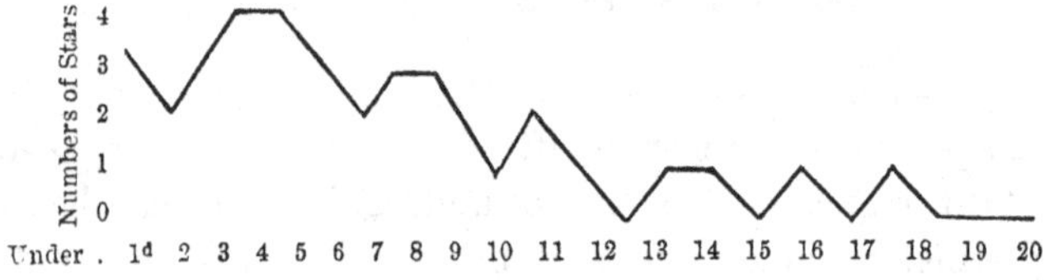

Fig. 17.—Distribution of all the periods of Variable Stars under twenty days.

only interrupt, although at perfectly regular intervals, the usually steadfast shining of certain stars. Of these two kinds, the former is conspicuously exemplified in β Lyræ—a star of which we have already made the acquaintance in connection with its gaseous spectrum—the latter in Algol.

The variations of β Lyræ, detected by Goodricke in 1784, were first completely investigated by Argelander in 1844.[1] They are of a somewhat complex nature, including two equal

Fig. 18.—Distribution of periods under twenty days, excluding those of Algol variables.

maxima separated by two unequal minima (see fig. 19). At its highest light the star is of 3·4, at its lowest, of 4·5 magnitude. The intervening minimum is usually at 3·9 magnitude, but at times it tends to become effaced, while at others it is scarcely less marked than the principal minimum. The intervals between all the four phases are approximately equal, but the curve of change is sharpest at the principal minimum,

[1] *De Stellâ β Lyræ Disquisitio.*

and it is then, accordingly, that the most exact observations are made. The entire cycle is traversed in twelve days, twenty-one hours, forty-seven minutes, and thirteen seconds, being some two hours and forty minutes more than the time occupied a century ago; and the retardation continues to progress, though by no means uniformly.

It is a singular fact that changes in *quality* are, in this star, combined with changes in *quantity* of light, yet in a perfectly independent fashion. The bright lines in its spectrum vary seemingly on their own account. Instead of flashing out towards maximum, as in long-period variables, they, if anything, fade, as if overpowered by the general increase of lustre. Mr. Maunder suspected the hydrogen rays to be actually dark, October 19, 1888, when the star was at a

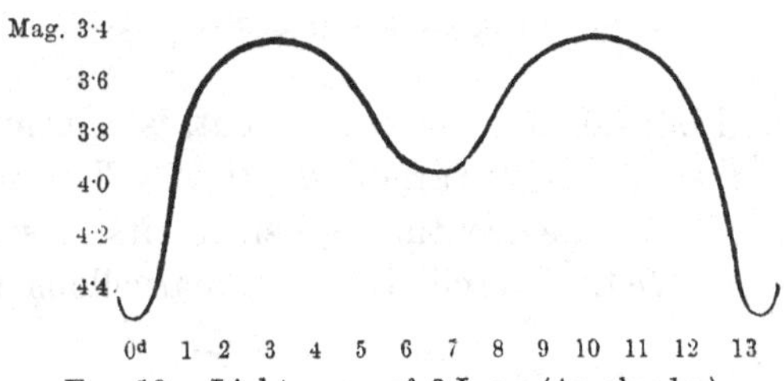

FIG. 19.—Light-curve of β Lyræ (Argelander).

maximum; while on the other hand, the only positive record of their showing by absorption was on September 5, 1882, the day after a secondary minimum. No rule, then, can be clearly made out, and an apparent anomaly has to be admitted. Beta Lyræ is the only short period variable known to give a gaseous spectrum.

Its double periodicity is reflected, though in a far less finished form, in R Sagittæ, a star fluctuating comparatively slowly and unpunctually from about the tenth to $8\frac{1}{2}$ magnitude. The period, normally of seventy days, shortens and lengthens alternately to the extent of some four days every ten years,[1] and includes two unequal maxima and two unequal minima (see fig. 20). The minima are subject to curious exchanges of intensity. They became equalised in 1873 with

[1] Baxendell, *Proc. Phil. Soc. of Manchester*, vol. xix. p. 120.

the result of cutting the period completely in two;[1] then the phase that had been subordinate grew to be the principal, while the principal declined to a subordinate rank; and there are signs that the original state of things will, after a time, be restored.

A pause in the decline, as if a second maximum were

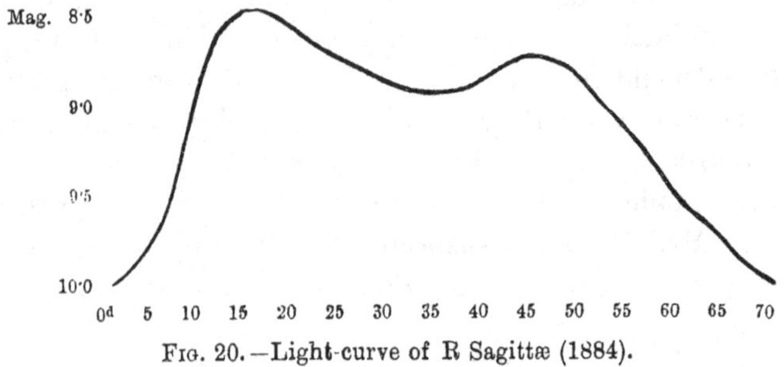

Fig. 20.—Light-curve of R Sagittæ (1884).

contemplated but failed to be carried out, is a common characteristic of short-period variables. It is well exemplified in δ Cephei, which has continued, since its discovery by Goodricke in 1784, to oscillate with marvellous regularity

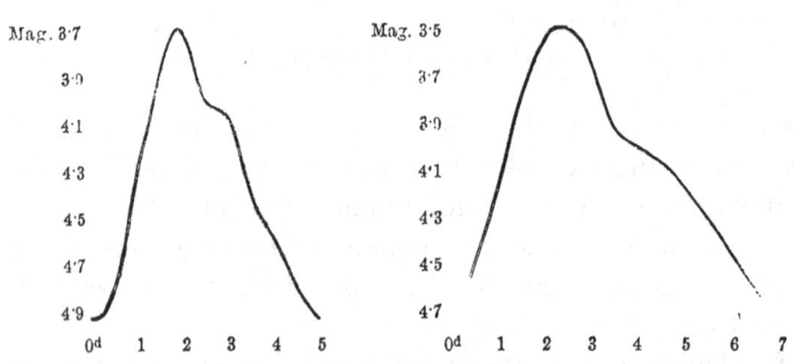

Fig. 21.—Light-curve of δ Cephei. Fig. 22.—Light-curve of η Aquilæ.

between 3·7 and 4·9 magnitudes in a period of five days, eight hours, forty-seven minutes, and forty seconds. A swift ascent is accomplished in $1^{d}\ 13{\cdot}6^{h}$; then, sixteen hours after maximum, occurs the halt marked by the *shoulder* of the curve in fig. 21. The spectrum of δ Cephei resembles that

[1] Schönfeld, *Vierteljahrsschrift Astr. Ges.* Jahrgang xxi. p. 301; Baxendell, *Proc. Phil. Soc. of Manchester*, vol. xxiv. p. 200.

of the sun in being crossed by a great number of fine lines. A wide double star, its full yellow colour contrasts effectively with the delicate blue of a smaller companion, perhaps physically related to, though not visibly circulating round it.

Like δ Cephei, η Aquilæ never sinks out of reach of the unarmed eye. At its faintest it is of 4·7, at its brightest of 3·5-magnitude. In the mode, no less than in the range of its fluctuations, it closely resembles the Cepheus variable (see its light-curve in fig. 22). Their period of seven days, four hours, fourteen minutes, and four seconds, has probably slightly expanded since Pigott first noticed them in 1784,[1] and is sometimes irregularly deviated from to the extent of several hours. The spectrum of η Aquilæ is of the solar type.

In 10 Sagittæ, a variable discovered by Mr. Gore in 1885, the typical pause in descent is strongly marked. The star

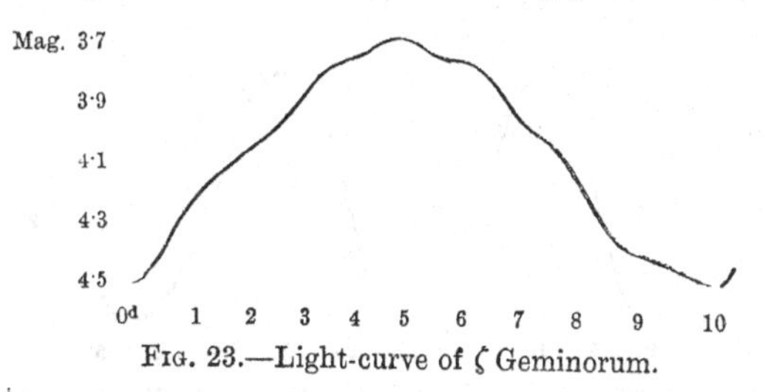

FIG. 23.—Light-curve of ζ Geminorum.

rises by eight-tenths of a magnitude in less than three days, then drops to its former level in five and a half.[2] In its dim phases it cannot be followed with ordinary sight.

The maximum of ζ Geminorum, noticed by Schmidt in 1847 to vary between 3·7 and 4·5 magnitudes, is, by a somewhat rare exception, symmetrically placed as regards the preceding and ensuing minima (see fig. 23). The period of ten days, three hours, forty-three minutes, and twelve seconds has lengthened by ten minutes since its definition forty-three years ago. In one southern star the rule of an ascent quicker than the descent is found to be inverted. This is R Trianguli Australis, discovered by Gould to vary from 6·6 to 8 magnitude, the change upward being accom-

[1] *Phil. Trans.* vol. lxxv. p. 127.

[2] Sawyer. *Astr. Jour.* No. 157; Chandler, *Astr. Nach.* No. 2749.

plished in two days, downward in less than a day and a half.[1] Another southern variable, also of Gould's detection, is remarkable for its excessively short period—the shortest with which we are acquainted except that of one Algol-variable. The light-curves of the two are placed in juxtaposition in fig. 24. R Muscæ increases from 7·4 to 6·6 magnitude in nine hours, and subsides back again in twelve hours, twenty minutes, so that the total length of its cycle is only twenty-one hours, twenty minutes. And its changes are rendered conspicuous by the circumstance that, as Dr. Gould remarks,

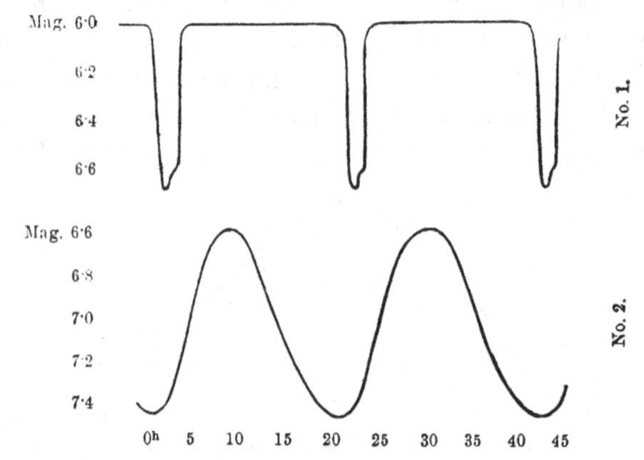

FIG. 24.—Light-curves of U Ophiuchi (No. 1), and R Muscæ (No. 2), the two variables of shortest known periods.

'its average brightness is so near the limit of ordinary visibility in a clear sky at Cordoba that the small, regular fluctuations of its light place it every few hours alternately within or beyond this limit.'[2]

The variability of stars like β Lyræ and η Aquilæ is almost more difficult to explain than the variability of stars like Mira Ceti. The sunspot analogy, it is true, survives in the similarity of the curves picturing the flow in time of both descriptions of change (compare figs. 16 and 21), but with profound differences. We are, accordingly, as good as destitute of a physical explanation of short-period variability. Mathematical theories on the subject have, indeed, been

[1] *Uran. Argentina*, p. 260. [2] *Ibid.* p. 258.

elaborated by Gyldén[1] and Pickering, but they scarcely profess to be more than devices for geometrically representing the progress of the phenomena. The combination is assumed in them of exceptional and barely possible modes of axial rotation with a highly elaborate arrangement of slag-covered, or otherwise darkened areas over the surfaces of stars, which, at least in Gyldén's view, must be mainly solid bodies. But spectrum analysis shows one of them (β Lyræ) to be of a gaseous constitution, and most of the others to resemble the sun; so that the existence on them of permanent patches of obscuration is out of the question. We are often called upon, in natural investigations, to admit what greatly widens experience, but we are always justified in rejecting what contradicts it.

The hypothesis of M. Plassmann[2] associates bodily tides, causing (it is assumed) a temporary increase of luminosity as they travel, with occultations by the tide-raising satellites. So artificial a combination seems little likely to be realised; yet we are reminded by Professor Pickering[3] that improbability must be expected in any theory of phenomena so antecedently improbable as those of variable stars. By Mr. Lockyer, on the other hand, they are considered due to recurring meteoric infalls; and both views, we believe, will prove to be correct in ascribing a compound nature to the objects in question. Their vicissitudes represent, *in some way*, there is reason to suppose, the effects of orbital movement; in what precise way, must for the present remain an open question. Electrical changes are no doubt concerned, but our ideas as to what these imply are so vague, that to appeal to them seems like catching at the easiest expedient for avoiding a confession of total ignorance.

Stars of the Algol type, forming Pickering's fifth class of variables, stand in many respects apart from all other objects of the kind. Their peculiarities are so marked, and, on the whole, so rigidly maintained, as to invite close scrutiny

[1] *Acta Societatis Fennicæ*, t. xi. p. 345.
[2] *Die veränderlichen Sterne*, p. 54.
[3] *Proc. Amer. Acad.* vol. viii. (1881), p. 259.

and minute comparison with theory. Variability is in them by short accesses, and consists always in a temporary loss of light. They undergo in fact what are now known to be real eclipses at stated intervals, while shining, for the most part, as steadily as ordinary stars. Their detection is for this reason so difficult that their scarcity may be more apparent than real. Up to the present, acquaintance has been made with only nine, the designations of which follow in the order of their discovery.

Name	Discoverer	Period				Amount and duration of change					
		D.	H.	M.	S.	MAG.		MAG.		H.	M.
Algol . .	Montanari, 1669	2	20	48	52	2·2	to	3·7	in	10	0
S Cancri . .	Hind, 1848	9	11	37	45	8·2	to	9·8	in	21	30
λ Tauri . .	Baxendell, 1848	3	22	50	53	3·4	to	4·2	in	10	0
δ Libræ . .	Schmidt, 1859	2	7	51	20	4·9	to	6·1	in	12	0
U Coronæ . .	Winnecke, 1869	3	10	51	15	7·6	to	8·8	in	9	42
U Cephei . .	Ceraski, 1880	2	11	47	30	7·2	to	9·4	in	10	0
U Ophiuchi .	Sawyer, 1881		20	7	42	6·0	to	6·7	in	4	30
Y Cygni . .	Chandler, 1886	1	11	58	31	7·1	to	7·9	in	6	0
R Canis Majoris	Sawyer, 1887	1	2	43	12	5·9	to	6·7	in	5	0

The first star on the list is the most accurate in its changes, and has been the most accurately observed of any star in the heavens. Their extraordinary character was determined, and an explanation of them by interpositions of a dark satellite suggested, by Goodricke in 1783, since when, some 13,600 minima have occurred in a manner perfectly consistent with the hypothesis. It became then of great interest to test its absolute truth, and the first means of doing so were afforded by Professor Pickering's strict inquiry into the conditions o the supposed recurring eclipse.[1] They proved to be in Algol all but perfectly complied with. Outside of the ten hours during which it parts with and regains three-fifths of its light, the star displays the required uniform lustre. The oscillation is of stereotyped pattern, the same in duration and extent now that it was fifty years ago, and that it will be probably fifty years hence. The precision of its performance seemed to correspond far better with the results of geometrical rule and measure than with those of the complex interaction

[1] *Proc. Amer. Acad.* vol. viii. (1881), p. 17; *Observatory*, vol. iv. p. 116.

of physical causes. The spectroscope testified in the same sense by showing the surviving light at minimum to be of unchanged quality. It is dimmed, as if in large measure cut off, but betrays no symptom of intrinsic modification. These singular correspondences have not proved deceptive. The postulated eclipses actually take place.

The manner in which their genuineness has been established illustrates the extraordinary versatility of modern methods of research. No problem in which distant light-sources are concerned seems beyond their capacity to grapple with. The received explanation of Algol's changes evidently involved the mutual revolution, in a period identical with theirs, of the eclipsed and eclipsing bodies. And since their orbits, to admit of a transit of the satellite over the primary, should lie almost edgewise to our sight, practically the whole of their velocity should, in the course of each revolution, be directed alternately straight away from, and straight towards the earth. Here, accordingly, spectroscopic measures, recommended by Professor Pickering,[1] were clearly applicable, and after being tried visually with some promise of success at Greenwich,[2] were decisively brought to bear, photographically, by Professor Vogel in 1888–9.[3] Their result was one of the most remarkable verifications of theory on record.

Before each minimum Algol was found to be moving away from the sun (independently of a continuous translation *towards* him of 2⅓ miles a second) at the rate of 26⅓ English miles per second; after each minimum, to be approaching with an equal speed; while at intermediate times the imprinted lines, by resuming their normal positions in the spectrum, showed the star to be then moving perpendicularly to the visual ray. Multiplying this velocity (of 26⅓ miles) by the number of seconds in Algol's period (247,732) we get an orbital circumference corresponding to a diameter of (in round numbers) two million miles. Moreover, since the proportionate dimensions of the bright and dark bodies are shown

[1] *Proc. Amer. Acad.* vol. viii. p. 34. [2] *Observatory*, vol. xi. p. 108.
[3] *Astr. Nach.* No. 2947.

by the amount of obscuration of one by the other to be very nearly as 100 to 83, their relative masses would also be known, if we could be sure that they are of the same mean density. The assumption as regards a mass shining with great brilliancy and one almost totally dark is certainly a hazardous one, but it receives some warrant from the example of the sun and Jupiter. By its aid Professor Vogel arrived at the following provisional data for the system of Algol.

Diameter of Algol . . .	1,061,000 English miles
,, satellite . . .	830,300 ,,
Distance from centre to centre .	3,230,000 ,,
Orbital velocity of Algol . .	26·3 miles per sec.
,, ,, satellite . .	55·4 ,,
Mass of Algol	$\frac{4}{9}$ solar mass
,, satellite	$\frac{2}{9}$,,

In the accompanying diagram C marks the centre of gravity round which both stars revolve with velocities inversely proportional to their masses. Thus, Algol travels in an orbit of only half the compass of that of its companion, because possessed of twice its attractive force. It is easy to see, too, that the duration of the eclipse compared with the length of the period gives the relation between the diameter of the occulted body and the diameter of the orbit of the occulting body; whence the absolute dimensions of one becoming known, those of the other follow.

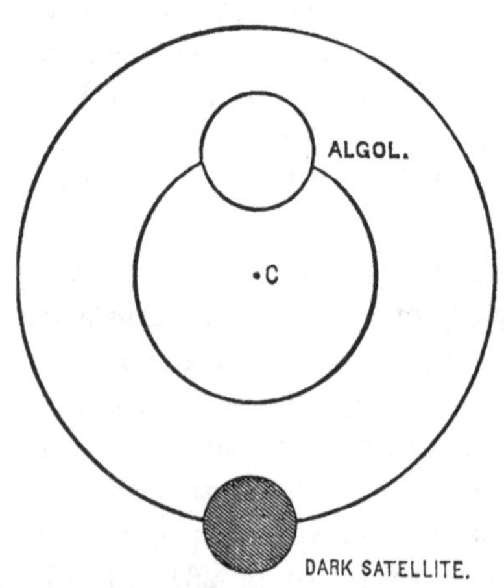

Fig. 25.—Algol during an Eclipse.

The density alike of Algol and of its satellite is less than a quarter that of the sun, or 0·38 that of water. They must both then be completely gaseous, and Professor Vogel finds evidence, in the early and later phases of eclipse, of the possession by each of very profound atmospheres. Spectroscopic confirmation, however, is totally wanting to the view that any part of the darkening is due to atmospheric interposition. A slight want of symmetry in the curve representing the light changes before and after minimum is associated by Dr. Wilsing[1] with the ellipticity of the satellite's path, causing it to slacken its course in moving off the disc. The deviation nevertheless from circularity must be of trifling amount, since otherwise the two bodies would come into contact at their nearest approach. As it is, the stability of the extraordinarily close system formed by them comes only just within the range of theoretical possibility. Although the period of Algol is now six seconds shorter than it was in Goodricke's time, and is still diminishing, it is nearly certain that compensation will eventually take place. The perturbation gives us perhaps the only hint that is ever likely to reach us of the association with Algol of further unseen companions.

It needs no argument to prove that the eclipse-theory of the variable in the head of Medusa must apply to all other members of the same sharply characterised class. Many of them, however, present anomalies which are the more deserving of careful study that they may one day throw an important light on the circumstances under which combinations of the indicated kind exist.

The light-change of S Cancri, the second of the Algol variables, was discovered by Mr. Hind in 1848, and its peculiar nature ascertained by Argelander in 1852.[2] The star remains steady during thirteen-fourteenths of its period, then declines, in eight hours and a half, to less than one quarter of its usual brightness, which it recovers in the course of thirteen hours more. Besides this wide inequality in the times of sinking and rising, the latter process is interrupted soon after it has begun by a marked pause,[3] represented graphically from Schönfeld's observations in fig. 26. A large irregularity, besides, has once been detected in the compass of this star's change. On April 14, 1882, Schmidt observed at Athens a

[1] *Astr. Nach.* No. 2960. [2] *Ibid.* Nos. 796, 804, 806.
[3] *Vierteljahrsschrift Astr. Ges.* Bd. ix. p. 230.

minimum nearly two magnitudes fainter than any he had seen before. During one hour the star remained sunken nearly to the *twelfth* magnitude.[1] The period of S Cancri is subject to a perturbation with a range of about forty minutes, and embracing rather more than three hundred minima.[2]

Inequalities of this kind, which in Algol sum up to a few seconds in a century, and grow to many minutes in S Cancri, are in λ Tauri counted by hours.[3] The task is a formidable one of explaining them on gravitational principles in a system of which the average period is under four days. The same star affords another example of an accelerated decrease, as compared with the increase of its light, which is also believed to

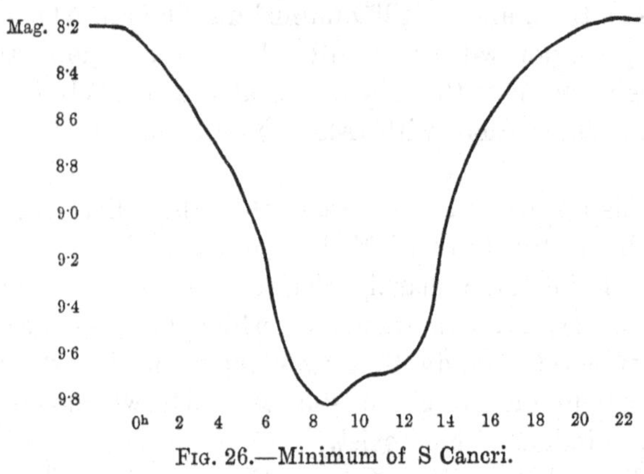

Fig. 26.—Minimum of S Cancri.

be slightly variable outside the regular accesses of change.[4] But although never descending much below the fourth magnitude, it has been comparatively little observed.

The period (about $2\frac{1}{3}$ days) of δ Libræ is shorter, while its time of oscillation is two hours longer than that of Algol. It follows that the supposed eclipsing body circulates in still closer contiguity to its primary than the Algol-satellite. The gap of space from surface to surface can not, in fact, much exceed one-third of the interval from centre to centre. The mean density of the components, too, must be so far below

[1] *Astr. Nach.* No. 2491.

[2] Argelander, *Bonner Beob.* Bd. viii. p. 397 ; Schönfeld, *Sirius*, Bd. x. p. 68.

[3] Schönfeld, *Jahresbericht*, Mannheim, Bd. xl. p. 76.

[4] Plassmann, *Die veränderlichen Sterne*, p. 42.

that of Algol,[1] that their subsistence as globes under the disruptive strain of their mutual gravitation might well appear incredible. There is even a further aggravating circumstance. The recurring drop of the star from 4·9 to 6·1 magnitude, occupies $5\frac{1}{2}$, the compensatory rise $6\frac{1}{2}$ hours; so that a circular orbit seems precluded, while a sensibly eccentric one would infallibly bring about an immediate collision! Inequalities of $5\frac{1}{2}$ seconds, comprised within a cycle of about nine years,[2] modify the periodicity of this interesting object.

The fifth Algol-variable, U Coronæ, declines from 7·6 to 8·8 magnitude in 4h. 30m.; regaining its normal brightness in 5h. 12m.; so here again we meet the relatively slow resumption of brightness prevalent in this class of objects. U Coronæ appears to be slightly variable independently of its systematic changes.[3] The disturbances of its period have been studied by Mr. Chandler.[4]

The variations of U Cephei, first recognised by M. Ceraski at Moscow, June 23, 1880, are more rapid and extensive than those of any of its congeners. In four and a half hours, the star is reduced to about one-ninth its ordinary lustre, losing light, at one stage of its decline, at the astonishing rate of more than one magnitude an hour! The obscurity lasts an hour and a half, but not with entire uniformity. The lowest point is touched at first,[5] and a pause in the ascent, like that inflecting the light-curve of S Cancri (see fig. 26), is indicated. Some complicated irregularities of period have further been ascertained by Mr. Chandler. The white rays of U Cephei turn ruddy at minimum, or rather, if we adopt Professor Pickering's suggestion,[6] the light of an enormous but imperfectly luminous satellite is substituted for that of the star. For the eclipse is in this case supposed to be total during the entire 'stationary period' of an hour and a half; *was* supposed we should say, for Mr. Chandler's observations of a checked

[1] See Mr. Maxwell Hall's computation as regards Algol, *Observatory*, vol. ix. p. 225.

[2] Schönfeld, *Jahresbericht*, Bd. xl. p. 95.

[3] *Ibid.* p. 96.

[4] Chandler, *Astr. Jour.* No. 205.

[5] *Ibid.* No. 199, p. 53.

[6] *Proc. Amer. Acad.* vol. viii. (1881), p. 389.

attempt at increase during the time assigned to totality are, if confirmed, inconsistent with its prolongation.

A star of the sixth magnitude in Ophiuchus, detected as an Algol-variable by Mr. Edwin Sawyer, of Cambridge-port, Massachusetts,[1] had its true period of twenty hours, seven minutes, and forty-two seconds, the shortest yet met with in any star, fixed by Mr. S. C. Chandler.[2] The obscuration lasts four and a half hours, and is about equally divided between a fall and a rise of seven-tenths of a magnitude.[3] But the return of light does not proceed uniformly (see fig. 24, No. 1). It is arrested, about half an hour after minimum, by a 'stand-still' of some fifteen minutes, independently recorded by Sawyer and Chandler.[4] Moreover, the satellite producing this anomalous eclipse must, it would seem, revolve quite close to, if not actually within the limit of distance shown by Roche of Montpellier to be the least at which an attendant globe could maintain its integrity against the tremendous strain of tidal forces. The average density of these conjoined bodies comes out less than one-fifth that of the sun, or 0·27 that of water. A remarkable observation of U Ophiuchi was made by Schjellerup, June 9, 1863.[5] He noted it with surprise (being aware that other observers had seen it much brighter) as of 7·7 magnitude, or one full magnitude below the lowest it has been known to touch in recent years.

The period of Y Cygni, added to the list of Algol-variables by Mr. Chandler, December 9, 1886, averages about a day and a half,[6] but fluctuates to an extent unparalleled in this kind of star.[7] The retardation of its phases between 1887 and 1888 amounted to *seven hours*, totally disconcerting prediction, and the period is now shortening as rapidly as it lengthened before.[8] The actual change does not exceed half a second at each of the returns, but these are so numerous, that the accumulating errors sum up in a short time to a startling aggregate. The deviations of Y Cygni, in fact (as

[1] *Astr. Nach.* No. 2412. [2] *Astr. Jour.* No. 161.
[3] *Astr. Nach.* No. 2484. [4] *Astr. Jour.* Nos. 162, 177.
[5] *Catalogue of* 10,000 *Stars*, Copenhagen, 1864, No. 6162, and *note.*
[6] Chandler, *Astr. Jour.* No. 163.
[7] Chandler, *ibid.* No. 185. [8] *Ibid.* No. 204.

Mr. Chandler points out), exceed those of Algol a hundred, those of U Ophiuchi a thousand fold.

The twenty-seven hours' period of R Canis Majoris is divided between apparently symmetrical oscillations lasting five hours, and a steady maximum of twenty-two hours' duration. Its individual peculiarities have yet to be determined; but the conditions of change are visibly such as can only be realised in a surprisingly close system.

Our survey of the Algol-variables has thus disclosed the following significant peculiarities. 1. In six out of nine cases, the eclipse deepens more rapidly than it lightens; in *no instance* is this relation inverted. 2. All the systems considered are made up of bodies much more tenuous than water, one member of each pair being nevertheless sensibly obscure,[1] while the other is brilliantly luminous. 3. All (so

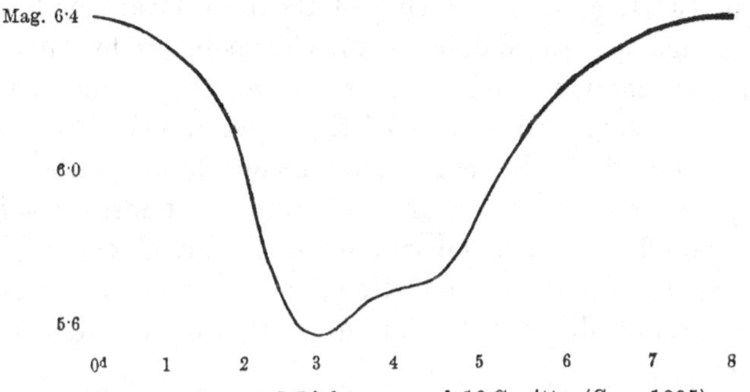

Fig. 27.—Inverted Light-curve of 10 Sagittæ (Gore 1885).

far as is known) are subject to periodical disturbances from invisible attracting masses. 4. Two of these stars (S Cancri and U Ophiuchi) have been observed to undergo at minimum a loss of light far exceeding the usual proportion. 5. Several Algol-variables *invert* most curiously the typical light-curves of such stars as δ Cephei and η Aquilæ (figs. 21, 22). To the quick advance of darkening in the Algol-stars corresponds a quick gain of light in δ Cephei and its allies; the pause in the ascent of the one class is represented with strange fidelity

[1] Vogel calculates that the companion of Algol must be, at any rate, eighty times less luminous; since otherwise a second minimum would be perceptible corresponding to its occultation by the primary star.

by a halt in the decline of the other. We give in fig. 27 the reversed light-curve of 10 Sagittæ, discovered by Mr. Gore in 1885 as variable in 8½ days. The resemblance to that of S Cancri (fig. 26) is unmistakable. We seem to have before us the characteristic minimum of an Algol star. It is difficult to believe in the total dissimilarity of the causes producing effects opposite indeed, yet displaying in their oppositeness so remarkable an analogy. The additional remark may be worth making that 'eclipse-stars' seem to agree in showing a Sirian spectrum; unless indeed R Canis Minoris prove an exception to the rule.[1] The solar type, on the other hand, predominates among ordinary short-period variables. Variable stars of all classes are probably at enormous distances—even on the celestial scale—from the earth. There is no sign that any of them are included among the stars in our comparative vicinity. One of the best means of forming a rough general judgment on this point is by amount of apparent motion; and variables remain in general nearly fixed in the sky. Mira, one of the most mobile, shifts its position indeed to the not wholly inconsiderable extent of twenty-five seconds of arc in a century; but measures for parallax would be much embarrassed by its changes of magnitude, and have not yet been attempted. We are thus absolutely without direct information as to the remoteness of variable stars.

Their distribution over the sphere presents some noticeable peculiarities. Contrary to what might have been expected, short-period variables, although on the whole much brighter objects than those of long periods, tend much more decidedly to concentration in the Milky Way, while the preference for its plane among those not variable belongs chiefly to faint stars. In Algol-stars, though present, it is less strong than in periodical stars of Pickering's fourth class. These lie, for the most part, along a great circle nearly, but not quite coincident with the medial line of the galaxy.[2] It is remarkable that

[1] See Pickering, *Henry Draper Memorial, Fourth Annual Report*, p. 5.

[2] The northern pole of this circle, according to Pickering, is situated in R.A. 13h., Dec. +20°. That of the Milky Way is in R.A. 12h. 40m., Dec. +28°.

their condensation-level (as is shown by its being projected into a *great* circle) passes through the sun. Within the zone itself, there is an evident disposition towards clustering. Where the Milky Way divides in Cygnus, the variables follow its southern branch, and they are thickly sown over the whole sky-region from Lyra to Sagittarius.[1] Indications, indeed, abound that the conditions of variability, and even of particular kinds of variability, are localised in space. Thus in Sagittarius, no less than four stars fluctuate in periods of six to seven days, and several others are subject to slower vicissitudes. Two adjacent stars in the Southern Triangle vary in unusually short periods. R follows S Coronæ Australis nearly in the same parallel, after an interval of only forty-six seconds of time. The new star which appeared in Scorpio in 1860 marked the centre of a group of six objects, all widely variable irregularly or in long periods. Five stars of a similar nature, including two virtually extinct Novæ, are collected in a small section of Ophiuchus; and in general the sites of temporary stellar apparitions are more or less closely dotted round with variables. There is reason to suppose that the circumstances favouring instability of light do not exist anywhere in the neighbourhood of the sun.

[1] Chandler, *Astr. Jour.* No. 193; Plassmann, *Die veränderlichen Sterne*, p. 85.

CHAPTER X.

THE COLOURS OF THE STARS.

THE stars differ obviously in colour. Three or four among the brightest strike the eye by their ardent glow, others are tinged with yellow, and the white light of several has a bluish gleam like that of polished steel. Reddish tints are, however, in the few cases in which they affect lucid stars, the most noticeable, and were the only ones noticed by the ancients.

Ptolemy designates as 'fiery red' (ὑπόκιῤῥοι) the following six stars: Aldebaran, Arcturus, Betelgeux, Antares, Pollux, and —*mirabile dictu*—Sirius! all the rest being indiscriminately classed as 'yellow' (ξάνθοι). Now Pollux at present, though by no means red, is at least yellowish, but Sirius is undeniably white with a cast of blue. A marked change in its colour since the Alexandrian epoch might thus at first sight appear certain, the more so that Seneca makes express mention of the dog-star as being 'redder than Mars;'[1] Horace has 'rubra Canicula' as typical of the heat of summer;[2] and Cicero, in his translation of Aratus, speaks of its 'ruddy light.' Nevertheless the case is doubtful. The questionable epithet, in all probability, crept into the 'Almagest' by a transcriber's error, Ptolemy not being responsible for it. In the early Arabic versions of that work it evidently did not occur, for Arab astronomers of the tenth and subsequent centuries ignored the imputation of colour to the dog-star, and Albategnius stated the number of Ptolemy's red stars as five.[3] Among the Latin writers, the misapprehension (if misapprehension there were) originated with Cicero, who was much more a rhetori-

[1] *Quæst. Nat.* I. i. [2] Sat. ii. 5, 39.
[3] W. T. Lynn, *Observatory*, vol. x. p. 104.

cian than a natural philosopher, and it became practically extinct with the perhaps unverified assertion of Seneca. It is, however, curious to find 'the fiery Sirius' coming up again in the verses of so close and original an observer as Tennyson.

There is better reason to believe that Algol has either really blanched with time, or may be subject to temporary suffusions of colour. It is now purely white, with a spectrum of the Sirian pattern, but appeared red in the tenth century to the Persian astronomer Al Sûfi.[1] His authority is considerable, and his only other addition to the ruddy stars of Ptolemy is fully justified by the ardent glow of 'Cor Hydræ.' It is worth recording, too, that Schmidt noticed in 1841 the 'demon-star' in Perseus as yellowish red, although he never in later years saw it otherwise than white.[2]

The same observer was amazed, March 21, 1852, to perceive Arcturus without a trace of the strong colour familiar to him in it during eleven previous years. In comparison with its paleness, Capella seemed bright yellow, Mars and Betelgeux glowed almost like fire.[3] It was some years before the star resumed its original hue, and the reality of the change, admitted by Argelander, was certified by the observations of Kaiser at Leyden.[4]

The periodical variations in colour of α Ursæ Majoris, the 'Pointer' next the pole, announced by Klein in 1867,[5] were long disbelieved in. Yet they appear actually to take place, though perhaps with much fitful irregularity. A series of observations with Zöllner's polarising colorimeter, executed in 1881 by M. Kövesligethy, of the O Gyalla observatory in Hungary, gave clear evidence of alternating fluctuations between red and yellow in a period of 54½ days.[6] The star is under no suspicion of varying in light.

The colours of the stars visible to the naked eye are faint and pale compared with those disclosed by the telescope. The

[1] Schjellerup, *Description des Étoiles Fixes*, p. 25.
[2] *Astr. Nach.* No. 1099.
[3] *Ibid.* No. 999.
[4] *De Sterrenhemel verklaart*, Pt. i. p. 597.
[5] *Astr. Nach.* Nos. 1663, 2111.
[6] *Sirius*, Bd. xiv. p. 2?3.

real gems of the sky are found low down in the scale of brightness. To some extent this is only what might be expected. Intense tints result from strong selective absorption in the atmospheres of the stars they distinguish, and strong absorption implies large loss of light. Stars shine with the rays that have survived transmission through the glowing vapours in their neighbourhood, and the more nearly those rays are limited to one particular part of the spectrum, the purer and clearer the resulting tint will be. A true prismatic hue could accordingly be produced only through an enormous reduction of brightness; but true prismatic hues do not exist among the stars,[1] the colours of which are always more or less copiously diluted with white light.

The science of star-colours has hitherto made little satisfactory progress. Attempts to set up a standard chromatic scale have not been successful,[2] and instrumental devices for ensuring just and equable judgments may sometimes induce larger errors than they avert.[3] Simple visual estimations, on the other hand, must be treated with great reserve, since 'personal equation' in this matter often assumes enormous proportions. The extreme of colour-blindness is reached by comparatively few, but endless minor individualities of perception vitiate the greater part of an accumulated mass of evidence which might otherwise justify inferences of real change. From the complex bundle of rays forming the image of a star, each retina picks out and accentuates those to which it is most highly sensitive, precluding the possibility of agreement as to delicate tints between many different observers. With both the Herschels, for instance, the equilibrium of colour was shifted towards the red end of the spectrum. Sir William's 'garnet' star (μ Cephei), Sir John's 'ruby' stars, have appeared merely reddish yellow to subsequent observers. Struve's assistant, Knorre, saw all stars indiscriminately white; Admiral Smyth, on the contrary,

[1] Struve, *Mensuræ Micrometricæ*, p. lxxxvi.

[2] See the system proposed by Franks, *Monthly Notices*, vol. xlvii. p. 269.

[3] See the results given by Kövesligethy, *Ueber eine neue Methode de Farbenbestimmung der Sterne*, Halle, 1887.

discriminated between shades of colour altogether inappreciable to most of those who have profited by his 'Cycle of Celestial Objects.'

Even of the same observer the impressions do not always agree. Fatigue and advancing years modify the colour-sense; and M. Safarik states that stars invariably appear redder to his left than to his right eye.[1] Atmospheric conditions, too, are powerfully operative. Misty air blots out faint tints and alters strong ones, azure visibly turning green through its influence. Height above the horizon is another circumstance to be taken into account before any useful comparisons can be made, while instrumental causes tend further to perplex their upshot. Large apertures in themselves help to bring out colour, especially in small stars; but the colour-correction of great refractors is always imperfect, and the outstanding blue fringe usually conspicuous in them must by contrast give a reddish tone to the image.[2] Reflectors produce a similar effect through absorption of many of the higher rays by the silvered glass or speculum metal forming their mirrors. Indeed, Mr. Franks, an assiduous cultivator of stellar chromatics, thinks there is not much to choose in this respect between reflecting and *perfectly* achromatic telescopes,[3] although their size and adjustment make a very great difference. Above all, only medium powers should be used in colour observations, since with high magnification all hues merge, or tend to merge, into yellow.

In this department, then, there is often little connection between appearances and realities. Discrepant statements are far from necessarily implying actual variation. The former abound; instances of the latter are to be met with, but can only be admitted with extreme caution.

The study of star-colours divides naturally into two branches—one concerned with isolated, the other with compound objects. Inquiries in the first case are simplified by

[1] *Vierteljahrsschrift Astr. Ges.* Jahrg. xiv. p. 378.

[2] Webb, *Student*, vol. v. p. 487.

[3] *Jour. Liv. Astr. Soc.* vol. v. p. 88. Mr. Franks considers that, on the whole, colour *pales* with increase of light. Struve held the opposite opinion. See *Mens. Microm.* p. lxxxvi.

the curious and unexplained fact that single stars are never markedly tinged except with red or yellow. Vega makes the nearest approach in the northern hemisphere to an independently blue star; γ Toucanæ, α Eridani, and ϵ Pavonis are the 'pale sapphires' of the southern sky. But they are very pale indeed, so pale as to produce no definite impression of colour upon ordinary eyes. Nor is the 'emerald' tinge of β Libræ more decided. We have accordingly to deal just at present only with 'red stars.'

The earliest list of thirty-three of them was drawn up by Lalande in 1805.[1] 'Ces étoiles,' Von Zach remarked in republishing it in 1822, 'annoncent toujours quelque chose de particulier; or toute particularité mérite d'être observée.'[2] We have to a great extent got rid of the notion which presented itself to John Michell in 1767,[3] that what they 'announce' is the impending extinction of their own fires, but their peculiarities have become, on that very account, all the more worthy of attention. Red stars are commonly variable both in light and colour; the display of colonnaded and zoned spectra belongs exclusively to them; and they are strongly characterised by atmospheric extent and instability. Few of them can be watched long and attentively without being caught in some singular phase of change.

Their systematic study began with the publication in 1866 of the late M. Schjellerup's Catalogue of 280 red stars;[4] ten years later, Mr. Birmingham of Tuam completed a similar work comprising 658 entries,[5] and Mr. Chambers laid before the Royal Astronomical Society, April 6, 1887, a catalogue founded on his personal observations during seventeen years.[6] Of 711 nominally 'red' stars in both hemispheres, he had examined 589, being virtually all those visible in England, with the result of finding the colour of most exaggerated. 'Orange' was to his eye the tint prevailing among them;

[1] *Connoissance des Tems pour l'an* 1806.

[2] *Corresp. Astr.* t. vii. p. 296.

[3] *Phil. Trans.* vol. lvii. p. 238.

[4] *Astr. Nach.* No. 1591; reprinted with numerous additions in *Vierteljahrsschrift Astr. Ges.* Jahrg. ix. p. 252.

[5] *Memoirs R. Irish Acad.* vol. xxvi. p. 249.

[6] *Monthly Notices*, vol. xlvii. p. 348.

true 'reds' were scarce; of stars meriting to be qualified as 'carmine' or 'ruby' he had not met above a dozen.

Observers in this line have indeed usually expressed disappointment with the tints described by their predecessors. Sir John Herschel's seventy-six 'ruby-coloured objects,'[1] fall short, by general admission, of the expectations raised by that epithet. Schjellerup found most of Schmidt's red stars nearly colourless, Birmingham made the same remark as to many of Schjellerup's, Chambers is dissatisfied with Birmingham's. Birmingham did not, however, claim much depth of colour for most of the objects he enumerated; he fully admitted their 'redness' to be, with few exceptions, largely adulterated with orange, and Schmidt has pointed out that there is absolutely no star of the pure and perfect crimson of a solar prominence. This is necessarily the case, since no stellar atmosphere is entirely opaque to rays in the upper part of the spectrum.

It would, nevertheless, be a mistake to suppose that red stars are not striking telescopic objects. Their light, even in the less distinguished specimens, has a lurid glow which at once marks them out from ordinary stars, and those of deeper tints shine with an ardour recalling the wrathful intensity of a stormy sunset. The contrast between a red and a white star in the same field of view is sometimes most beautiful and effective. Thus, in the southern constellation Grus, π^1 and π^2 show like little burnished discs of copper and silver respectively, seen under strong illumination.

Among conspicuous stars, Antares, in the heart of the Scorpion, is the ruddiest, Betelgeux comes next, and both are suspected to vary in tint. Aldebaran and Arcturus have figured immemorially in the short list of visibly fiery objects, to which Al Sûfi added α Hydræ, and Father Noël γ Crucis. But their colours are mere pale shades compared with those instrumentally brought into notice. 'Hind's crimson star,' otherwise known as 'R Leporis,' appeared to its discoverer in 1845 like 'a drop of blood on a black field.' As with most other variables, however, increase of light brings with it a paling of colour, intense redness giving place to a coppery

[1] *Cape Observations*, p. 448.

hue as it rises above the eighth magnitude. Its spectrum is of the fourth type, with particularly strong absorption of the blue rays, a very small proportion of which penetrate the veil of dense vapours surrounding this remarkable object.

A similar one, now known as V Hydræ, is No. 16 of Lalande's, No. 136 of Schjellerup's Red Stars, and was recorded by Dr. Copeland at Dunsink, March 22, 1876, as 'brown red,' and of 7·2 magnitude.[1] But three years later, Dr. Dreyer found it risen to the sixth magnitude, and of a 'most magnificent copper red,' while Birmingham observed it in 1874 as of the eighth, Dunér in 1884, down to 9·5 magnitude. Its fluctuations of light are thus very considerable.

Close to one of the gems of the Southern Cross, an eighth magnitude star was described by Sir John Herschel as of 'the fullest and deepest maroon-red, the most intense blood-red of any star I have seen. It is like a drop of blood when contrasted with the whiteness of β Crucis.'[2] Among other southern stars remarkable for colour, are R Sculptoris, no less 'intensely scarlet' now than when Gould saw it at Cordoba; R Doradûs, glowing like a live coal out of the darkness of space, and L_2 Puppis; all of them noted variables.

In the northern hemisphere, V Cygni bears the palm for depth of tint, which, however, pales somewhat towards its maxima; and not far inferior to it are R Cassiopeiæ, R Leonis, R Crateris, and Mira, with U Cygni and U Cassiopeiæ, both splendidly set off by the vicinity of blue attendants. Crimson, indeed, verges, in these and other periodical stars, more and more towards orange in their brightening phases; yet they always remain technically 'red.' A few cases of complete, if temporary change of colour have, however, been recorded. Thus, a seventh magnitude star in the Lynx (90 Schjellerup) noted by Struve as 'rubra,' by Secchi as 'bella gialla,' seemed blue, or bluish white to Birmingham January 13, 1874, a confirmatory and nearly contemporaneous observation being made at Greenwich.[3] A star of 8½ magnitude (148 Schjellerup)

[1] *Dunsink Observations*, vol. iv. p. 55. [2] *Cape Observations*, p. 448.
[3] *Memoirs R. Irish Acad.* vol. xxvi. p. 269.

called 'scarlet' by Lord Rosse in 1861, 'dark red' by d'Arrest December 8, 1866, showed no colour to Birmingham May 8, 1874. Dunér found it, nevertheless, of a deep orange red in 1884, and it is characterised by a fine colonnaded spectrum. Again, Schjellerup was struck with the redness of a star in Aquila[1] in 1863, which after an interval of ten years, appeared to Birmingham actually *blue*; and similarly, a bluish-white object occupied the place, November 14, 1850, of a star in Taurus marked 'very red' by Mr. Hind, September 3, 1848.[2] One further instance may be mentioned, in the hope of attracting the attention of southern observers to an object of particular interest. A fifth magnitude star in Argo, known as *r* Velorum,[3] was notoriously red during Gould's stay at Cordoba. But it seemed to the present writer at the Cape in the autumn of 1888, perfectly white, or tinged, if tinged at all, with blue; and a spectrum to correspond, continuous, save for a probable dark F, was derived from it. We have unluckily no knowledge as to the nature of its spectrum in its antecedent red stage.

The following list collects a few of the best authenticated examples of colour-change among the red stars.

Name	Magnitude	Change
1 Schjellerup	8·7	Ruby to slight tinge of red.
4 Schjellerup	8·8	Deep to pale red.
5 Schjellerup	7–8 (var. ?)	Garnet to white.
8 Schjellerup	7·5–8	Very red to white.
Copenhagen Catalogue 3282	8·0	Red to blue.
63 Schjellerup . . .	7·3	Red to bluish white.
90 Schjellerup . . .	7·0	Red to bluish white.
148 Schjellerup . . .	8·5–9·5 (var. ?)	Scarlet to white.
168 Schjellerup . . .	7·7	Orange to white.
214 Schjellerup . . .	7·5	Red to blue.
222b Schjellerup. . .	7·5–8·0	Red to white.

The changes of colour visible in temporary stars have generally been in an opposite direction to those of ordinary

[1] No. 6803 of the *Copenhagen Catalogue* = No. 214 of Schjellerup's *Red Stars*.

[2] *Monthly Notices*, vol. xi. p. 46.

[3] The place for 1875 of the star is R.A. 10h. 16m. 48s., D. −41° 1·3′. There is another *r* Argûs of the same magnitude.

variables. Their sanguine tints faded, instead of deepening with the decline of their light. Thus, Tycho's star, though it passed through an intermediate stage of redness, was of a leaden white when it disappeared. T Coronæ ran nearly the same course. Nova Ophiuchi (1848) and Nova Andromedæ were ruddy at first, colourless later. Nova Cygni from orange turned bluish. The corresponding colour phases of a star a little to the north of a nebula in Leo (New Gen. Cat., 3107) seem to mark it as of peculiar constitution. It fluctuates from about eighth to ninth magnitude, and simultaneously from decided orange to slight blue.[1] One of the very few periodical stars which lose colour and light together is T Geminorum. Below the ninth magnitude it has scarcely a tinge of yellow; at maximum it flashes deep orange or red.[2]

Red stars are very unequally distributed. Certain wide tracts of the sky are nearly destitute of them; in some, they occur profusely. The Milky Way between Aquila, Lyra, and Cygnus was called by Birmingham the 'Red Region';[3] but other galactic constellations, such as Perseus and Cassiopeia, are all but exclusively formed of white stars.[4]

Evidently, however, real partialities of colour-distribution must be to a great extent masked by the projection, to the eye, of objects at indefinite distances from each other upon the same portion of the sphere. Hence extensive local collections of similar stars may be so confused with overlying and underlying aggregations as to be completely unrecognisable. Smaller groupings are more readily detected. It is by no accident that, in the immediate neighbourhood of one red star, others are so apt to be met with; and the 'brick red' and 'ruby' pairs included in Herschel's Cape list, may with confidence be assumed to be severally in some sort of physical connection. Red stars, it was remarked by the same authority, are conspicuous in many clusters both by brightness and situation; and Father Secchi was struck with the critical

[1] *Dunsink Observations*, vol. iv. p. 54.

[2] Hind, *Astr. Nach.* No. 839.

[3] *Memoirs R. Irish Acad.* vol. xxvi. p. 255.

[4] Franks, *Monthly Notices*, vol. xlvi. p. 343; see also Osthoff, *Wochenschrift für Astr.* Bd. xix. (1876), p. 326.

positions of such objects as regards spiral or radiated stellar arrangements in the Milky Way.[1]

The principle of colour by association is barely indicated in clusters, and chiefly in a 'jewelled' one about κ Crucis, as to which we shall say a few words by-and-by.[2] It is in double stars carried out to the highest perfection. Nature is inexhaustible in her display among them of harmonies, contrasts, and delicate gradations of hue. They not only vividly sparkle in green and gold, azure and crimson, but shine in the sober radiance of fawn and olive, lilac, deep purple, and ashen gray. Chalcedony, aquamarine, chrysolite, agate, and onyx have counterparts in the heavens as well as rubies and emeralds, sards, sapphires, and topazes. These beautiful tints do not occur at random. We can partially discern some 'law of order' governing their development; but empirically as yet, and without any true insight into its cause.

Mariotte of Dijon, a physicist, but no astronomer, was the first to speak of *blue* stars. 'Les étoiles qui paraissent bleues,' he wrote in 1681, 'ont une lumière faible, mais pure et sans mélange d'exhalaisons.'[3] But he gave no examples, and it is not easy to divine what class of objects he alluded to. The chromatic observation of double stars was really begun by Father Christian Mayer, at Mannheim in 1776; the interest of his preliminary efforts was, however, completely absorbed in the splendour of Herschel's similar, but vastly more extensive and assured results. He not only discovered a great number of exquisitely tinted couples, but by his success emphasised the importance of systematic attention to colour in double stars.

His example was followed by F. G. W. Struve, who in 1837 classified from this point of view 596 of the brightest known stellar pairs. The upshot was to prove agreement in colour the rule, contrast the exception.[4] Just half, or 295 of the objects examined had both components white; 118 had both yellow or reddish with slight differences of intensity; sixty-

[1] *Atti dei Nuovi Lincei*, t. vii. p. 72. [2] See *infr.* p. 240.

[3] *Oeuvres*, t. i. p. 287.

[4] *Étoiles Doubles*, pp. 33-4; *Mensuræ Microm.* p. lxxxii.

three were tinged with blue. The instances of genuine contrast numbered 120, and in *all* of these the small star was called blue. The rule is, moreover, without exception, that no primary member of a dissimilarly tinted pair is blue.

The reality of chromatic contrasts in double stars was established by the persistence of colour in satellites during the obliteration of their primaries by an interposed wire or bar; and besides, as Struve remarked, optically produced tints should be invariably complementary, which is far from being the case. A curious proof of this independence was afforded by a double occultation of Antares and its companion, observed by Dawes in 1856. The small star, emerging first from behind the moon, seemed as perfectly green viewed thus alone, as when half lost in the glare of the great red star it is attached to.[1] The same phenomenon was re-observed in 1878.

The connection between inequality of brightness and inequality of colour in coupled stars did not escape Struve's notice.[2] He found a mean difference of less than half a magnitude between the exactly similar members of 375 pairs; of over one magnitude for 101 stars showing varied shades of the same colour; and of nearly two magnitudes in 120 cases of contrasted tints. Professor Holden, taking into account all known physical pairs and none others, reached, in 1880, an analogous result.[3] Where there was identity of colour the average difference of lustre proved to be only half a magnitude; where there was diversity, the luminous inequality mounted to two and a half magnitudes. One hundred and twenty-two of the stars considered belonged to the first class, forty to the second. Now markedly unequal are generally wide pairs,[4] so that disparity of hue is seen to prevail in systems formed by a large star and a comparatively small and remote companion; while genuine twin suns, of not very different radiative power, and of similar radiative quality,

[1] *Monthly Notices*, vol. xvi. p. 143; Niesten, *Ciel et Terre*, t. ii. p. 96; Webb, *Cel. Objects*, p. 386.

[2] *Mens. Microm.* p. lxxxii.

[3] *Amer. Jour. of Science*, vol. xix. p. 467.

[4] Doberck, *Astr. Nach.* No. 2278.

circulate as a rule rapidly and closely round their common centre of gravity. Why this is so we cannot tell; the bare fact is before us.

A good many beautifully coloured stars are, nevertheless, ascertained to be in mutual revolution. The yellow and rose-red components of η Cassiopeiæ finish their circuit in about two hundred years; those of ε Boötis, shining chrome yellow and sea-water blue, in probably upwards of twelve hundred; ξ Boötis and π Cephei, orange and purple, o Cephei and τ Cygni, golden and azure pairs, are all in undoubted orbital movement. A good many richly tinted stars, on the other hand, appear stationary, doubtless because their distances apart are so considerable as to make their revolutions inordinately slow. Thus the emerald-green companion of α Herculis has preserved, during a century, an invariable position with regard to the ruddy star it depends upon; and Antares forms with its sea-green satellite a somewhat similar and equally rigid combination. The fixed pair β Cygni (Albireo) shining with 'yellow topaz' and 'aquacœlestis blue' light, present perhaps the most lovely effect of colour in the heavens, nearly matched, however, by the variable δ Cephei and its cœrulean companion. Among numerous other examples of contrasted or harmonising tints in double stars, may be mentioned: γ Andromedæ, orange and green; γ Delphini, yellow and pale emerald; η Persei, golden and azure; 24 Comæ Berenices, orange and lilac; 12 Canum Venaticorum, pale yellow and fawn; ν Serpentis, sea-green and lilac; a pair in Cassiopeia (Σ 163), copper colour and blue; 17 Virginis, light rose and dusky red; o Draconis, orange and emerald.

Bright white stars have not unfrequently small blue ones in their vicinity. A distant companion of Regulus seems as if steeped in indigo; Rigel has an azure attendant; λ Geminorum, one of an amethystine shade; 84 Ceti and 62 Eridani are made up each of a white and a lilac star; while the sapphire 1 Boötis is grouped closely with one, more loosely with two other subordinate blue objects.[1]

[1] Flammarion, *Catalogue des Étoiles Doubles*, p. 76.

Two questions at once suggest themselves about the colours of double stars. To the first, Are they real? a decisively affirmative answer can be given; but the second, Are they permanent? cannot be disposed of with such promptitude. The subtlety of hues resulting from a highly complex set of retinal impressions renders them peculiarly liable to *subjective* variation. As evidence of *objective* variation, then, random notes of colour are of little or no use. Only the estimates of skilled observers, trained to the needful precautions, furnished with suitable instruments, above all, owning normal eyesight, are worth weighing and comparing. Under this rule of exclusiveness, the testimony requiring the admission of real change shrinks surprisingly in compass, but it does not wholly disappear. Colour-variables are to be found among compound, no less than among single stars.

Owing partly to instrumental, partly to personal causes, the elder Struve perceived as purely white many stars seen by Herschel with a tinge of red or yellow. Disagreements in the opposite direction merited then particular attention, and such disagreements were especially marked in two cases.[1] The components of the splendid couple γ Leonis were described by Herschel in 1784 as both white, the smaller inclining slightly to pale red.[2] But Struve saw them in 1837 golden yellow and 'reddish' green, Admiral Smyth 'bright orange and greenish yellow;' and such (allowing for some inevitable inconsistencies both in the definition and perception of colour) they still remain. Here then we have a strong presumption of genuine change, which in the companion instance of γ Delphini, is raised almost to certainty. These last stars, noted by Herschel in 1779 as both perfectly white, showed golden yellow and bluish green to Struve's scrutiny. The progress of alteration may perhaps be marked by the younger Herschel's and South's record of them as white and yellowish in 1824.[3] Their contrasted tints of orange and green now strike the eye at the first glance with the smallest telescope.[4]

[1] *Mensuræ Microm.* p. lxxxvii.

[2] *Phil. Trans.* vol. lxxv. p. 48.

[3] *Ibid.* vol. cxiv. p. 363.

[4] Noble, *Hours with a Three-inch Telescope*, p. 111.

Another pair famous for colour-fluctuations is 95 Herculis, composed of two equal stars of $5\frac{1}{2}$ magnitude, planted (to appearance) immovably within 6″ of each other. Familiar with them as vividly tinted objects, Professor Piazzi Smyth was astonished, on pointing his telescope towards them from the Peak of Teneriffe, July 29, 1856, to perceive them both white.[1] In the following year, however, they shone as before in 'apple green and cherry red,' and were so observed by Admiral Smyth, Dawes, and others. Captain Higgens[2] actually watched these colours fade and revive in 1862–3, in the course of about a year, but no trace of them has been seen of late; the stars of 95 Herculis are of an identical palish yellow.[3] Their history goes back to 1780, when Herschel observed them as bluish white and white; J. Herschel and South called them 'bluish white and reddish' in 1824, Struve, 1828–32, greenish yellow and reddish yellow, in precise agreement with Pickering's appraisement in 1878.[4] Thus, the 'magnificent tints of orange and green' which Secchi admired in 1855, and Piazzi Smyth missed in 1856, were of a transitory character.

In the well-known binary, 70 Ophiuchi, there has been an equally undoubted change. Except an 'inclination to red' in the smaller, the elder Herschel perceived no colour in either of these stars; his son and Sir James South called them white and 'livid'; yet they were recorded by Struve as of an especially intense yellow and purple, by Admiral Smyth as 'pale topaz and violet.' They are now both yellow, very much as they were seen by Secchi in 1855, and by Franks in 1876; the companion was, however, marked 'purplish' at Harvard College in 1878, 'rose-coloured' by Flammarion in 1879.

The three stars of ζ Cancri are usually yellow, but Dembowski noticed them as all white 1854–6, the remoter component turning yellowish or olive in 1864–5.[5] This form of change is not very rare among revolving stars, both of which deepen and lighten, for the most part simultaneously,

[1] Smyth, *Sidereal Chromatics*, pp. 35, 78.
[2] *Ibid.* p. 80.
[3] Noble, *Op. cit.* p. 105.
[4] *Harvard Annals*, vol. xi. p. 150.
[5] *Astr. Nach.* Nos. 1110, 1574.

through various shades of primrose and cowslip; while the development in other couples of the more vivid hues of the spectrum tends towards the production of contrast. It often happens, too, that one component only varies in colour, in which case the change *always* affects the satellite star. The attendant for instance of δ Herculis has appeared by turns ashen, 'grape-red,' blue, and bluish green; that of δ Cygni was observed by Struve as grey in 1826–33, but conspicuously red in 1836, blue by Dawes in 1839–41, alternately red, blue, and violet by Secchi in 1856–7, grey once more by Dembowski in 1862–3, red by Engelmann in 1865. Since then it has commonly seemed light blue;[1] but showed nevertheless with the great Nice refractor in 1883 and 1886 as yellow or orange.[2] Again, the multiple star σ Orionis includes one if not two colour-variables; the distant companion of γ Leporis changed from pale green in 1832 to garnet in 1851 and 1874; and the satellite of ν Serpentis from lilac in 1832 to 'native copper' in 1851.[3]

M. Niesten of Brussels attempted in 1879 to assimilate colour-variations in revolving stars to the supposed production of sunspots through planetary influence.[4] A tendency he thought could be traced to the emission by both components of white or yellowish light near 'periastron,' or their nearest approach to each other, deeper or different tints becoming more and more prominent with their increasing distance. But we need only recall, in proof of the insufficiency of this hypothesis, that two of the most strikingly changeable couples—95 Herculis and γ Delphini—have *no* perceptible relative movement.

The truth is, that we are entirely ignorant as to the causes of the phenomena such objects present, and that the only kind of data by which there is much hope of their being elucidated have not yet begun to be collected. Eye-estimates of colour, however trustworthy, do not reach below the surface; they are mere indications, which the spectroscope and the

[1] Engelmann, *Astr. Nach.* No. 1676.

[2] *Annales de l'Observatoire de Nice*, t. ii. (Perrotin).

[3] Smyth, *Sid. Chromatics*, p. 29; Webb, *Cel. Objects*, p. 389.

[4] *Bull. de l'Acad.* Bruxelles, t. xlvii. p. 50.

'spectrograph' (a photographic spectral apparatus) can alone help us to interpret. The foundation of a science of chromatic variability will be laid by the first definite statement of a change of spectrum in connection with a change of colour. Until this is forthcoming, there can be no tangible ground for so much as a conjecture on the subject. Similarly, the condition *sine quâ non* of redness in single stars can only be ascertained by learning what are the prismatic alterations supervening upon its disappearance. We should naturally expect them to consist in a marked falling-off in the effects of absorption; but it is also well to remember that the changes in colour of temporary stars have been accompanied by a shifting of the *emissive centre of gravity*, the red fading out

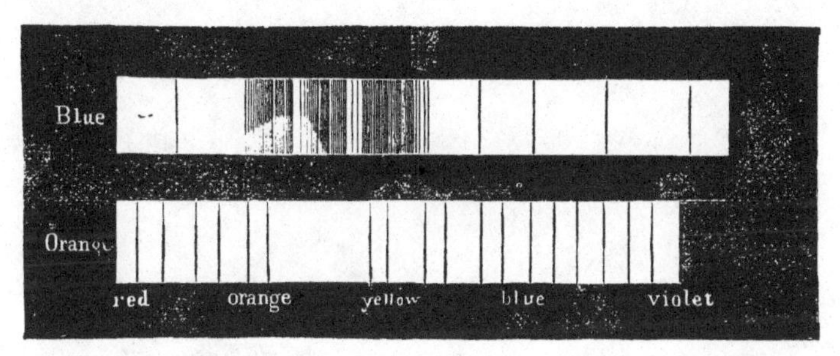

Fig. 28.—Spectra of the Component Stars of β Cygni (Huggins).

of the spectrum of Nova Cygni, for instance, long before the yellow and green.

The spectra of permanently coloured double stars are marked by what we may call 'complementary absorption.' Dr. Huggins showed in 1864[1] that in β Cygni the yellow rays of the larger star escape paying toll to its enveloping vapours, while in its companion the blue rays are comparatively exempt from their encroachments (see fig. 28). A corresponding result was obtained for *a* Herculis. It appears then certain that the contrasted colours of coupled stars are not simply an effect (as has sometimes been supposed) of their having reached different stages of growth, but arise from a profound modifi-

[1] *Phil. Trans.* vol. cliv. p. 431.

cation of their physical state, due in some unexplained way to their mutual influence. The sky-blue, lapis lazuli, and violet tints of numerous secondary stars, unrivalled to the eye among solitary objects, are proved by light-analysis to owe their origin to exceptional atmospheric conditions.

CHAPTER XI.

DOUBLE STARS.

A DOUBLE star is one that divides into two with the help of a more or less powerful telescope. The effect is a strange, and might have appeared beforehand a most unlikely one. Yet it is of quite ordinary occurrence. Double stars are no freak of nature, but part of her settled plan; or rather, they enter systematically into the design of the Mind which is in and above nature.

The first recognised specimen of the class was ζ Ursæ Majoris, the middle 'horse' of the Plough, called by the Arabs 'Mizar,' which Riccioli found at Bologna, in 1650, to consist of a $2\frac{1}{2}$ and a 4 magnitude star within fourteen seconds of arc of each other. Both are radiantly white, and they make a glorious object even in a very small telescope. The accident of a bright comet passing, on February 8, 1665, close to γ Arietis ('Mesarthim') led to the discovery of its duplex nature by Robert Hooke in the course of his observations on the comet. The components, each of the fourth magnitude, and eight seconds apart, are perfectly alike both in light and colour. Meanwhile Huygens had seen θ Orionis —perceived to be quadruple in 1684—as triple in 1656; α Crucis, in the southern hemisphere, was divided by some Jesuit missionaries sent by Louis XIV. to Siam in 1685, and α Centauri by Richaud at Pondicherry in 1689;[1] making in all five double stars detected during the seventeenth century. Four more—γ Virginis, Castor, 61 Cygni, and β Cygni—were picked up in the course of the ensuing fifty-five years; and in 1776, Father Christian Mayer began at Mannheim a deliberate search for stellar couples. His thirty-three discoveries

[1] Flammarion, *L'Astronomie*, t. ii. p. 164.

in two years might be described as the preliminary washings from the rich lode struck a few months later by Sir William Herschel.[1]

The plentifulness of double stars was in itself an irresistible argument for their reality. That any two unconnected bright stars should be projected closely side by side upon the sphere was improbable; that such a contingency could be repeated blindly hundreds of times was what no sane man ought to have been capable of believing. But human credulity is nowhere more conspicuous than in what it is prepared to attribute to chance; and it needed such clear evidence of mutually circling movements as Herschel was able to produce in 1803, to establish the conviction of the *physical* existence of double stars.

The fact is one at which we can never cease to wonder. It brings us face to face with a state of things entirely unfamiliar, and of which the purpose lies beyond the scope of our limited understandings. So accustomed are we to the 'sole dominion' of our own great star, that the presence of *two* suns in one sphere might well at first sight appear incredible. Yet there are many things 'undreamt of in our philosophy' which are nevertheless true. Every drop of stagnant water is a world of uninterpreted mysteries; what we choose to call the 'order of nature' is violated at every instant, inexplicably, by our own volition; and if that order be attended by anomalies upon the earth, how much less shall we venture to prescribe its course in the heavens?

The term 'double star' is obviously quite indefinite, apart from some agreement as to its meaning; and it was in fact used by early observers in a far wider sense than it is now usually considered to bear. Many of the small and remote attendants upon brighter stars recorded by the Herschels could scarcely be presumed to have any real connection with them; 32″ was fixed by Struve as the maximum interval between the components of a genuine double star, or 16″, unless both were brighter than the ninth magnitude; the younger

[1] See the writer's *History of Astronomy*, p. 21, 2nd ed.

Struve's 'Pulkowa Catalogue' included no stars beyond the narrower limit; and Mr. Burnham rejects all pairs below the eighth magnitude above 5″ apart. This progressive restriction has almost necessarily accompanied the improvement of telescopes. With the powerful and perfect refractors now in use, really close pairs accumulate faster than they can continue to be observed; and the further collection of the innumerable loosely associated stellar points they reveal would be mere inane waste of time.

Already some twelve thousand double stars, in the Herschelian sense, have been registered, of which nearly half correspond, by the closeness of their combination, to strict ideas of what a double star should be; about 1,400 are separated by 2″ or less, and between 600 and 700 are visibly revolving. These last interesting cases multiply at present with especial rapidity. They are most apt to occur, as might be expected, among stars at the shortest apparent distances from each other, and requiring accordingly the highest optical powers for their detection. Our acquaintance with most of these is for that reason quite recent, and their movements are only coming to be recognised as one pair after another is re-measured after a few years' interval.

The singular profusion with which stars are planted side by side with a bare *hairbreadth* of sky between, was made apparent by Mr. Burnham's discoveries at Chicago, while he still pursued the profession of a stenographer. His thousand new pairs included 743 at an average distance of 1″·58.[1] This means that the total interval from centre to centre of these objects was just equal to the width of a human hair held thirty-six feet from the eye. About one-tenth of that distance is the minimum at which, even with the great Lick telescope, stars can be divided, but by no means the minimum at which they can separately exist. It is quite certain that numberless stars which must always, either through their distance from ourselves, or the closeness of their companions, remain optically single, are nevertheless

[1] *Memoirs R. Astr. Soc.* vol. xlvii. p. 317.

compound; hence of any given star, as of a chemical 'element,' we can say, not that it is indivisible, but only that it has never been divided.

Such stellar pairs as are known to be in orbital movement are called 'binary stars,' to signify that they form real dual systems. The finest specimen of this kind in the northern heavens is Castor, or *α* Geminorum, composed of a second and a third magnitude star 5″·18 apart. They are both white with a greenish tinge, and can be divided with a very moderate telescope, so that the sight of this brilliant and suggestive object is not reserved for the inner circle of astronomers. Now it happens that Bradley observed the relative situation of these stars in 1719, and the comparison of his record with measures of the present day shows that they have shifted in the interim to the extent of 125°, or more than a third of a revolution. To complete an entire one, then, they would need at the same rate about 420 years. The end of that term, however, will pretty certainly find them not more than half way round. And the reason is obvious. Throughout the last century they were travelling much faster than they are travelling now; for their orbit is so eccentric that their greatest is twice their least distance, and 'periastron' occurred in 1750. During three and a half centuries to come they will still continue to slacken speed, until at last the furthest point of mutual withdrawal is reached, and the slow process of approach entered upon. The circuit cannot be completed, it has been calculated, in much less than a thousand years.

The mass of the conjoined stars of Castor could easily, with the help of Kepler's laws, be deduced from the elements of their orbit, if only their distance from the earth were known. But on this point we have unluckily no authentic information.[1] It seems probable, however, that they are extremely remote. For we can scarcely suppose the two luminaries taken together to contain less matter than the sun, and unless they do, they

[1] Johnson's parallax of 0″·1984 is almost certainly erroneous. If it were exact, the joint mass of the stars of Castor would be less than $\frac{1}{18}$ that of the sun.

must be at or beyond a distance from which light would spend forty-four years in reaching our eyes.[1] But if the sun were thus vastly removed into space, we should receive from it only $\frac{1}{100}$ part of the light we actually receive from Castor, which is hence likely to be a more massive and distant body than we have experimentally assumed it to be.[2] There is no doubt, however, of its possessing much more than the solar luminosity in proportion to mass.

The brightest is also (with one exception) the widest pair of revolving stars in the sky, and a third distinction—that of being nearer to us than any other known sidereal object—accounts for the first two. In *a* Centauri are combined two stars so brilliant that the lesser, though emitting only one-third as much light as its neighbour, is still fully entitled to rank as of the first magnitude. Sir John Herschel found them both yellow, the second even brownish-yellow, but they are now undeniably white, and the companion-star must have gained considerably in lustre during the present century unless Feuillée, Lacaille, Brisbane, and Dunlop erred egregiously together in calling it of fourth magnitude.[3] Since they were observed by the Franciscan monk, Louis Feuillée, at Lima in 1709, these stars have completed two revolutions and entered upon a third, yet there is still some uncertainty as to the length of their period. It seems, however, impossible that it should fall short of eighty-five, and unlikely that it should exceed eighty-eight years.[4] The orbit is about as much elongated as that in which Faye's comet travels round the sun, and carries the stars, accordingly, at 'apastron' to more than twice their 'periastron' distance. They are now about 16″ apart, and are separating fast, having in 1875 swept through their point of nearest approach. The 'mean radius,' or half the major axis of the computed ellipse, if seen

[1] Corresponding to a parallax of 0″·075.

[2] The mass of a star moving in an orbit of known angular dimensions grows with the *third* power, its light with the *second* power of the distance ascribed to it. Mass accordingly *gains upon* light as the star is pushed farther and farther back into space.

[3] See Flammarion's *Catalogue*, p. 81.

[4] Powell, *Monthly Notices*, vol. xlv. p. 18.

square from the earth, would subtend an angle of $18\frac{1}{2}''$, corresponding at the star's distance of twenty-five billion miles to an actual span of (in round numbers) one thousand million miles; so that these lustrous objects are sometimes as close together as Jupiter is to the sun, and never as far off as Uranus. Their mass (computed from Doberck's elements including a period of eighty-eight years) is just twice, their light about $2\frac{1}{3}$ times, that of the sun.

The spectacle is beyond doubt an amazing one of two such bodies united thus *organically* into a single majestic system. That it includes many other members may be taken for granted, although we may never succeed in observing them, and are unable, even in imagination, to bestow or arrange them satisfactorily. Evidently, no planetary scheme or schemes at all resembling our own can depend upon the stars of α Centauri. A Mercury or a Vulcan, at the most, might find shelter in the close vicinity of one from the disturbing power of the other, its possible inhabitants enjoying the combined or alternating radiance of a greater and a lesser sun. Comets entering these precincts must be perplexed to decide between the two potentates claiming their allegiance, and perhaps on occasions pay their court to each in turns, throwing out tails, as they do so, in all sorts of anomalous and contradictory directions. It has, however, been suggested that the clients of double stars circulate about both simultaneously, in orbits wide enough to keep them beyond the reach of dangerous perturbations from either. This is, of course, conceivable, if somewhat unlikely, nor is it impossible that the two kinds of scheme may be combined and harmonised into one highly complex system.

The stars of 61 Cygni, like those of α Centauri, share a rapid onward movement through space. They are among our nearest stellar neighbours, and there is nearly the same amount of inequality between them. They are, however, very much fainter than the southern luminaries they in some respects resemble, one falling somewhat short of the fifth, the other of the sixth magnitude. They are both, the lesser especially, deeply tinged with yellow.

Although they have been under continuous scrutiny since 1753, when Bradley noted the differences in their times of transit, it is only within the last few years that the curvature of their path has become perceptible. While marching alongside of it with what is called a 'common proper movement,' the smaller also shifts its place sensibly as regards the larger star. But for a century and upward the shifting appeared to take place along a straight line. If this had really been the case, the fact would have abolished the presumption of their binary character, and compelled the belief, which was actually adopted by Captain Jacob in 1858,[1] that the stars would eventually part company and cease to have even an apparent connection. It is now, however, evident that they have not the slightest intention of doing so. The first symptoms of a bend in their track were noticed by Mr. Wilson in 1875;[2] and Dr. C. F. W. Peters's investigation in 1885[3] not only showed them to be revolving, but set forth, at least provisionally, the circumstances of their revolution.

In a period of 783 years the companion-star describes round its primary (supposed for the purposes of calculation to be at rest) an orbit of larger angular or apparent dimensions than those assigned to any other stellar path. Its 'mean radius' measures $29\frac{1}{2}''$, and this, at the distance of 61 Cygni (about forty billion miles), is equivalent to an interval just $65\frac{1}{2}$ times that between the earth and sun. These stars then circulate in an orbit more than twice as wide as that of the planet Neptune, and nearly as eccentric as that of the planet Mercury. But in the solar system a body revolving at $65\frac{1}{2}$ times the earth's distance from the sun would finish its rounds in 530 years, while the components of 61 Cygni need for theirs close upon eight centuries. Hence their mutual attraction is clearly less than the power swaying the planetary movements. Their united mass, in fact, is less than half that of the sun. But since the combined surface of two approximately equal stars, together containing 0·45 of the sun's mass, amounts to 0·73 of the sun's surface, they

[1] *Edinburgh New Phil. Jour.* vol. vii. p. 107.

[2] *Monthly Notices*, vol. xxxv. p. 326.

[3] *Astr. Nach.* Nos. 2708-9.

ought, if equally luminous, to give 0·73 of the sun's light. The proportion of it, however, actually emitted by them is only 0·14. They are less luminous, then, per unit of surface (if of the same mean density) in the ratio of 14 to 73, or, say, five times; and, indeed, the sun in their situation would shine as of fully third magnitude, while they barely reach the fifth. These stars are hence very likely to be far advanced in condensation, and may be drawing near the close of their career as light-givers.

The conditions of this system would seem much more propitious to tranquil planetary circling than those of α Centauri. There is more room for it, to begin with; the stars of 61 Cygni are six times farther apart, and their orbit is much more nearly circular. The perturbing influence of each upon the dependents of the other would accordingly be much less formidable than where two powerful orbs contend, as it were at close quarters, for gravitational pre-eminence. The planets of the smaller stars, too, may, on account of the mild quality of their radiations, nestle quite close to them without receiving an excessive amount of heat. If the earth, for example, were suddenly set revolving round one of the pair in the Swan at the mean distance of Mercury from the sun, we should be conscious of no appreciable rise of temperature.

A pair (No. 190 of Herschel and South's catalogue), strongly resembling 61 Cygni, is found in the constellation Libra. A rapid voyage in concert is in them, too, complicated by relative displacements which, though apparently rectilinear, are doubtless in reality a small section of a wide curve.[1] Comparatively little attention has, however, as yet been paid to them, nor are we acquainted with their distance from the earth. They are both yellow, and there is a disparity of about one magnitude between them; hence the analogy with 61 Cygni is physical as well as geometrical.

Two other allied systems are formed by the coloured pairs 70 Ophiuchi and η Cassiopeiæ. Chromatic similarity is, indeed, at present impaired by the substitution of yellow for

[1] Flammarion, *Catalogue*, p. 85; Burnham, *Publications Washburn Observatory*, vol. i. p. 130.

the contrasting rosy hue of the companion-star in 70 Ophiuchi, but it may be expected to become re-established. Both couples (as was pointed out by Mr. Sadler)[1] progress through space at about the same rate, and both are at nearly the same distance of twenty 'light years' from the earth; they show spectra of an identical type (the solar), and the light-power of each relative to mass appears to be very nearly that of the sun. Both, too, have proved singularly recalcitrant to computation. The orbit of 70 Ophiuchi, though one of the earliest experimented upon, can still only be regarded as provisionally determined. The stars have, indeed, hitherto so persistently refused to keep to their predicted places, that disturbances by invisible members of their system have sometimes been brought in to account for their anomalous behaviour. The period of revolution included in the most recent set of elements (Mr. Gore's) is eighty-eight years; the mass of the two stars is nearly three times that of the sun; and their orbit is so considerably eccentric that their distance apart ranges from fourteen to forty-two times the radius of the earth's orbit.

The still more elongated path of η Cassiopeiæ, traversed in 167 years (according to Mr. J. B. Coit's elements), is of a far ampler sweep. Its mean radius is 56½ times that of the terrestrial, nearly twice that of Neptune's orbit. The stars are, nevertheless, at their nearest approaches, not much farther away from each other than Uranus from the sun, and since they together contain more than six times the solar quantity of matter, perturbations of great intensity must at such times affect any trains of attendant bodies they may separately possess.

Laplace's conjecture that space might hold as many dark as bright masses has received some countenance from the phenomena of double stars. For among them are reckoned effects of the attraction of unseen upon the movements of seen bodies, while in one case the detection of an imperfectly luminous object has, like the discovery of Neptune, ensued upon the theoretical indication of its place.

[1] *English Mechanic*, vols. xli. p. 410, xliv. p. 322.

From the nature of the proper motion of Sirius, Bessel inferred in 1844 that it did not travel alone. The line traced out by it must, were it solitary, have been straight, whereas it undulated markedly and regularly once in about half a century. Revolution in that period round an obscure companion was indicated; the elements of the hypothetical Sirian system were computed by Peters and Auwers, and the precise position of the Sirian satellite was assigned by Safford in September 1861. On January 31 following, it was found just in the right spot by Alvan G. Clark, of Cambridgeport, Massachusetts.

The companion of Sirius is a dull yellow star of 8·5 magnitude, almost lost in the glittering radiance of its great neighbour. Their apparent distance having diminished from 10″ in 1862 to 5″ in 1889, it is now an object of such extreme difficulty that only a few of the best telescopes in the world can show it, and the only very recent observations have been made by Mr. Burnham with the thirty-six Lick refractor.[1] As the periastron passage is timed to occur in 1896, it will be some years before the pair again begins to open out.

For many years astronomers did not feel justified in admitting that a body so little luminous as Clark's companion should still be massive enough to sway the onward march of Sirius visibly to and fro. The real source of attractive power, it was thought, still remained to be discovered. But what seemed improbable has, with time, become fully established. The little star picked up at Cambridgeport has now, during twenty-eight years, conformed with such general fidelity to the theoretical orbit of the disturbing body, as to leave no doubt of their identity. The system thus constituted is a very remarkable one. Its chief member is a body extremely bright in proportion to its mass; its secondary member is a body abnormally massive proportionately to its light. Sirius shines like ten thousand, it gravitates like two of its companions. There must hence be an enormous disparity of temperature between them, with a probably corresponding difference of mean density. The smaller body may thus

[1] *Astr. Nach.* No. 2884.

have already advanced far on the road towards planetary solidity and obscurity.

At the distance of Sirius (about fifty billion miles) the sun would appear as a star of the third magnitude. An accumulation of sixty-three suns would then barely supply the emissions of that brilliant orb, the attractive energy of which is, nevertheless, little more than twice that governing the solar system. The revolutions of its satellite are completed in 58½ years (adopting Mr. Gore's recent orbit), at a mean distance twenty-two times that of the earth from the sun, with excursions, at apastron, a hundred millions of miles beyond that of Neptune. Now to control motion so swift in so spacious an orbit, 3¼ times the amount of matter contained in the sun must be present; of which one-third belongs to the satellite itself, constituting it a body rather more ponderous than the sun, though giving no more than $\frac{1}{160}$ of its light. Thus the contrast between the components of this binary star could scarcely among *visible* objects be more pronounced. But a further point of strangeness is reached in the pairing of completely invisible with brilliantly lustrous bodies. This seems to be the case with Procyon, the lesser dog-star.

Since its motion is disturbed in precisely the same way as the motion of Sirius, there can be no doubt that it forms part of a binary combination. But the second member of that combination has not been discovered, and we may fairly say, Mr. Burnham's search for it in the clear air of Mount Hamilton with the great Lick refractor having proved fruitless, cannot be discovered with any optical means now available. Professor Otto Struve's illusory detection of it March 19, 1873, is a curious example of the tricks chance now and again plays upon the most wary observers. The false image seen (produced by reflections between the lenses of the equatoreal) was always situated about 10″ from the genuine image of the large star, and in a horizontal line from it; but it happened that the varying positions of Procyon with regard to the meridian at the times of the successive observations, gave rise, by an extraordinary series of coincidences, to just such an effect of revolution in the satellite as was anticipated by theory.[1] The orbit described by Procyon round the common centre of gravity of itself and its companion, is rather smaller than the orbit of Jupiter. It is nearly circular, and is traversed in a period of forty years. But our ignorance as to the distance apart of the conjoined bodies entails, of course, ignorance as to their mass; hence the real circumstances of the system still remain an enigma. Procyon is farther off from the earth than Sirius, though not in the proportion of its inferiority in magnitude. Its actual light-emissions are about one-third those of the greater dog-star.

The detection of partially obscure stellar schemes opens a wide field to conjecture. They may be much more numerous than is supposed, for the undulatory movements betraying their existence should escape notice in any but the nearest stars by their minuteness, and even in these if the plane they lay in made any considerable angle with the line of sight. Here, however, the spectroscope comes to our assistance by providing the means of investigating variable motion *in* the line of sight.[2]

Nor can we avoid probing with our thoughts the relation obviously existing between stars like Procyon and stars like Algol. The common peculiarity of being attended by dark satellites affects motion sensibly in one case, light in the other. But the distinction may be more apparent than real. It is conceivable that the satellite of Procyon may have originated in a situation resembling that now occupied by the satellite of Algol, from which it was gradually pushed outward by the reactive effects of tidal friction. The eclipses of Procyon could never, however, under any circumstances, have been visible from the earth, for its orbit lies square to our visual ray, and an interposing body must needs travel edgewise towards and from us.

There are other criteria besides that of visible revolution in an orbit, by which physical can be distinguished from optical double stars. Since 1812, when Bessel pointed out

[1] *Bull. de l'Acad. Impériale*, t. xviii. p. 564; *Monthly Notices*, vol. xxxvii. p. 193.

[2] See *infra*, p. 212.

the conclusiveness of the argument for real connection implied in the advance together of the stars of 61 Cygni,[1] 'common proper motion' has been universally admitted as a proof of genuine association. Thus the lustrous pair γ Arietis has continued relatively fixed since Bradley measured it in 1755; yet its members are fellow-travellers through space, and assuredly keep mutually circling as they go, although so slowly that a century and a quarter count almost for nothing in the majestic cycle of their revolutions. Again, the brightest star in the Southern Cross is made up of two stars of the second magnitude 5″ apart, the situations of which have not perceptibly changed since Sir John Herschel determined them in 1834. This, however, could scarcely be the case unless both shared the small movement attributed to the compound object. And even independently of this positive test, the probabilities are enormously against the accidental close juxtaposition of two stars so brilliant and so nearly equal as those of α Crucis.

The circumstance testifies strongly to the prevalence of physical connection between stellar pairs, that the average difference of brightness between them grows steadily with their distance apart, approximately equal being thus usually contiguous objects. Every degree of inequality is indeed found in undoubted systems; still the chances of optical association must obviously increase enormously, even at the same distances, with increase of optical disparity.

The background of the sky is so thickly strewn with small stars, that we cannot be surprised if some of them happen to occupy critical situations. There is, indeed, more reason for surprise at finding that certain remote satellites of bright stars seem indissolubly united to them. Regulus, for example, carries with it, as it pursues its way across the sphere, a star between the eighth and ninth magnitude, of an ultramarine colour, and discovered by Winlock at Washington to be itself closely double. The interval between the pair and its governing body amounts to nearly three minutes of arc. Castor, too, has attached to it a tenth magnitude star at 74″; and Aldebaran forms with a minute object at 31″ what seems to be a

[1] *Monat. Correspondenz*, Bd. xxvi. p. 160.

permanent combination resembling in its effect to the eye that of Mars with his outer satellite.[1]

Where two close stars seem fixed, relatively and absolutely, the case for their physical union must depend upon circumstantial evidence alone. But this is sometimes of overwhelming strength. Contrast of colour, for instance, may afford grounds for a persuasion amounting almost to a certainty of real relationship. Such tints as blue, green, and violet, only occur among mutually associated stars; nor can we possibly suppose the association upon which they depend for their production to be merely apparent. The topaz and azure components of β Cygni, for instance, have no appreciable motion of any kind, and they are separated by a gap of 34″, exceeding the limit of distance of real double stars as defined by Struve. But it is impossible to doubt that their brilliant hues are truly expressive of the systemic union from which they in some unknown way result. The same may be said of δ Cephei and its bright blue attendant, and of the much closer and nearly equal stars 95 Herculis, the inference in their case being strengthened by the concerted changes of colour recorded of them. We might be sure, too, of the dependent status of the emerald satellites of the red stars α Herculis and Antares, even if it were not independently proved by a community with their primaries in very slow progressive movement.

Nevertheless, some highly coloured pairs have been concluded by M. Flammarion, from a careful study of their relative displacements, to be optical.[2] Among these are o Draconis, which might be called a replica, subdued in lustre, of β Cygni; σ^2 Piscium, golden and blue; 42 Piscium, yellow and green; a gold and purple pair in Auriga (OΣ 154); two nearly equal stars in Perseus (Σ 434) showing lovely tints of golden yellow and azure, with several besides.

If these colours be inherent, it is difficult to believe that the stars distinguished by them are simply thrown together by perspective. Before venturing to pronounce, however, we

[1] Burnham, *Astr. Nach.* Nos. 2189, 2875.

[2] *Comptes Rendus*, t. lxxxvii. pp. 836, 872.

must wait, and let their motions develop. By rashly anticipating nature, we often only display our ignorance of her hidden workings. Meanwhile, experiments might be made with a view to detect any lurking influence of contrast upon the hues of these particular couples, which, however regarded, present points of curious interest.

The display of similar goes quite as far as the display of dissimilar colours of an unusual kind, towards proving a physical union between adjacent stars. Strikingly red pairs, for instance, even when pretty widely separated, can hardly be the result of chance. Several of them are known, but their fixed character does not invite frequent observation. Variability of light supplies another valuable indication of relationship. When common to both members of a pair, it leaves no room for doubt on the subject. We shall recur to this topic in the next chapter.

Stars with ascertained proper motions characterise of themselves the nature of their companionship. For either they keep on together, or they show signs of incipient separation, and so slowly but surely discriminate between a lasting union and mere temporary contiguity. In the latter case the movement of one of the stars referred to the other necessarily proceeds along a straight line, so that rectilinear displacement is an infallible, and the only mark of an optical couple. One curiously close (Σ 1516) occurs in the constellation Draco. Two stars, of 7 and 7·5 magnitude, passed in 1856 so near to one another by the hazard of their paths nearly intersecting, as to present the effect of two points of light, one inch apart, at a furlong's distance from the eye. Their angular separation, then only 2″·6, is now 13″, and it will continue to grow indefinitely. Their absolute disconnection has been confirmed by direct measurements showing them to differ enormously in remoteness from the earth. The larger of the two, by one of those singularities which abound in the heavens, forms a genuine pair with a star very much fainter than its spurious companion.

From what has been said, it is clear that a good deal of patience is needed for the investigation of double stars. The facts about them must often be allowed to ripen for a long time before they can be turned to account. Sooner or later, however, their fruit cannot fail to appear. There is, perhaps, no other branch of science in which industry is so sure to be rewarded with definite results. The first step is to separate clearly perspective from physical couples; and this can only be done by the persistent repetition of exact measures. The next is to detect circulatory movements in the latter as they begin to be apparent, or to keep watch on them as they progress. Their careful comparison with theory may at any time bring surprising novelties to light. For each stellar system is in effect a world by itself, original in its design, varied in its relationships, teeming with details of high significance. But at present only an imperfectly traced outline of the construction of some three score among thousands of them, is before us; their multitude, in fact, distracting attention. Yet it would be better to make intimate acquaintance with one than to know hundreds superficially. All the resources of modern inventiveness should be enlisted in these inquiries. Not only the revolutions, distances, and masses of double stars, their movements *across* and *in* the line of sight, should be determined with ever-increasing precision, but their colours and magnitudes, and above all, their separate spectra, both visual and photographic, should be recorded. By such means as these, real knowledge will be augmented far more than by the most brilliant success in the telescopic detection of new pairs. This has its own interest and value, but the recesses of sidereal structure must be otherwise explored.

CHAPTER XII.

VARIABLE DOUBLE STARS.

THE light-changes of double stars are commonly of a fitful and indecisive kind. They may affect one or both members of stationary pairs; but visibly revolving stars, as a rule, conspire to vary, if they vary at all. The alternating fluctuations of γ Virginis, discoverable only by close attention to the swaying balance of lustre between the components, are in this respect typical. Each may be described as normally of the third magnitude, and each in turn declines by about half a magnitude and recovers within a few days, yet so that the general preponderance during a cycle of several years, remains to the same star. The existence of this double periodicity was recognised in 1851 by M. Otto Struve, who, however, despaired of investigating it with success in a latitude where the stars in question never rise more than 30° above the horizon.[1]

Their circulation is in the most eccentric of ascertained stellar orbits (see fig. 30). The ellipse traversed by γ Virginis in 180 years is, in fact, proportionately somewhat narrower than the path round the sun of Encke's comet, so that the stars will in 1926 be separated by fully seventeen times the interval of space between them in 1836, when they merged into a single telescopic object. Their inequalities of light seem to have developed as they approached each other; at least, they first began to be noticed by Struve in 1818, and they at present tend to become obliterated, whether to revive with regained proximity towards the close of the twentieth century, future observations must decide. A spectrum of the

[1] *Observations de Poulkowa*, t. ix. p. 122.

Sirian pattern is combined with a perceptible tinge of yellow in their light.

Relative variability is in 44 Boötis still more marked than in γ Virginis. But here a fundamental disparity between the components is seldom and temporarily abolished. Noted by Herschel as considerably unequal in 1781, they appeared to him perfectly matched in 1787. And it may be noted that they had in the interim passed periastron. Struve observed, June 16, 1819, a difference between them of two magnitudes, which had sunk to half a magnitude in 1833. Argelander found them precisely equal June 6, 1830; Dawes perceived, April 27, 1841, a slight advantage on the side of the usually smaller star; while the superiority of its companion was recorded by M. Dunér at Lund as ranging, during the years 1869 to 1875, from 0·4 to 1·3 magnitudes.[1] Since their changes are often simultaneous, though not always in the same direction, their combined variability has never been conspicuous. The stars of 44 Boötis complete their rounds in a highly eccentric orbit in a period of 261 years.[2] Their tints, varying from yellow and sky-blue to white and dull grey, cannot be without influence upon their photographic magnitudes, which were determined at Paris in 1886 to be 5·3 and 6. Their joint light, though of the same spectroscopic quality, has then only one-twelfth the intensity of that of γ Virginis.

The component stars of ζ Boötis when photometrically measured at Harvard College in 1883 were of 4·4 and 4·8 magnitudes, but the order of their brightness has been at least three times reversed during a century of observation.[3] Their period of revolution must be of enormous length. From 1796 to 1841 they appeared fixed; then a very slow wheeling movement became perceptible, accompanied by a diminution of distance, and it now taxes the powers of the best telescopes to divide them.[4] Their spectrum is of the Sirian type.

[1] *Lund Observations*, 1876, p. 74. [2] Doberck's elements.

[3] *Harvard Annals*, vol. xiv. p. 458; *Observations de Poulkowa*, t. ix. p. 143; Dunér, *Mésures Micrométriques*, p. 68.

[4] Crossley, *Handbook of Double Stars*, p. 299; Tarrant, *Jour Liverpool Astr. Society*, vol. v. p. 77.

An analogous object is α Piscium, made up of a fourth and a fifth magnitude star at 3″ distance, and revolving in a period unlikely to be much less than two thousand years. The larger certainly varies in light, and perhaps also in colour, the smaller certainly in colour, and perhaps also in light.[1]

An observation made by Mr. Tebbutt in New South Wales, August 22, 1887, gave a unique proof of the relative variability of a close double star in Virgo (OΣ 256). At its occultation by the moon on that night, the chief part of the light went out with the disappearance of the reputed lesser star, the component which had of late passed for its primary remaining still for a few moments separately but dimly visible.[2] Similar but less marked reversals had already been noticed by O. Struve and Dembowski in this slowly circulating pair.[3]

Variability, as we have said, affects both or neither of two stars so intimately united that their orbital movements have become apparent after a comparatively short lapse of time. A possible exception, however, to this rule is met with in δ Cygni. This beautiful and delicate pair was discovered by Sir William Herschel in 1783, but in 1802 and 1804 he totally failed, with improved optical means, to see the eighth-magnitude companion. His son was equally unsuccessful under the best atmospheric conditions in 1823, and Sir James South and Gambart in 1825.[4] It emerged to view, however, with Struve's nine-inch Fraunhofer in 1826, and has since been rarely missed. An occultation of one star by the other, postulated to account for the telescopic singleness of the pair between 1802 and 1826, was by their subsequent movements decisively shown not to have taken place, and the alternative hypothesis of a temporary loss of light in the small component was, almost of necessity, adopted. Yet it has received no strong countenance from recent observations. M. Dunér acquired the conviction from seven years of experience that the visibility of an object at all times difficult

[1] *Harvard Annals*, vols. xi. p. 112, xiv. p. 433; Flammarion, *Catalogue*, p. 12
[2] *Observatory*, vol. x. p. 391. [3] *Obs. de Poulkowa*, t. ix. p. 327.
[4] *Phil. Trans.* vols. cxiv. p. 339; cxvi. p. 376.

depends entirely upon the state of the air;[1] and Mr. Burnham seems to be of the same opinion.[2]

Changes of colour in this satellite star are nevertheless patent. Struve found it of an ashen shade from 1826 to 1833; in 1836 of a bright red.[3] It has since generally appeared blue; but Dunér saw it once olive, though otherwise always red; and intervals of greyness are also on record. The computed orbits have hitherto failed to represent the movements of the system with any degree of accuracy; but Mr. Gore's with a period of 377 years may prove more successful.

Relative variability has recently been detected by M. Flammarion in γ Arietis; it is, or has been, also present in θ Serpentis, 38 Geminorum, π Boötis, ϵ Arietis, and many other couples, most if not all of which give spectra of the Sirian type. Their agreement in the possession of this particular quality of light is the more remarkable from its being the badge in solitary stars of exceptional emissive stability. Every 'white star,' so far known to be variable, has proved also to be compound, and those of the Algol type are so far from making an exception to this rule, that they are among the most rapid of possibly existing binaries.

Besides these, we are acquainted with only two Sirian stars, δ Orionis and S Monocerotis, which have had periods of light-change assigned to them. The first is a wide double star of dubious variability; the second is the leading member of a straggling cluster, and was discovered by Winnecke in 1867 to change from 4·9 to 5·4 magnitude in 3d. 10h. 38m. Of two close attendants, the inner and brighter, at a distance of 2″·8, has been thought to be in slow circulation, but the point is still unsettled. The system has no appreciable proper motion.

A star situated near α Virginis is the only object characterised by a spectrum of the first type known to undergo extensive *intrinsic* fluctuations of light. From a comparison of observations going back to the tenth century, when Al Sûfi

[1] *Mésures Micrométriques*, p. 118.
[2] Westwood Oliver's *Astronomy for Amateurs*, chap. vii
[3] 'Color comitis egregius,' *Mens. Microm.* p. 297.

registered it as of fifth to sixth magnitude, Schmidt ascertained, in 1866, its irregular variability from the fifth to the eighth.[1] The anomaly of such changes in a Sirian star was brought more into harmony with other examples by Burnham's division of 'Y Virginis,' at Chicago in 1879, into two nearly equal components, less than half a second (0″·47) apart.[2] A subsequent observation gave no satisfactory evidence of alteration, either in brightness or position, during the intervening two years.[3] This is not surprising, since accesses of light-change in double stars are very generally followed by long intermissions. Nor could orbital motion, although presumably in progress, be expected to become so quickly apparent. The study of its laws, and of the varying magnitudes of the members of this singularly interesting system, ought not to be neglected by the possessors of great telescopes.

The fluctuations of U Tauri (observed by the late Mr. Baxendell, 1865 to 1871), like those of Y Virginis, seem for the present suspended. The twofold nature of this object, which is situated quite close to a variable nebula, was detected by Mr. Knott, December 4, 1867. Each component is of 9·7 magnitude, and they lie 4½″ asunder.[4] No recent observations of them that we are aware of have been made.

The variotinted pairs δ Cephei and β Cygni both belong to the class treated of in the present chapter. The former is the well-known short-period variable with which we have earlier become acquainted;[5] the latter changes slowly and almost imperceptibly between 3·3 and 3·9 magnitude.[6] The satellite is, in each case, exempt from the suspicion of instability. Not so the fifth magnitude attendant of α Herculis. The elder Struve considered that it declined occasionally to one-sixth its normal brightness;[7] and Father Secchi also perceived irregularities,[8] which have, however, for many years past ceased to be noticeable. The green hue of this star

[1] *Astr. Nach.* No. 1597; *Harvard Annals*, vol. xiv. p. 456.

[2] *Observatory*, vol. iii. p. 192.

[3] *Mems. R. Astr. Society*, vol. xlvii. p. 190. [4] *Ibid.* vol. xliii. p. 78.

[5] See *ante*, p. 132. [6] Klein, *Astr. Nach.* No. 1663.

[7] *Mensuræ Micrometricæ*, p. 97. [8] *Atti dell'Accad. Pont.* t. vii. p. 62.

(although to many eyes appearing blue) is in reality quite unchanging.

The variations from 6 to 6·8 magnitude of U Puppis, detected by Mr. Espin in 1883,[1] derive added interest from the strong probability that they integrate the changes of two close components. The star is the chief member of one of Struve's wide fixed pairs (Σ 1097); after being 'elongated' by Dembowski in 1865, it was fully resolved by Burnham in 1875 into two unequal objects at an interval of 0″·80. There is no other short-period variable of so obviously compound a nature. Its spectrum is of the solar type.

A corresponding long-period star is η Geminorum, perceived by Mr. Burnham, during a visit to Mount Hamilton in 1881, to form 'a splendid unequal pair, likely to prove an interesting system.'[2] Its revolutions will deserve the more attention that no star showing a banded spectrum has yet given perceptible signs of orbital movement. As a variable, η Geminorum may be described as an abortive specimen of the Mira class. Its phases, run through in a period of 229 days, are always ill-marked, and at times almost wholly suspended. The share of the companion in bringing about these partial effacements has yet to be determined. Its changes, for instance, may possibly be to some extent of a compensatory character; or its influence may tend to interrupt the regular progress of those of its primary. The latter seems the more likely alternative.

The presumption of sympathy in variability between closely-conjoined stars, although supported by many facts, has not yet been tested with any degree of strictness; but the converse proposition, that agreement in light-change implies physical connection, is of all but self-evident truth. Two variables in Cygnus,[3] for example, situated 24″ apart, may safely be assumed to constitute a system, their 'ruddy and cærulean' tints being a confirmatory circumstance. A still more striking combination is presented by U Cassiopeiæ and

[1] *Monthly Notices*, vol. xlvii. p. 432.

[2] *Ibid.* vol. xlvii. p. 204 Burnham's measures at Lick, in 1889, *sugges* orbital movement, see *Astr. Nach.* No. 2930.

[3] *h* 1470 = Lalande 38428.

a blue companion at 59″, with which its strong red glow contrasts at times splendidly. The principal star fluctuates irregularly from the sixth to below the ninth magnitude; the attendant from the eighth to the tenth. The probability of their being united by a special tie is overwhelming. Accordant variability of a conspicuous kind is an argument for its existence to the full as convincing as the possession of a common proper motion.

The crimson tint of U Cygni, discovered by Mr. Knott in 1871 to vary from above the eighth to below the eleventh magnitude in a period of 466 days, was described by Webb as 'one of the loveliest in the sky.' It is set off by the blue rays of a companion at 63″, which seems to fluctuate in colour, but little, if at all, in light. Their azure is, however, no mere optical effect of contrast, since (though capable of fading independently)[1] it survives without alteration the telescopic extinction of the adjacent red luminary. U Cygni is the only star belonging to the fourth spectral order open to a suspicion of being in systemic connection with a neighbour.

A good many variables have satellites as to which no such suspicion arises. Thus, the ninth magnitude star within 46″ of the beautiful 'carmine'-tinted object S Orionis is undistinguished either by colour or change, and they hence very likely form only a perspective couple. The same inference applies to three small stars contiguous respectively to R Cassiopeiæ, R Crateris, and Mira Ceti, detachment through the proper motion of the variable being, in the last case, visibly in progress.

The light-changes of connected stars are of great importance to the theory of stellar variability. For since they affect objects which the character of their spectra warrants us in saying would shine steadily if single, their mutual influence is manifestly concerned in generating fluctuations of a certain ill-pronounced type. This relation is rendered the more significant through the possibility, brought into view by Mr. Lockyer's meteoric theory, that variability of every

[1] Birmingham, *Trans. R. Irish Acad.* vol. xxvi. p. 300; Gemmill, *English Mechanic*, vol. xlvi. p 340.

kind may depend for its production upon external action by closely-circulating and to us invisible bodies. A test too may be furnished by fluctuating couples to the opinion that luminous instability belongs to a late stage of stellar existence. The contemporaneous origin and similar constitution of members of binary systems are indicated to our minds as highly probable. If this be so, development should proceed, other things being equal, quickest in the smallest masses, more slowly in the larger.[1] Hence, if it were true that variability accompanied decline, companion-stars should be far more unstable in light than their primaries. Facts, however, contradict this inference. Mr. Burnham is altogether incredulous as to the alleged disappearances of certain satellites through loss of light; and in every undoubted case of variability in one member only of a pair, the member it distinguishes is the principal star.[2] This circumstance gives a pregnant hint on the obscure but eminently interesting subject of the origin, history, and mutual relations of conjoined stars.

[1] Lockyer, *Proc. R. Society*, vol. xliv. p. 90.

[2] For a list of variable Double Stars, see Appendix, Table II.

CHAPTER XIII.

STELLAR ORBITS.

THE strong presumption that the law of gravitation would prove truly universal has been fully borne out by investigations of stellar orbits. Binary stars circulate, it can be unhesitatingly asserted, under the influence of the identical force by which the sun sways the movements of the planets, the earth the movements of the moon. It is true that this does not admit of mathematical demonstration, but the overwhelming improbability of any other supposition amounts practically to the same thing.[1] The revolutions of the stars are hence calculable, because conducted on familiar principles; their velocities have the same relation to mass, their perturbations may lead to similar inferences as in the solar system.

Observations, however, must precede calculations; and they are rendered arduous in double stars by the extreme minuteness of the intervals to be measured. Many revolving pairs never separate to the apparent extent of a single second of arc; yet this fraction of a second may represent, in abridgment, a span of some thousands of millions of miles. Infinitesimal errors, magnified in this proportion, become of enormous importance, and often impenetrably disguise the real aspect of the facts.

For determining the relative situations of adjacent stars, two kinds of measurement are evidently needed. The first gives their distance apart, the second the direction of the line joining them as regards some fixed line of reference. That selected is the 'hour-circle,' or great circle passing through the pole and the larger star, and the angle made with it by the line of junction of the pair is called their 'position-angle.'

[1] Tisserand, *Bulletin Astronomique*, t. iv. p. 13.

It is counted from 0° to 360°, in a direction opposite to that of the movement of watch-hands; and a star is hence said to be in direct revolution if its circuit is from north to south through east, but in retrograde revolution, if it is oppositely pursued.

Now the successive places, from year to year, of the moving star, obtained in this way with absolute accuracy, would fall into a perfect ellipse, the foreshortened delineation of the real ellipse traversed in space. For the star is seen by us projected against the sky, or rather upon the plane touching the sphere at that point, while the actual orbit may lie in any one of an infinite number of planes. The two curves, nevertheless, have relations by which one can be deduced from the other with geometrical certainty. Both are ellipses, and in both the 'radius vector,' or line drawn from the satellite to the primary, describes equal areas in equal times. But the position of the chief star at the focus of the real ellipse is not maintained in its perspective representation, in which the 'projected focus' is often quite unsymmetrically placed. Once then the *seeming* orbit of a binary star is thoroughly ascertained, the problem of determining its *actual* orbit is as good as solved, the transition from one to the other being effected by a mathematical operation of no considerable difficulty. Even when the seeming orbit is a straight line the process remains feasible, and in fact one of the most reliable stellar theories relates to a couple, the movements of which are conducted in a plane passing, it may be said, accurately through the earth.[1] The stars of 42 Comæ Berenices appear simply to oscillate to and fro in a period of somewhat less than twenty-six years, never diverging to a greater extent than about half a second, and occulting each other completely twice in the course of a revolution. Discovered by Struve in 1827, they have since five times presented an aspect of indivisible singleness. Other mutually occulting pairs are γ Coronæ Borealis with a period of ninety-five, 44 Boötis with one of 261 years, and a binary in Ophiu-

[1] O. Struve, *Monthly Notices*, vol. xxxv. p. 367; *Atti dell' Accad. Pont.* t. xix. p. 259.

chus (Σ 2173) of somewhat fluctuating brightness, revolving in forty-five years.

Nothing would at first sight seem easier than to lay down, from a sufficient store of data, the apparent path of a star. Yet the task is often a most embarrassing one. Owing to the excessive minuteness of the quantities concerned, the best observations can give only loose approximations to the actual facts. The margin of uncertainty, always very wide, at times exceeds any reasonable limit, and computers are hence obliged, as a rule, to reject some part of the materials before them as misleading and incompatible with the rest. But the exercise of discretion leads to diversity of results, and totally different orbits can thus be derived from the same set of observations, by varying their treatment so as to distribute differently their inherent errors. Where only a moderate arc of the orbit has been seen to be described, the problem of ideally completing it admits, from the indeterminateness of its conditions, of no rigid solution. When the companion of Sirius, for instance, had been eighteen years under scrutiny, it was still impossible to decide whether it would return to its starting-point within the half century allotted to it on grounds independent of its visible movements, or depart on a remote excursion from its primary, demanding some hundreds of years for its accomplishment.[1]

In no department of astronomy is the mischief of 'personal equation' so sensible as in the measurement of double stars. Nearly all available data are prejudicially affected by it, and those emanating from different individual sources are thus often rendered exceedingly inharmonious. Much labour and ingenuity have been spent in determining its direction and amount for various observers, with a view to freeing their results from its effects; and after all, it remains a question whether the observations so elaborately corrected are not more misleading than in their 'raw' state.

All these complications can be at once swept away by substituting the camera for the micrometer. The photographic method leaves no room for systematic, very little for

[1] Plummer, *Monthly Notices*, vol. xlii. p. 63.

accidental errors. G. P. Bond, of Cambridge (U. S.), showed in 1857, long before the introduction of the modern 'dry plates,' its wonderful capabilities for the accurate registration of the varying relative situations of double stars;[1] and those capabilities were more fully realised in 1886 by the skill of the MM. Henry. We are permitted to give in fig. 29 some specimens of the Paris photographs of double stars. The repeated impressions shown of each pair were obtained by allowing free play to the diurnal motion during certain definite intervals between successive short exposures. Thus the line of displacement of the stars traces out part of a circle of

Fig. 29.—Four Double Stars photographed at Paris. (From Mouchez's 'Photographie Astronomique.')

declination, and their angles of position are directly measurable from plates, embodying the data for their own orientation. The exactitude of determinations from them proved very remarkable; for ζ Ursæ Majoris the 'mean error' of single measures of distance amounts to only 0″·077, of angle to 0·°55.[2] And this is no illusory precision, undermined by evasive uncertainties, but the statement of a fact hard enough not to crumble in the handling. Stellar orbits will then really be known when they come to be calculated solely from photographic measures.

[1] *Astr. Nach.* No. 1129.

[2] Mouchez, *La Photographie Astronomique*, p. 44.

Their application is, however, at present restricted to such pairs as are neither very unequal nor very close. The diffusive brightness of Sirius, for instance, leaves no possibility of getting a separate print of its companion; nor could even the much lesser disparity between the stars of δ Cygni be made compatible with the distinct self-portraiture of both. Again, the minimum interval at which even perfectly equal couples have hitherto been successfully photographed exceeds two seconds, closer objects running together on the plate. As experience and invention progress, these limitations may be removed, but they are as yet effectual.

The systematic adoption of the new method promises to bring material reinforcement to the resources of the science of compound stars. Observation will be enabled by it to keep pace with discovery, and stars need no longer be found only to be forgotten. The exposure of a sensitive plate during a few seconds yearly to the rays of each couple will supply, after a time, a stock of facts impaired in value by no perplexing inconsistencies. The apparent orbits of revolving stars will be virtually inscribed, for the benefit of computers, in each tolerably complete collection of negatives. In 'smoothing the curve,' a process always inspiring distrust, little of arbitrary choice will remain; and the representative ellipse, instead of threading its way amid a straggling crowd of outlying observations, will pass, not, indeed, actually *through* (which would imply the annihilation of error), but very close to all the given places.

About sixty stellar orbits have so far been computed by Dr. Doberck, Mr. Gore, MM. Glasenapp, Dunér, Celoria, and some others. But for the most part tentatively only, nor always with success. Predictions are often very far astray, the moving stars showing themselves totally regardless of the trammels of theory. Thus, the equal members of the lucid pair γ Coronæ Australis had, in 1887, run ahead of anticipation to the extent of twelve degrees; and the period of 6 Eridani, another fine southern binary, has had, owing to grave discrepancies with the Sydney observations of 1871 to 1881, to be lengthened from 117, first to 224, then to 302 years. Nor can we feel much confidence that the path of 61 Cygni is really that traced out by Peters, or that Castor will duly complete the millennial course prescribed to it by Thiele and Doberck. But, as we have said before, the cause of these uncertainties is to be found in the limits placed upon visual accuracy by the conditions of our existence, not at all in any want of ability on the part of the computers.

Satisfactory acquaintance has hardly yet been made with the movements of more than half-a-dozen stellar pairs. Those of the quicker kind naturally exhibit their nature soonest, and indeed revolutions require to be finished, or nearly finished, before they can be said to be ascertained. Among the best stellar theories extant is that of ξ Ursæ Majoris, one of two fourth magnitude stars marking the hindmost paw of the Great Bear. Divided by Herschel in 1780, this couple was made by Savary, in 1828,[1] the subject of the first experiment in the extension of Newtonian principles to the sidereal universe. It succeeded; for the stars were found to describe, as nearly as could be expected, the orbit calculated for them on the supposition that their mutual gravitation was the influence binding them together into a moving system; and the validity of Newton's law wherever matter exists has never since been open to serious question.

The path of ξ Ursæ has of late been several times re-investigated, and with results so concordant as to give a strong assurance of their approximate accuracy. It is a considerably elongated ellipse, the eccentricity being expressed by the fraction $\frac{2}{5}$, which is just twice that of the orbit of Mercury, half that of the orbit of Encke's comet. The period of traversing it is $60\frac{1}{2}$ years; its semi-major axis would subtend, if seen without foreshortening, an angle of $2\frac{1}{2}''$, and it lies in a plane inclined 56° or 57° to what we may call the ground-plane of the heavens—the tangent-plane, that is, to the sphere at that point.[2]

We are ignorant of the mass of this system because we

[1] *Conn. des Temps*, 1830, pp. 56, 163.

[2] Data on these several heads, together with others defining the situation of periastron, and of the line of intersection of the orbital and reference planes, constitute what are called the 'elements' of a star's movements.

are ignorant of its distance from the earth; but if we were to assume, by way of illustration, that it is at what Struve somewhat precariously estimated to be the 'average distance' of a fourth magnitude star,[1] it would follow, from the vast scale of its construction combined with the rapidity of its movements, that the gravitational power residing in it exceeds 194 times that of the sun. There is, indeed, a strong likelihood that these figures exaggerate both the remoteness and the massiveness of ξ Ursæ, but whatever its distance and whatever its mass, we can say with certainty that it is an intensely luminous body. If of the same mean density, it must be, square mile for square mile of surface, of about two and a half times the solar brightness.

Determinations of the distances of binary stars are of special interest from their leading to a knowledge of their masses. The connection is easily explained. Angular measurements, which are the only ones possible to be got of objects out of tangible reach, are convertible into definite linear values when the radius of the sphere they refer to becomes known—in other words, when the interval of space between the eye and the objects measured is ascertained. So that the dimensions, in seconds of arc, of the orbits of stars at measured distances give at once their dimensions in millions of miles, whence, with the help of the periods of the revolving objects, their masses easily follow. For by the law of gravity, the attractive power of any system is proportionate to the cube of the mean distance of the bodies composing it, divided by the square of their period. Employing, then, as a unit of space in this little calculation, the distance of the earth from the sun, and the year as our unit of time, we get the mass of each pair of revolving stars in terms of the sun's mass. It comes out, of course, larger in the ratio of the cube of the distance for the same period, and smaller in the ratio of the square of the period for the same distance. Swiftly-moving and spacious systems contain accordingly great quantities of matter; sluggish ones comparatively little.

[1] *Étoiles Doubles*, p. 46. From Mädler's elements, Struve deduced a value for the mass of ξ Ursæ equal to 159 times the solar mass.

The quickest of known binaries is δ Equulei, divided at Pulkowa in 1852 into two stars of about 4½ magnitude, set so close together that only the very best instruments can show them separately. Nor can even they do so at all times. The stars move in a plane almost coincident with the visual ray, and seem consequently to perform little vibrations, carrying them apart, at intervals of approximately seven years, to the extent of 0″·44. During some intervening years they are optically merged into one. The difficulty of rightly interpreting such inconspicuous appearances is obviously very great, and has not yet been overcome, recent measurements by Mr. Burnham showing Wrublewsky's period of 11½ years[1] to be decidedly too short.[2] The true period of δ Equulei will probably be found not to differ much from fourteen years.

The pair most nearly approaching its rapidity is β Delphini, discovered by Burnham in 1873, and completing its revolutions, by the recent investigations of M. Celoria, in seventeen years; while ζ Sagittarii, first divided by Professor Winlock in 1867, comes third with a period of 18½ years. The 'occulting stars' in Coma Berenices (Flamsteed's No. 42) circulate in 25½, ζ Herculis and an interesting couple in Leo (Σ 3121) each in a few months less than thirty-five years. Among stars at known distances from the earth, 85 Pegasi has the shortest period (twenty-two years), and its mass proves to be eleven times that of the sun. These stars, as Mr. Gore remarks, present rather the appearance of a sun and planet than of two suns. The primary centuples the light emitted by its satellite, and there is just the disparity between them that would be presented by the sun and Jupiter if of the same intrinsic brilliancy. These would, on the condition supposed, constitute, at the distance of sixty light-years (attributed to 85 Pegasi), a pair of respectively 7½ and 12½ magnitude, never above 0″·28 asunder. The utmost powers of the great Lick refractor would, however, scarcely be adequate for their separation.

In all, eleven star-couples have had periods assigned to them of less than fifty years, fifteen of less than a hundred but more than fifty. The slowest of *computed* binaries is

[1] Glasenapp, *Astr. Nach.* No. 2771. [2] *Ibid.* No. 2875.

ζ Aquarii, a bright, nearly equal pair just in the equator, needing, according to Dr. Doberck, 1578 years to finish its circuit. But there is no certainty on this point. Half a dozen totally different orbits could probably be accommodated to the arc of 45° described since 1779, equally well with that provisionally fixed upon.

But even these leisurely movements are swift compared with others which, after the lapse of upwards of a century, seem barely nascent or even non-existent. The apparent fixity, indeed, of stars at the considerable intervals separating the components of β Cygni, δ Cephei, θ Serpentis, and others, accords with what we know of the prodigious scale of sidereal construction, but the indication for moderately close pairs of periods ranging up to or beyond twenty thousand years is startling. Such are 95 Herculis, α in the same constellation, and γ Arietis; the movements of ζ Ursæ Majoris, causing a change of angle of five degrees in 135 years, *suggest* their completion in about 10,000, and 4,000 *may* suffice for those of γ Delphini. This strange inactivity shows that the systems it characterises are either of exceedingly small mass, or else inconceivably remote from the earth.[1]

We have positive knowledge of the masses only of such stellar couples as have had both their parallaxes (the equivalents of their distances from the earth) and their orbits determined. They are seven in number—namely, Sirius, α Centauri, 61 Cygni, 70 Ophiuchi, η Cassiopeiæ, 85 Pegasi, and o^2 Eridani; and their total mass proves to be that of thirty suns.[2] On an average, then, each of these systems contains nearly four and a half times as much attractive energy as the solar system, each individual star being more than equal in this respect to a pair of suns like ours. Were the extension of this mean conclusion legitimate, the distances of all stars revolving in ascertained orbits might be inferred from their assumed massiveness (since the relation between distance and mass is convertible), and upon this principle Mädler derived what he called the 'hypothetical parallaxes' of binaries,[3]

[1] Mädler, *Fixsternsysteme*, p. 10. [2] See Appendix, Table V.
[3] *Der Fixsternhimmel*, p. 82.

reckoning, however, the mass of each pair to be only that of a single sun. This estimate is now seen to be much too small, and the distances founded upon it to fall proportionately short of the truth.[1] But, indeed, no general conclusions of the kind are fit for application to individual cases. The range of variety is so great that only simulated knowledge can be obtained in this way. Collective inferences, however, are not therefore worthless. Thus, from averaging the masses of only seven binaries, we have already gathered plausible grounds for believing our sun to occupy a low rank as a centre of attraction. It may be, nevertheless, that the swifter binaries, which can at present alone figure on such a list, give too high an average mass.

Calculations based upon the orbital elements of revolving stars tell nothing of their *relative* masses. They apply only to the common stock of matter in each system, leaving its apportionment to be otherwise tested. This cannot be done except through the due apportionment of movement between the members of the system—an arduous task, hardly yet begun to be grappled with.

There is no such thing in nature as a stationary body round which other bodies circulate. Answering motion there must always be, though on a scale reduced just in the same proportion that the mass is increased. Thus, the earth describes, under the influence of the moon, an ellipse exactly similar to that described by the moon under the influence of the earth, but eighty-one times smaller. And the sun corresponds in the same way to the revolutions of every one of the planets, notwithstanding that the centre of his movement as regards each of them, with the single exception of Jupiter, lies far beneath his own surface. Binary stars, however, are often probably almost equally massive, and therefore almost equally mobile bodies. The fixity of one member of each pair is purely conventional—an indispensable fiction without

[1] The masses of revolving stars vary, *cæteris paribus*, as the cubes of their distances from the earth. Of systems identical in period and apparent movement, one twice as remote as another will be eight times as massive, one three times as remote twenty-seven times as massive, and so on.

which measurements would be impracticable. Those actually made give the sum of the movements of both stars, and an orbit computed from them represents the sum of their distinct orbits. Identical in shape and position with the true ellipses, it differs from them only in size, its linear dimensions in any direction being equal to both theirs taken together.

The genuine centre of movement of two mutually circling stars is their common centre of gravity, which lies on a straight line drawn from one to the other, at a distance from each inversely proportional to its mass. The precisely similar orbits traversed by each are then spacious in the same inverted ratio. The larger star performs the smaller circuit, and *vice versâ*. In the case of their equality, their orbits must intersect if elliptical, but coincide if circular, when the stars will pursue each other along the same track, while occupying in it,

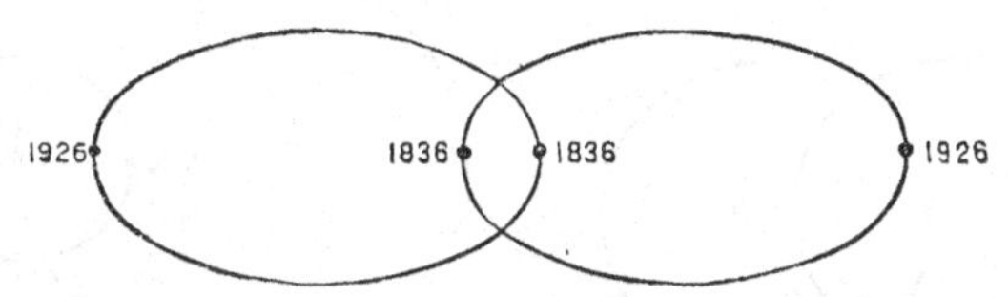

Fig. 30.—Orbits of the Components of γ Virginis.

at each successive moment, diametrically opposite positions, nor could either, during an eternity of undisturbed revolution, gain a hair's breadth upon the other. Circular movement is, however, rarely even approximated to in stellar systems, the members of which usually follow highly eccentric paths.

We may take γ Virginis as an example of a pair moving in equal ellipses, the relations of which are shown in fig. 30. They have, it will be seen, a common focus, the seat of the centre of gravity, from which the stars (being of equal mass) must always be equally distant. Neither can approach to or recede from this point of origin of the force acting upon them without the other simultaneously doing the same; the two must be in periastron, and retire towards apastron together, losing, and subsequently regaining, velocity by the same gradations. The stars of α Centauri are also believed to travel in equal orbits, but in much less elongated ones than those of γ Virginis (see fig. 31).

The movements of unequal stars are similarly conducted. That is to say, the *proportion* of their respective distances from the common focus is invariable. They are accordingly always found in corresponding parts of their orbits, and at opposite ends of a right line passing through the focus. The manner of their revolutions can be realised by a glance at fig. 32, representing the orbits of Sirius and its companion, the small ellipse belonging of course to the bright star.

Obviously, from what has been said, knowledge of the relative masses of binary stars would ensue upon knowledge of the relative dimensions of their separate orbits. But for its attainment, a prolonged series of most delicate micrometrical

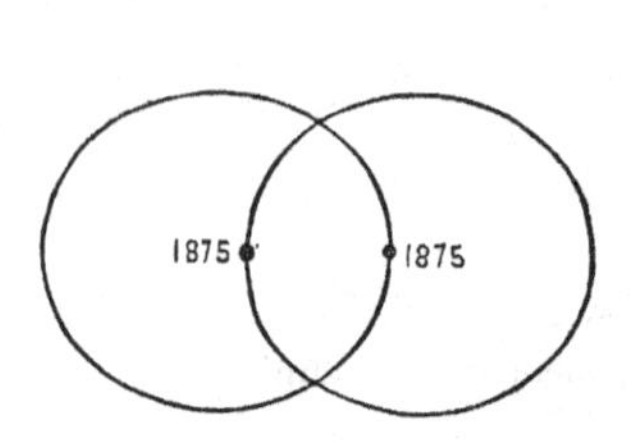

Fig. 31.—Orbits of the Components of α Centauri.

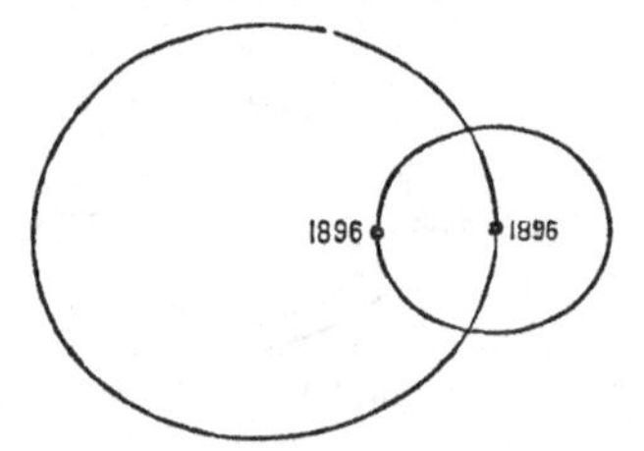

Fig. 32.—Orbits of Sirius and its companion.

measurements between each of the revolving objects and some neighbouring star chosen as a reference-point, would in general be necessary.[1] Such measurements could now be executed with comparative facility by photographic means, and their results can scarcely fail to afford valuable information regarding the physical condition of stars. The relations of mass to light-power, for instance, could thus be investigated with some definiteness. All that we know about them at present is that they vary largely, and often unaccountably.

The proper motions of three binary systems have afforded some information as to the distribution of matter in them; for the track pursued by each component is necessarily a sinuous one, like that of the moon round the sun, while the

[1] Pickering, *Proc. Amer. Acad.* vol. viii. p. 6 (1881).

centre of gravity of the two advances uniformly in a straight line. Now this neutral point was found by Mr. Stone[1] to be situated about midway between the stars of α Centauri, showing them to be not far from equally massive; it is just half as distant from Sirius as from its companion, and lies, according to M. Ludwig Struve,[2] nearly five times closer to the large than to the small star of η Cassiopeiæ. The latter object contains then $\frac{1}{5}$ as much matter as its primary, yet it emits only $\frac{1}{28}$ of its light, so that it is notably deficient in brightness. The same peculiarity, in an enhanced degree, characterises, as we have seen, the Sirian satellite; and its appearance in the secondary member of the great pair in the Centaur is the more remarkable that their equality in point of mass gives no good ground for the supposition of their having reached different stages of evolution. The key to the puzzle may be found in spectroscopic observations.

From them, indeed, we may hope before long to get, not only interpretations of such results as we are now in search of, but the results themselves. For this purpose the principle of the change in the refrangibility of light through motions of recession or approach can be made available. Altered refrangibility makes itself perceptible in displacements of known spectral lines, and these can now be measured on photographic plates[3] with such exactitude as to give perfectly trustworthy information respecting the radial velocities of the objects the displaced lines are derived from. If the objects be the members of a binary system, the determination of their relative velocities in line of sight will suffice to fix their relative masses, since the proportion of their orbital movement, directed at any given instant towards or from the earth, must always be the same for each component. The inverted correspondence between mobility and mass thus holds good for radial measurements.

These are facilitated by the circumstance that the simultaneous 'end-on' motions of mutually revolving stars are, by the necessity of the case, oppositely directed. They reach a

[1] *Monthly Notices*, vol. xxxvi. p. 258.
[2] *Copernicus*, vol. i. p. 199.
[3] See *ante*, p. 31.

maximum when one is in the ascending, the other in the descending node of their orbits—that is, when one star is crossing the plane of projection on its hither course, the other on its further excursion; they disappear on the stars reaching two intermediate positions, when the whole speed of circulation flows transversely to the line of sight.[1] These alternating changes make it possible, by repeated observations at sufficient intervals of time, to eliminate the constant effect due to the proper motion of the system. The variable residue will then be the element sought.

Its determination promises to be fruitful in several ways. The total orbital velocity of each pair for which it was ascertained could easily be calculated, and the dimensions of the orbit thus implied would disclose the absolute mass and distance from the earth of the combined stars.[2] Still a further piece of information would be conveyed. Simple visual observations are helpless to discriminate between the nearer and the further sections of a stellar orbit. The effect to the eye is the same whether the moving star be approaching on the left-hand side and receding on the right, or approaching on the right and receding on the left of its apparent path. In other words, the ascending and descending nodes are indistinguishable, and the computed ellipse may indifferently lie in the plane assigned to it, or in one equally tilted in the opposite direction.[3] This ambiguity might be removed by a single spectroscopic observation at either node, telling whether the star was advancing or retiring.

A great number of binary couples will doubtless respond satisfactorily to this last simple test; but only those most favourably situated can be expected to prove susceptible of the more refined measurements leading to detailed results. The spectroscopic method has naturally no bearing except on stars moving in known orbits; nor is it likely to be successfully applied unless the planes of those orbits are more nearly level than perpendicular to the visual line, and the objects traversing

[1] Niven, *Monthly Notices*, vol. xxxiv. p. 339; Palisa, *Astr. Nach.* No. 2941.
[2] Rambaut, *Proc. R. Irish Acad.* vol. iv. ser. ii. p. 663.
[3] R. Falb, *Sirius*, Bd. vi. p. 121.

them are bright enough, and sufficiently far apart, to show distinct spectra. Their remoteness, provided these conditions be fulfilled, is of no consequence,[1] spectroscopic motion-displacements being absolutely independent of distance; and it may accordingly be possible to determine in this way stellar parallaxes so minute as to be otherwise insensible.

The pair on the whole best suited for experiments of this nature is *α* Centauri. In April 1879 the radial part of the velocity relative one to the other of the components amounted (according to Mr. Rambaut's calculation)[2] to eight miles a second—a quantity easily measurable with the Potsdam photographic apparatus. The same element for Sirius will reach a value, in September 1890, of thirteen miles a second; but here the spectrum of the secondary star is much too faint to be communicative as regards motion. Among couples promising to become so are 70 Ophiuchi, λ Ophiuchi, and ξ Ursæ Majoris; their treatment, to be successful, must, however, be exceptionally skilful, patient, and delicate.

A remarkable circumstance connected with the last-named system gives a special interest to the spectroscopic determination of its translatory movement. The stars composing it appear to move in a progressively widening orbit—a visual effect, it is conjectured, of their continuous approach to the earth.[3] Only the approach should be improbably rapid to account for the observations, and it has not yet been shown to be present at all. Theoretically, every stellar pair must be in course of separation or closing-up according as it is advancing or receding along the line of sight, and its parallax could even (in the abstract) be derived from the proportion borne by the change to the rate of motion causing it. But modifications produced in this way are not likely, for an indefinite time, to enter into the domain of practical science.

Another indirect method, scarcely less remote from realisation, of establishing the distances from the earth of

[1] Fox Talbot, *Report Brit. Assoc.* 1871, p. 34. Pt. ii.
[2] *Proc. R. Irish Acad.* vol. iv. p. 669.
[3] Birkenmajer, *Sitzungsberichte*, Wien, Bd. xciii. ii. p. 786.

computed binaries, was proposed by Savary in 1827.[1] It aims at learning the real dimensions of stellar orbits from the time taken by light to cross them; and this would become known through the delay in the visibility of the more remote component (simultaneous observations giving non-simultaneous places),[2] if only acquaintance with stellar revolutions were indefinitely more accurate than it is. Thus the 'equation of light' for 61 Cygni, the circuit performed by which is the most spacious we are cognisant of, is only nine hours, an interval during which the orbital motion of the pair is utterly lost in the uncertainties of calculation.

In an amendment to Savary's plan, Houzeau attempted in 1844[3] to turn to account for the same purpose certain optical inequalities produced in the onward motions of connected stars by the progressive transmission of light. But their interest is almost purely speculative. Villarceau showed that the only possibility (and that a remote one) of their being rendered apparent depended upon the preliminary ascertainment of the relative masses of the stars concerned.[4] It is then safe to predict that this species of 'light-aberration' will not prove instrumental for the advancement of definite knowledge.

Professor Pickering[5] and Mr. Monck[6] of Dublin have separately perceived the existence of a relation between the movements and magnitudes of binaries, rendering it possible to determine their comparative superficial luminous power quite independently of their distances. It is necessary, however, to assume either that the components of each pair are of similar quality in this respect, or else that one of them is of negligeable mass; nor does the formula distinguish between extent of surface and intensity of shining. The degree in which matter is concentrated in the bodies

[1] *Conn. des Temps*, 1830, p. 169; Arago, *Annuaire*, Paris, 1834, p. 286; Struve, *Mens. Microm.* p. clxxii.

[2] Birkenmajer, *loc. cit.* p. 717.

[3] *Astr. Nach.* No. 496.

[4] *Conn. des Temps*, 1878, *Additions*, p. 3; *Comptes Rendus*, t. xxxiv. p. 353; Seeliger, *Vierteljahrsschrift Astr. Ges.* Jahrg. xxi. p. 285.

[5] *Proc. Am. Acad.* vol. viii. p. 14 (1881).

[6] *Observatory*, vol. x. p. 96; *Knowledge*, vol. xii. p. 141.

considered is left untouched by it; their mean density may have any assignable value.

The result of these inquiries is to confirm the prevalence of astonishing varieties in the emissive powers of different stars. Thus, γ Leonis is three hundred times more brilliant than the sun, if its density be the same; or, if we suppose its brilliancy the same but its density inferior, then the substance of these globes must be seven times rarer than atmospheric air at standard pressure, in order to give the required bulk.[1] It is possible, however, that their movements, when better known, may disclose their possession of more gravitating power than Dr. Doberck's orbit implies. Between their discovery by Herschel in 1782, and 1878, they traversed an arc of only 35°, leaving obviously much room for uncertainty as to their future course. Their spectrum, which is of the solar type, in no way accounts for their abnormal splendour.

In the same way, the brilliancy of δ Cygni comes out (with Behrmann's elements) one hundred times that of the sun; δ Equulei and ζ Sagittarii emit fully thirty times, γ Virginis and 25 Canum Venaticorum more than sixteen times, Castor nearly fifty times the solar light relative to mean density. Indeed, our solitary star is so generally and so far surpassed in luminosity by the members of binary systems, that Professor Pickering inclined to think that injustice had been done to it in current estimations. Stars are nevertheless to be found standing vastly beneath it as light-givers. We have already met with one notable instance to this effect in the companion of Sirius, and there are several others. Thus about half the sun's brilliancy belongs to 61 Cygni, and no more than $\frac{1}{72}$ part of it is claimed by a pair of revolving ninth-magnitude stars forming part of the triple combination known as 40 (otherwise o^2) Eridani. The singular contrast between the brightest and duskiest of known star-couples was pointed out by Mr. Monck. 'If γ Leonis,' he remarks, 'were replaced by a star of equal mass but similar in character to 40 Eridani, its light would be reduced to $\frac{1}{20000}$ of its present

[1] *Proc. Amer. Acad.* vol. viii. p. 14.

intensity, while if the fainter pair of 40 Eridani were replaced by a star of equal mass whose brilliancy was equal to that of γ Leonis, its light would be increased twenty thousand fold, and it would outshine every star in the sky except Sirius.'[1]

The most striking general peculiarity of stellar orbits is their high eccentricity. Nearly all of them are greatly more oval than the planetary paths round the sun, and a large proportion approach to cometary shape. This remarkable fact indicates, in Professor Kirkwood's opinion,[2] the division of the parent nebulæ of double stars before the acceleration by contraction of their rotatory movement had made much progress. As regards the development of tangential velocity, these systems were thus in a sense abortive, and their members, beginning to revolve under the nearly unbalanced influence of their mutual gravitation, necessarily pursued very much elongated tracks. There is no tendency to an agreement between them as to the direction in which these tracks are pursued. The revolutions of binary stars are indifferently retrograde or direct. Whether they to any extent affect a common plane is a question that has yet to be decided. If they do, their preference is not for the level of the Milky Way, but for that at right angles to it.[3] Nothing like proof, however, has as yet been given of the regulation, by any fixed principle, of the inclinations of stellar orbits. The existence of a fundamental plane of movement would be of high significance as regards the history and relations of the sidereal world; yet the whole drift of modern research suggests, rather than the close and rigid union between its parts which it would indicate, a loose connection destined to be extensively modified by time.

[1] *Journal Liv. Astr. Soc.* vol. v. p. 174.

[2] *Amer. Jour. of Science*, vol. xxxvii. p. 233 (1864).

[3] Gore, *Eng. Mechanic*, vol. xlvi. p. 296.

CHAPTER XIV.

MULTIPLE STARS.

THE further resolvability of a great many double stars is perhaps the most curious result of modern improvements in the optical means of observing them. With every addition to the defining power of telescopes, the visible complexity of stellar systems has increased so rapidly as to inspire a suspicion that simple binary combinations may be an exception rather than the rule. The frequency with which what appeared to be such have yielded to the disintegrating scrutiny of Mr. Burnham and others, suggests at any rate the presence of an innate tendency, and seems to show that the duplicity of stars is no accident of nebular condensation, but belongs essentially to the primitive design of their organisation. Although we can never become fully acquainted with all the detailed arrangements of stellar systems, we are then led to suppose them far more elaborate and varied than appears at first sight. Each, we cannot doubt, is adapted by exquisite contrivances to its special end, reflecting, in its untold harmonies of adjustment, the Supreme Wisdom from which they emanate.

The continuance of the process of *optical dissociation*, begun by the splitting-up of an apparently simple star, sometimes shows the primary, sometimes the satellite, not unfrequently both primary and satellite, to be very closely double. Ternary systems are accordingly of two kinds. In one, the smaller star consists of two in mutual circulation, and concurrent revolution round a single governing body; in the other, an intimately conjoined pair guides the movements of an unattended attendant. The planetary type of construction is uncommon or unknown. No star has been *ascertained* to possess two or more companions circulating (so to speak) co-ordinately. Groups possibly indicating such a disposition of parts exist, but perspective may have a share in producing them. The variable S Monocerotis, with its two client-stars, is an example.

Among the most interesting triple stars of the double-satellite description is the brilliantly coloured γ Andromedæ. The original components, of third and fifth magnitudes 10″ apart, remain *in statu quo* since they were seen by Father Mayer in 1777, but their secular journey together over an arc of 6″ establishes the genuineness of their relationship. In 1842 the sea-green companion was found to be itself double. With the fifteen-inch Pulkowa refractor, Otto Struve caught sight of a thin black line (representing probably a gap of some thousands of millions of miles) dividing it into two stars, of respectively 5½ and 7 magnitudes, since perceived to be revolving at a rate which would carry them completely round their orbit in about 500 years. They have already progressed so far as to obliterate the 'thin black line' testifying to their separate existence. Even Mr. Burnham's keen eyesight, aided by the utmost powers of the Lick thirty-six inch, is no longer adequate to distinguish them. All that can be seen is that the blended object they constitute, instead of being perfectly round, like a truly single star, is slightly 'oblong' in one direction. When the pair was more open than it is now, a difference of tint between the components was noticed by several observers,[1] and it is not improbable that the emerald effect of their light results from the merging together of actually distinct yellowish-green and blue radiations.

A pair in some respects similar, but much fainter, is attached at about the same apparent interval to the lustrous white star Rigel. An excessively difficult object at the time of its detection, it has of late become impossible. Mr. Burnham has for some years failed to extract from it the slightest sign of duplicity.[2] Yet its elongation was suspected both by him and Mr. Herbert Sadler in 1871,[3] and was

[1] Gledhill, *Observatory*, vol. ii. p. 269; Barneby, *ib.* p. 229.

[2] *Astr. Nach.* No. 2929.

[3] *Astr. Register*, vol. xviii. p. 15.

laboriously verified, first at Chicago in 1877, then from Mount Hamilton in 1879. Change, under these circumstances, seems much more probable than error, and great interest will attach to future observations of the tiny sapphire star joined in disparate union with the chief luminary of Orion.[1]

A ternary group, corresponding to these two in plan, but greatly enlarged in optical scale, has already been cursorily noticed.[2] It consists of a 4·5 magnitude star, designated by Flamsteed 40, by Bayer o^2 Eridani, with a faint and far-away double satellite, all three discovered by Herschel in 1783. The physical association of the pair with the large star at an interval of 82″ would be improbable, were it not certified by their possession in common of an exceptionally swift proper motion. An advance during the last century over a space nearly equal to a quarter of the moon's diameter has modified their relations only by a trifling approach to their primary of the dependent stars, due perhaps to slow circulation round it in an orbit presented edgeways to our sight.

They have, in the mean time, almost finished a circuit of each other, and will have completely finished it, by Mr. Gore's calculation, within 139 years from the date of their detection. And since their distance from the earth has been measured, the real size of their orbit and their joint mass are also known. We find then that the average interval between them is thirty-six times that separating the earth from the sun, so that (their path being only moderately eccentric) they never approach as near to each other as Neptune does to our central orb, which they together surpass two and a half times in gravitative power. But in their place, from which light reaches us in nineteen years, the sun would shine as a fifth-magnitude star, while they combine into one of only ninth magnitude. Their feeble luminosity thus once more forces itself upon our attention, and compels us to reflect upon the possibility of whole systems existing in unimpaired mechanical perfection, but wrapt in perennial darkness. For what purpose existing, who can tell? The flight of our

[1] Rigel, although called β Orionis, decidedly surpasses α Orionis in brightness.
[2] See *ante*, p. 203.

thoughts is short, and the ultimate aims of the Maker are remote. To attempt to compass them is to invite palpable failure.

Among double primaries Castor holds the first place, through the lustre of its components and the vastness of the scheme completed by the captive star borne along in their train. Another specimen is ε Equulei, one of Herschel's pairs, the larger member of which was again divided by Struve in 1835. The feat had become possible through the progress of orbital motion, the continuance of which has since rendered it easy. Signs of circulation in the 7·5 magnitude star at 11″ perhaps exist,[1] but inconspicuously. It is, however, an undoubted satellite of the close couple.

Physical, too, almost certainly is the group constituting 12 Lyncis. Here a white star of the sixth magnitude has an immediate neighbour nearly as bright, but reddish, the two describing in 486 years orbits slightly less eccentric than that of the minor planet Pallas.[2] Unless their joint mass be less than the solar, they must be so far off that their light takes *at least* 120 years to reach our eyes. Their brightness relative to mass is twice that of the sun. Considering the length of their period, the fixity of a bluish attendant at nearly five times the mean interval between them is not surprising.

The movements of the third star (of 7·2 magnitude) in the ternary combination ξ Scorpii are of a somewhat perplexing nature. They suggest interruptions, due possibly to disturbances by an unseen agency. Mädler conjectured for them a period of 1469 years;[3] a couple of centuries may, however, elapse before they develop sufficiently to be computed. They *seem*, too, to progress in an opposite direction from those of the close double star which controls them at an apparent distance of 7″; but we can learn only from the spectroscope whether this is really the case.

The primary in this system consists of a fourth and a fifth magnitude star at 1″·3, just separable, accordingly, with a good four-inch telescope. The orbit assigned to them by

[1] Flammarion, *Catalogue*, p. 139. [2] Gore, *Astr. Nach.* No. 2808.

[3] Tarrant, *Journal Liverpool Astr. Society*, vol. v. p. 205; Crossley, *Handbook*, 320; Schorr, *Observatory*, August, 1890, p. 281.

Dr. Doberck approximates, in an unusual degree, to a circle, and is traversed in 96 years. The 'mass-brightness' of these objects is twelve times that of the sun, and they, no less than their attendant, are believed to be slightly variable. Their spectrum is of the Sirian type.

The relations of the three stars of β Scorpii are different, but equally genuine. Two, of respectively third and fifth magnitudes, but almost certainly variable, were first observed by Herschel in 1779; the third, discovered by Burnham a century later, makes with the primary an exceedingly difficult, unequal pair at $0''\cdot96$.[1] No *relative* movements have yet been perceived in this system, which nevertheless asserts its organic unity by the harmony of its advance through space.

One of the most curiously interesting of all the stellar systems known to us is ternary from a visual, quaternary from a physical point of view. It is composed of three bright members and an obscure one, all in comparatively rapid mutual circulation. The division of ζ Cancri, by Tobias Mayer in 1756, into a fifth and a sixth magnitude star about $5\frac{1}{2}''$ asunder was the preliminary to Herschel's further analysis. 'If I do not see extremely ill this morning,' he wrote on November 21, 1781, 'the large star consists of two.'[2] This was the earliest example of the decomposition of a double into a triple star.

The next distinct view of these close objects (called for convenience A and B, the remoter star C) was obtained by Sir James South at Passy in 1825, but Struve's nine-inch Fraunhofer showed them easily, and they have never since been lost sight of. Re-observation at once rendered patent their swift movement of revolution. Before the close of 1840 they had, by resuming the positions in which they were originally observed, authoritatively declared their period to be not far from sixty years. And their orbit lies in a plane so nearly square to the line of sight, that foreshortening takes little effect upon it, and occultations are hence impossible. Although the maximum interval between the stars scarcely exceeds one second, and the minimum interval is no more

[1] *Monthly Notices*, vol. xl. p. 100. [2] Crossley, *Handbook*, p. 247.

than 0″·2; they never close up beyond the dividing powers of first-class instruments.

But the orbital movements of the couple A B make only part of a complex scheme of displacements. 'This star,' Sir John Herschel remarked in 1826, 'presents the hitherto unique combination of three individuals, forming, if not a system connected by the agency of attractive forces, at least one in which all the parts are in a state of relative motion.'[1] And he added that, if really ternary, its perturbations must present 'one of the most intricate problems in physical astronomy'—a forecast which bids fair to be fully verified.

The star C apparently retrogrades round A B at an average rate of half a degree a year, indicating (if maintained with approximate uniformity) revolution in a period of 600 or 700 years. But this average rate is subject to very remarkable irregularities. The path traced out in the sky, far from being a smooth curve, is looped into a series of epicycles, in traversing which the star alternately quickens and slackens, or even altogether desists from its advance, while increasing or diminishing, by proportionate amounts, its distance from the centre of motion. This anomalous behaviour, detected by M. Flammarion in 1873,[2] was both detected and interpreted by M. Otto Struve in 1874.[3] The vagaries of the third component of ζ Cancri proved, from his investigation, to be very far from unmethodical. The accelerations which they included were shown to be perfectly compensated by retardations, and to be accompanied unfailingly by expansions outward of the parts of the track where they occurred, while contractions inward attended slackened movements. An explanation too was hazarded, the substantial truth of which was amply attested by M. Seeliger's subsequent researches.[4]

It seems then that the star C is merely a satellite to a dark body round which it describes, in 17½ years, a little

[1] *Phil. Trans.* vol. cxvi. p. 326.

[2] *Catalogue*, p. 49.

[3] *Comptes Rendus*, t. lxxix. p. 1463.

[4] *Sitzungsberichte*, Wien, Bd. lxxxiii. Abth. 2, p. 1018; *Denkschriften*, Munich, Bd. xvii. Abth. i. 1889; Harzer, *Astr. Nach.* No. 2764; *Observatory*, vol. xii. p. 116.

ellipse with a mean radius of one-fifth of a second. Together this singular pair circuits, or, more probably, is circuited by A B, the invisible disturbing body being, quite possibly, the most massive of the system. If this be the case, it is also, of course, the most nearly stationary, and should be regarded as the centre round which the lucent trio revolve—an arrangement hinting to us that the collocation in the same orb, familiar to us in the solar system, of the functions of rule and light-giving may, on occasions, be dispensed with. An anti-Copernican system, at any rate, appears to be to some extent exemplified by ζ Cancri. Here a cool, dark globe, clothed possibly with the vegetation appropriate to those strange climes, and plentifully stocked, it may be, with living things, is waited on, for the supply of their needs, by three vagrant suns, the motions of which it controls, while maintaining the dignity of its own comparative rest, or rather of its lesser degree of movement. For the preponderance of this unseen body cannot approach that of a sun over its planets; hence its central position is by no means undisturbed.

Another lesson already learnt from the stars of *a* Centauri [1] is emphasised by the relations as to mass and luminosity of the components of ζ Cancri. It is, that such relations are not merely prescribed by the inevitable progress of cooling. For, if they were, the most ponderous should, among bodies of contemporaneous origin, be invariably the least advanced; it should still be a distributor of light and heat after all its companions had sunk to extinction. But in ζ Cancri the largest reservoir has been the soonest exhausted; the globe containing most matter has far outstript its associates, and reached the planetary stage while they are still in their meridian glory as Sirian suns. The contrast is heightened in the close pair, by their possession (according to Professor Pickering's calculation) of nine times the solar emissive power relatively to mass; and all the three visible components are in this respect most likely of all but identical quality. Their real differences of magnitude, too, seem to be slight, although at times exaggerated by relative variability. The entire group

[1] See *ante*, p. 199.

is transported across the sphere at the rate of 15″ a century, but its distance from the earth is unknown.

A quadruple system of an unique kind is perhaps formed by ζ Ursæ Majoris with three variously related attendants. Already mentioned as a slowly revolving double star,[1] it is besides escorted on its indefinite journey onward by the fifth-magnitude star Alcor, the two making the combination popularly designated the 'Horse and Rider.' Since the interval between them is of 11′30″, they can easily be distinguished with the naked eye; nevertheless Alcor, totally overlooked by the Greeks, was regarded as a test-object for keen eyesight by the Arabs. Its gradual brightening is thus strongly suggested.[2] The probability that Mizar and Alcor mutually revolve is strong, but not overwhelming; their connection *might* be otherwise explained.[3] If they do, their *annus magnus* must be of enormous, to our ideas of interminable length. An estimate for it of 190,000 years is, however, purely conjectural.

The fourth member of the group has betrayed its existence after an absolutely unprecedented fashion. On the Harvard photographs of the spectrum of Mizar, the lines contained in it come out *periodically doubled.*[4] This can only mean that two approximately equal stars are united into a single telescopic object so closely as to be distinguishable only through the line-displacements due to their orbital revolution. Since one component must always retire from, as the other approaches the earth, the lines in their respective spectra are at such times pushed in opposite directions, and imprint themselves consequently in duplicate upon the sensitive plate; when, on the other hand, the stars are moving *across* the visual ray, the lines cease to be shifted and appear single. This naturally occurs twice in each revolution, and those of ζ Ursæ are in this way found to be accomplished in 104 days, with a velocity, oppositely directed for each star, of about fifty miles a second. They are hence 143 millions of miles apart, or about as far as Mars is from the sun, and their total mass equals that of *forty* suns.

[1] See *ante*, pp. 163, 195.

[2] Flammarion, *Catalogue*, p. 75.

[3] See *infra*, p. 347.

[4] *Fourth Annual Report*, p. 6; Pickering, *American Jour. of Science*, vol. xxxix. p. 46. See Monck's anticipation of this result in *Jour. Liv. Astr. Soc.* vol. vi. p. 115.

Unless, indeed, their orbit be inclined to the visual ray from the earth; in which case, both its size and their mass should be increased proportionately to the amount of the inclination.

This unexpected discovery has been followed up by the announcement that the spectral lines of β Aurigæ are doubled every alternate night, implying orbital revolution in a period of four days. We have then plainly arrived at the threshold of a new era as regards the investigation of close stellar systems, and can only await in silent wonder the development of their strange peculiarities. The varied possibilities meanwhile of stellar companionship are exemplified by Vogel's spectrographic discovery that Spica (α Virginis) revolves at a minimum rate of fifty-six miles a second, in an orbit with a radius of three million miles, round the common centre of gravity of itself and an obscure, or partially obscure, companion.[1] His suspicion that Rigel is similarly coupled may have been confirmed by the time these lines meet the eye of the reader.

Real quaternary stars are often self-discriminating; their arrangement into two adjacent couples is more characteristic of physical connection than any possible distribution of three stars can be. And in effect, several perspective groups of a single star with a genuine pair, such as δ Equulei, 85 Pegasi, and β Delphini, are visibly in course of being dissolved by proper motion, while no 'double-double' combination has yet given signs of breaking up.

A representative specimen of the latter class offers itself in ε Lyræ, a star of the fourth magnitude, a little to the north-east of Vega. Exceptionally keen eyes show it as double, and one of the brilliant surprises provided by the heavens for Sir William Herschel was that of finding each component further divisible. The discovery, though beautiful and interesting, was easy; all the four stars can be seen with a good three-inch telescope. The 'preceding' pair, or that which crosses the meridian first, is distinguished as ε^1, the 'following' pair as ε^2 Lyræ; and Flamsteed attached the numbers 4 and 5 to them respectively. The former consists of a fifth and a sixth magnitude star 3″ asunder; the constituent

[1] *Sitzungsberichte*, Berlin, April 24, 1890. Cf. Maunder, *Observatory*, July 1890, p. 238.

stars of ε^2 are nearly equal (5·3 and 5·5 magnitudes), and are set a little closer together (at 2″·45). Their revolutions, too, appear to be performed about twice as quickly as those of the neighbouring couple. From the shifting of their relative situations since 1779 to the extent of nearly half a right angle, their period may be estimated at about one thousand years; while that of ε^1 Lyræ can hardly fall short of two thousand. The practicability of computing either orbit is still remote.

The small common proper motion (9″ a century) of these bright couples affords positive evidence of their union into one vast system. At their unmeasured, perhaps immeasurable distance, the gap between them of 3½″ may well stand for a chasm costing light itself some months to bridge; yet the stress of their mutual gravity reaches across it, compelling their circulation in orbits so spacious that a single round of them must occupy an era of no insignificant duration, even in the life of a star. The four stars of ε Lyræ give a spectrum of the first type, combined, in the first two, with a decidedly yellowish colour. But this is often the case with double stars.

'A miniature of ε Lyræ'[1] is offered to our regard in ν Scorpii. This is perhaps the most beautiful quadruple group in the heavens, from the narrow limits within which the brilliant objects composing it are crowded. As a wide double it was noticed by C. Mayer in 1776; after seventy years the smaller star was divided by Mitchel at Cincinnati; and the larger one of fourth magnitude yielded similarly, in 1874, to the insistence of Burnham. These last very close stars are certainly revolving,[2] and both pairs, at an interval of 43″, share a slight drift through space.

The sixth-magnitude star 86 Virginis may be said to consist of a double primary with a double satellite at 27″. Acquaintance with the group in its true aspect was made through Burnham's analysis of one of Struve's wider pairs. The movements doubtless progressing in it have not yet become perceptible.

Eighteen 'double-double' combinations (one (Σ 2435) with a total extent of no more than 15″) were enumerated by

[1] Flammarion, *Catalogue*, p. 96.

[2] Burnham, *Memoirs R. Astr. Society*, vol. xlvii. p. 288.

Burnham in 1882,[1] and two were subsequently discovered by Professor G. W. Hough.[2] They perhaps exist more numerously than we have as yet any idea of.

A 'double-treble' star, so called by Herschel, has been the subject of numerous successive discoveries. With the slightest optical assistance σ Orionis, a star of 3·7 magnitude, just beneath the middle star in the belt of Orion, separates into two wide and unequal components, each of which was, October 7, 1779, perceived by Herschel to be triple.[3] As usual in such cases the process of resolution was continued, and the assemblage was described by Barlow as 'double-quadruple, with two very fine stars between the sets.'[4] These last, however, are not unlikely to be mere optical associates. To this intricate group Burnham has added a further element of complexity by detecting, in the autumn of 1888, its leading member to be formed of a fourth and a sixth magnitude star, a quarter of a second apart.[5] The discovery, like some others, raises a question as to the point where stellar subdivision can really be said to cease. Certainly not where visual limitations interfere with our recognition of it.

The essential character of σ Orionis is that of being made up of two distinct, yet evidently connected *knots* of stars, and the same knot (Σ 672)[6] contains all the four brightest components. These differ, and perhaps vary in colour, and their influence may be assumed to predominate in a system, which, however, gives no sensible evidence of movement, whether of the circulatory or the advancing kind.

The multiple star 45 Leporis is organised on a plan less markedly definite than that governing the structure of σ Orionis. It consists of four principal and five subordinate members, the last successively discovered by Sir John Herschel, Burnham, and Hall.[7] The star B, of eighth magnitude, stands out through its ruddy colour from its white companions.[8]

[1] *Observatory*, vol. iv. p. 176. [2] *Astr. Nach.* No. 2778.
[3] *Phil. Trans.* vol. lxxii. p. 124.
[4] Smyth, *Cycle of Cel. Objects*, ed. 1881, p. 156. [5] *Astr. Nach.* No. 2875.
[6] Struve, *Mens. Microm.* pp. 149, 245.
[7] Burnham, *Memoirs R. Astr. Soc.* vol. xliv. p. 238; *Astr. Nach.* No. 2062. *Observatory*, vol. iv. p. 177. [8] G. Knott, *ibid.* pp. 184, 212.

Measurements of the groups are still so recent that their repetition can hardly be expected for some time to come to give evidence of motion. Blended into a single object, the nine stars, covering an extent of 125″, are just visible to the naked eye.

The nebular relations of double and multiple stars were noticed with surprise by Sir John Herschel at the outset of his career.[1] Although admitting without hesitation their physical character, he was without the means of establishing it since made available, and could support his conviction only by the utter improbability of such collocations as he pointed out being fortuitous. Thus, a close, minute stellar couple is planted at the exact centre of a faint round nebula in Leo (New Gen. Cat., No. 3230); and the same kind of coincidence recurs twice in the southern constellation Dorado (N. G. C., Nos. 1732, 1951). Two pairs in Sagittarius, each set in the midst of a nebula (N. G. C., Nos. 6589, 6590), may from their contiguity be suspected to constitute one system; and two ninth-magnitude stars at 15″, marking very nearly the foci of an elliptical nebula in the same region (N. G. C., No. 6595), are certainly not accidentally projected upon it. 'One of the most curious objects in the heavens' (according to Sir John Herschel),[2] is a trio of stars arranged in an equilateral triangle, the sides of which measure 4″, and relieved upon a shield of milky light (N. G. C., No. 1931); an analogue, possibly, of two adjacent stellar trios observed at Harvard College in 1860, in the places of two small but bright nebulæ recorded by Bond and Auwers in 1853 (N. G. C., Nos. 2399, 2400). Mere differences of seeing might, it is true, account for these apparent substitutions; but the alternative supposition that the stars came more plainly into view through a loss of light in their nebulous surroundings deserves at least to be tested by renewed experiments, visual or (better still) photographic. But there are other still more noteworthy instances of the association of composite stars with nebulæ. The whole framework of the great nebulous structure in the sword of Orion seems to rest upon the stellar group designated θ, or rather θ' Orionis; for

[1] *Memoirs R. Astr. Soc.* vol. vi. p. 78. [2] *Ibid.* vol. iii. p. 54.

there is a second θ not far off, itself a wide double star, and the two together form, to the eye, one diffuse object, singly catalogued by Ptolemy, Tycho Brahe, and Hevelius. But it is with θ' exclusively that we are at present concerned.

On the very slightest telescopic persuasion, it allows itself to be seen as quadruple. The four stars into which it divides are severally of fifth, sixth, seventh, and eighth magnitudes, the greatest interval between any two of them not exceeding 21″. None of them is in visible subordination to any other; they stand, it might be said, on an equal footing, at the four corners of a rudely quadrilateral figure, or 'trapezium.' They maintain their places, too, both absolute and relative, with singular rigidity. After two and a half centuries of observation, no shifting of them can be detected. They are hence likely to be at a prodigious distance from the earth.

The rule that such groups seem more crowded as they are better seen, has not been infringed here. A fifth star of the eleventh magnitude was added to the company by Struve, November 11, 1826, and a sixth, then still fainter, but which has since probably gained somewhat in light, by Sir John Herschel, February 13, 1830. Both of these, though closely associated, each with one of the larger stars, share their apparent immobility.[1] Variability in light has often been ascribed, and as often denied to them. Burnham's experience is against it; yet the curious fact that Robert Hooke saw the fifth star in 1664 with a non-achromatic three-and-a-half-inch telescope,[2] is strongly indicative of temporary brightening. Professor Holden at least found it undiscernible at Washington with so much of the great refractor left uncovered;[3] nevertheless, all six of the trapezium-stars have, at favourable moments, been made out with both achromatics and reflectors of three to four inches aperture.[4] It can scarcely then be claimed for Hooke's observation that it demonstrates change.

[1] Burnham's measures seem decisive on this point. See *Memoirs R. Astr. Soc.* vols. xliv. pp. 203, 237, xlvii. p. 244; *Monthly Notices*, vol. xlix. p. 357.

[2] *Micrographia*, p. 242.

[3] Washington Observations for 1877, App. ii. p. 7.

[4] Webb, *Cel. Objects*, p. 367.

Further members of this group have, at various times, been half seen, half surmised; but their existence, always problematical, has been disproved through the application of the Lick thirty-six inch; for the three new stars perceived from Mount Hamilton by Alvan G. Clark and Barnard could certainly have been detected with no less powerful instrument. Two of them lie within the trapezium; the third, a double star of extraordinary minuteness and difficulty, just outside it.[1] Their positions are shown on the accompanying diagram (fig. 33), where the names of the observers are attached to the recently discovered stars.

About half a degree to the north and south respectively of θ Orionis are situated the double star C, and the triple ι

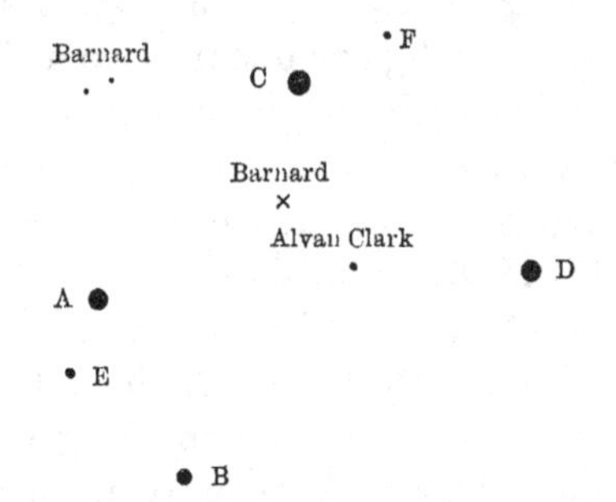

FIG. 33.—Stars of the Trapezium.

Orionis, each with a considerable encompassing nebula, in manifest dependence upon the great intermediate mass gathered about the trapezium. But if all three (as appears certain) belong to the same formation, the stellar groups constituting their nuclei must be connected, however remotely, into a single system. We cannot indeed regard them as performing mutual revolutions, which should be quickly arrested by the resistance of the gaseous stuff with which that whole region is filled; and we are equally unprepared to admit their abandonment to the influence of gravity issuing in catastrophic collisions. But we know of no third alternative. What then are we to conclude? Only, it seems to us, that it

[1] Burnham, *Monthly Notices*, vol. xlix. p. 352; *Astr. Nach.* No. 2930.

is premature to conclude anything. The hint, however, as to the activity in the sidereal system of forces totally alien to our experience should be carefully borne in mind.

In one other great nebula a multiple star is apparently dominant. The nuclear group in the trifid nebula (N. G. C., 6514) consists of a close quartette covering an angular extent of only 19″, with two extremely faint additional stars discovered by Professors Langley and Holden. Complete apparent fixity characterises the arrangement.

The frequent association of compound stars with nebulæ is no mere isolated fact. For they pass by insensible degrees into star-clusters, the fundamentally nebulous nature of which is rapidly, with the aid of photography, becoming established as an indisputable truth. The conjecture is even plausible that the formation of a multiple star in a great nebula represents the initial stage of the development from it of a crowded cluster, minor nebulæ giving rise to lesser groups; and if objects of the kind have not yet, so to speak, been turned out of the workshop, it is no wonder that fragments of their raw material still cling round them. Compositeness of structure may thus measure primitiveness of condition, illustrating, though to us dimly, the sequence of Divinely decreed changes by which cosmical order is gradually more and more fully disengaging itself from the 'loud misrule' of chaos.

CHAPTER XV.

THE PLEIADES.

From multiple stars the transition is easy to star-clusters. These seem to embody completely the idea contained in germ in the former class of objects. They are collections, often on the grandest scale, of sunlike bodies small and large, united in origin and history, acted upon by identical forces, tending towards closely related ends. The manner and measure of their aggregation, however, vary widely, and with them the cogency of the evidence as to their organic oneness. There are innumerable cases in which it absolutely excludes doubt; there are some in which it is rather persuasive than convincing. It is not then always easy to distinguish between a casual 'sprinkle' of stars and a genuine cluster. Nor can the movement-test, by which so many physical have been discriminated from optical double-stars, be here applied. Internal displacements of a circulatory character have not yet become apparent in any cluster, and there is only one with an ascertained common proper motion.

This is the immemorial group of the Pleiades, famous in legend, and instructive, above all others, to exact inquirers—the meeting-place in the skies of mythology and science. The vivid and picturesque aspect of these stars riveted, from the earliest ages, the attention of mankind; a peculiar sacredness attached to them, and their concern with human destinies was believed to be especially close and direct. Out of the dim reveries about them of untutored races, issued their association with the seven beneficent sky-spirits of the Vedas and the Zendavesta,[1] and the location among them of the centre of the universe and the abode of the Deity, of which

[1] Bunsen, *Die Pleiaden und der Thierkreis*, p. 434.

the tradition is still preserved by the Berbers and Dyaks.[1] With November, the 'Pleiad-month,' many primitive people began their year; [2] and on the day of the midnight-culmination of the Pleiades, November 17, no petition was presented in vain to the ancient kings of Persia; [3] the same event gave the signal at Busiris for the commencement of the feast of Isis, and regulated less immediately the celebration connected with the fifty-two-year cycle of the Mexicans. Savage Australian tribes to this day dance in honour of the 'Seven Stars,' because 'they are very good to the black fellows.' The Abipones of Brazil regard them with pride as their ancestors.[4] Elsewhere, the origin of fire and the knowledge of rice-culture are traced to them. They are the 'hoeing-stars' of South Africa,[5] take the place of a farming-calendar to the Solomon Islanders, and their last visible rising after sunset is, or has been, celebrated with rejoicings all over the southern hemisphere, as betokening the 'waking-up time' to agricultural activity.

To the Greeks of Hesiod's age their 'heliacal rising' (the first visible before sunrise) announced, each May, the opening of the season for navigation; and their name thus came to be interpreted (from *plein*, to sail) the 'sailing-stars.' But this etymology was doubtless—like the derivation of 'elf' and 'goblin' from *Guelf* and *Ghibelline*—an afterthought; and it may be confidently maintained that the word 'Pleiades,' bearing, like its Arabic and Hebrew equivalents, the essential signification of a 'cluster,' came from the Greek *pleîones*, many, or *pleiôs*, full.[6]

The similarity of the traditions respecting the swarm of celestial 'fireflies,'

Quæ septem dici, sex tamen esse solent,

is as surprising as their universality. That they 'were seven

[1] Haliburton, *Nature*, vol. xxv. pp. 100, 317; Van Sandiak, *L'Astronomie*, t. iv. p. 367.

[2] Haliburton, *Festival of the Dead*, p. 46. [3] *Ibid.* p. 13.

[4] Lubbock, *Origin of Civilisation*, p. 316, 4th ed.

[5] J. Hammond Tooke, in an interesting paper read in January 1889 before the S. African Philosophical Society.

[6] *Nature*, vol. xxxv. p. 608.

who now are six,' is asserted by almost all the nations of the earth, from Japan to Nigritia, and variants of the classical story of the 'lost Pleiad' are still repeated by sable legend-mongers in Victoria, by 'head-hunters' in Borneo, by fetish-worshippers amid the mangrove swamps of the Gold Coast. An impression thus widely diffused must either have spread from a common source, or originated in an obvious fact; and it is at least possible that the veiled face of the seventh Atlantid may typify a real loss of light in a prehistorically conspicuous star. Some members of the collection are at present, there is little doubt, slightly or slowly variable,[1] and progressive tendencies of the kind are in more than one case suggested to be present. Thus Alcyone, the chief of the collection, now of the third magnitude, and just twice as bright as the brightest of its companions, was either not one of the four Pleiades observed by Ptolemy, or was then much fainter than it has been from Tycho Brahe's time to our own. So at least Francis Baily concluded from a careful examination of the records,[2] and he knew better than most men how large an allowance has to be made for ancient inaccuracy. Abdurrahman Sûfi, too, the competent reviser of Ptolemy's observations, expressly states that the Alexandrian quartette appeared to him, in the tenth century, the most lustrous among the Pleiades.[3] Yet none of them can be identified with the actual *lucida*. A literal explanation of the old legend appears then feasible, and Professor Pickering's suggested identification of Pleione with the missing Atlantid has much to recommend it.[4] The display by this star of a gaseous spectrum resembling that of P Cygni makes it at least fully credible that, like P Cygni, it should have been noted for temporary brilliancy. It is now of 6·2 magnitude—that is, just beyond the range of ordinary eyesight.

The five stars ordinarily visible besides Alcyone (see

[1] C. Wolf, *Annales de l'Observatoire de Paris*, t. xiv. ii. p. 26; Lindemann, *Mémoires de l'Acad.*, St. Pétersbourg, t. xxxii. Sér. vii. No. 6. p. 29.

[2] *Memoirs R. Astr. Soc.* vol. xiii. p. 9.

[3] Schjellerup, *Description des Étoiles*, p. 132; Flammarion, *Les Étoiles*, p. 294.

[4] *Astr. Nach.* No. 2934.

Plate I. *Frontispiece*) are Electra and Atlas, each fluctuating slightly above and below 3·8 magnitude; Maia, now of the fourth, but on the rise; Merope and Taygeta, the inferiors of Maia by respectively a quarter and a half magnitude. Celæno, the seventh or concealed star, gives only about half as much light as Taygeta.

It can be seen, however, with many others, under favourable circumstances. Maestlin, the tutor of Kepler, perceived fourteen, and mapped eleven Pleiades previously to the invention of the telescope; Carrington and Denning counted fourteen,[1] Miss Airy marked the places of twelve with the naked eye.[2] The faintest of these fell but little short of the sixth, and there are twenty-five Pleiades down to the seventh magnitude,[3] each of which (with perhaps one or two exceptions) might be separately visible in a transparent sky or from an elevated station. But their crowded condition makes this impossible, and gives rise rather to the effect as described by Kazwini in the thirteenth century, of 'six bright stars with a number of dusky ones between.'[4]

With the use and increase of telescopic powers, the populousness of the cluster has been amazingly increased. An object-glass scarcely exceeding two inches diameter showed Robert Hooke in 1664 seventy-eight Pleiades,[5] and Michell's conjecture, in 1767, that there might be more than a thousand of them,[6] has been superabundantly verified by the results of modern labours. Over an area about Alcyone measuring 135′ × 90′, M. Wolf catalogued, at the Paris Observatory in 1876, 625 stars to the fourteenth magnitude; on the MM. Henry's sensitive plates, in 1885, 1,421 made their appearance in a smaller space, and the number was brought up to 2,326 by exposures of four hours in November and December 1887. The faintest objects thus registered, although nominally of the eighteenth, were probably, by strict photometric reckoning, of about the sixteenth magnitude.

How many of them really belong to the group, and how

[1] Webb, *Cel. Objects*, p. 393. [2] *Monthly Notices*, vol. xxiii. p. 175.
[3] *Harvard Annals*, vol. xiv. pt. ii. p. 398. [4] Ideler, *Sternnamen*, p. 147.
[5] *Micrographia*, p. 241. [6] *Phil. Trans.* vol. lvii. p. 259.

many are referred to it by perspective, can be determined with the help of time and patience. As regards some of the better-known stars, the process of discrimination has already begun.

Bessel's measurements of the places, relative to Alcyone, of 52 Pleiades,[1] executed with the Königsberg heliometer during the twelve years from 1829 to 1841, furnished a starting-point for investigations of their internal movements. The upshot of the first effective comparisons was to exhibit these as null. From a collodion-print of the cluster taken by Rutherfurd of New York in 1865, Dr. Gould redetermined nearly all Bessel's stars with such accuracy as to make it certain that no appreciable *interstitial* shiftings had occurred in the course of a quarter of a century.[2]

Now this seeming rigidity in effect implied a great deal. For the point of origin of the measures in question is not immovably fixed in the sky. The chief Atlantid has a secular proper motion (according to Newcomb) of 5″·8, the possession of which in common by the whole stellar band virtually demonstrated their effectual union. Where one among many objects is ascertained to be moving, relative fixity can only mean that all drift together; and so the unique phenomenon was brought to light of the transport in block across the sphere of a couple of thousand congregated suns. Even if the whole of this apparent displacement should prove to be, as it were, reflected from the solar advance, its significance of physical kinship among the objects affected by it would be nowise impaired. For the circumstance of their being similarly affected by it would suffice to locate them in the same region of space under the immediate influence of their constraining mutual gravity.

The establishment of a general unanimity of movement among the Pleiades was the first step towards investigating their relations; the next was to seek evidence of systematic change. This perhaps remains to be found; its highly recondite nature has at least been rendered unmistakable by the

[1] *Astr. Nach.* No. 430. [2] *Observatory*, vol. ii. p. 16.

labours of Wolf,[1] Pritchard,[2] and Elkin.[3] Displacements within the cluster, if they have been genuinely measured at all, are barely nascent. But with its *analysis* some progress has been made, as the result of Dr. Elkin's work at Yale College in 1884–5.

Leaving nothing to be desired in the way of skill and care, it was the more strictly comparable with Bessel's from having been executed, like his, with a heliometer, one of exquisite workmanship, completed in 1882 by the Messrs. Repsold, of Hamburg. Sixty-nine stars, down to 9·2 magnitude, were included in the survey, only one of Bessel's being omitted, while seventeen were added from the Bonn Durchmusterung. The close agreement, on the whole, between the places determined, after an interval of forty-five years, at Königsberg and Yale, enhanced the significance of some minute discrepancies, the most considerable of which were held, with a strong show of reason, to imply that six of the objects on Bessel's list were only apparent members of the cluster.[4] They should indeed be regarded as pseudo-Pleiades, intruders into a company from which they will eventually be expelled through the irresistible effects of incompatible movements. Exempt from the influence of the current bearing Alcyone and its true associates slowly towards the south-south-east, they remain almost absolutely stationary in the background of the sky, and are accordingly in course of being left behind.

The proper motion of Alcyone reverses, with approximate accuracy, the direction of the sun's progress through space. It may hence be regarded as parallactic, that is, transferred in appearance from our own. But the six stars lying apart from this perspective drift must (unless on the improbable supposition of their being our fellow-travellers) be so remote, that the path traversed by the solar system in forty-five years, estimated to be at least twenty thousand millions of miles

[1] *Annales de l'Observatoire*, t. xiv. ii.; *Comptes Rendus*, t. lxxxi. p. 6.

[2] *Monthly Notices*, vol. xliv. p. 357.

[3] *Trans. Yale College Observatory*, vol. i. pt. i. 1887.

[4] Dr. Elkin expressed this view with considerable reserve as regards four of the six stars.

in length, dwindles as viewed from them—certainly with some help from foreshortening—to evanescence! The sun seems stationary to them, as they seem stationary to us. Nor is their brightness such as to discredit the inference of their remoteness; they range in magnitude from 7·9 to 9·2.

Besides these half-dozen destined deserters, two stars (*s* and 25 Pleiadum), may, for an opposite reason, be presumed to stand aloof from the collection. Instead of lagging behind, they are hurrying on in front. They exhibit in excess the movement shared by the great majority of their seeming companions. They are then nearer to the sun, by perhaps one-third, than the veritable Pleiades, from which they, as well as the six 'fixed' objects visually intermixed with them, should henceforth be carefully distinguished.

The distance of Alcyone from the earth has never been measured; but it can be calculated, given the direction and velocity of the sun's translation, on the hypothesis that the proper motion of the star is simply a perspective effect of that translation. Now we know that the sun is travelling towards a point in the constellation Hercules not far from the star λ; and the rate of its journey is unlikely to fall short of fifteen, or to exceed twenty miles a second. Adopting the lower estimate, we find the distance of the Pleiades to be nearly 1,500 billions of miles, or 250 light-years (parallax $= 0''\cdot013$). And this may be considered a minimum value.

Our own sun, thus prodigiously remote, would shrink to a star below the tenth magnitude, and much fainter, accordingly, than any of those measured by Dr. Elkin. There can be little doubt, in fact, that the solar brilliancy is surpassed by sixty to seventy of the Pleiades. And it must be, in some cases, enormously surpassed; by Alcyone 1,000, by Electra 480, by Maia nearly 400 times. Sirius itself takes a subordinate rank when compared with the five most brilliant members of a group, the real magnificence of which we can thus in some degree apprehend.

The scale of its construction is no less imposing. No judgment can of course be formed as to the interval of space separating any two of the stars belonging to it. All of them are seen projected indiscriminately upon the same plane, without regard to the directions in which they lie one from the other. The line joining Maia, for instance, with Alcyone, may be foreshortened to any extent, or not at all. No criterion is at hand which we can apply. Of the dimensions, however, of the cluster as a whole, some notion can be gathered. For its shape—irrespective of some outlying streams of small stars—may be taken to be rudely globular; and since a circle described from Alcyone as a centre with a radius of 48′ includes all the principal stars, sixty of Elkin's sixty-nine, fifty-two of Bessel's fifty-three falling within it, the apparent diameter of the denser part of the aggregation cannot differ much from 96′. But the proportion of the radius to the distance of a globe of known angular dimensions is easily arrived at, and here comes out (in round numbers) as one to seventy-one; so that the bodies situated close to its surface are seventy-one times nearer to their central luminary than their central luminary is to us. If they revolve round it, it is at the stupendous interval of (at least) twenty-one billion miles, costing light three and a half years to cross; and the period of their circulation may well be reckoned by millions of years. Upon these dependent orbs, Alcyone shines with eighty-three times the lustre of Sirius in terrestrial skies; yet the presence of 67,000 Alcyones would only just compensate for the withdrawal of even such a diminished sun as brightens the firmament of Neptune. From stars more centrally placed, the chief of the cluster doubtless appears a veritable sun, although it may not be to all the primary light-giver. An assemblage like the Pleiades distributed round our sun would extend *compactly* three-quarters of the way to α Centauri, its feelers and appendages indefinitely farther. Hence there would be ample room in it for secondary systems and particular associations of luminous bodies. And, in point of fact, the actual cluster contains several of Burnham's close double stars, presumably in mutual revolution, to say nothing of the doubtful companion of Atlas, which, distinctly visible only once to Struve in 1827, gave nevertheless some sign of its presence during an occultation by the moon, January 6, 1876.[1]

All that is certain about the movements just beginning to be perceptible among the Pleiades, is that they are very small. Dr. Elkin concluded from the minuteness of the displacements brought to light by his measures, that 'the hopes of obtaining any clue to the internal mechanism of the cluster seem not likely to be realised in an immediate future;' and remarked the especial immobility of the brighter stars.[2] Electra, alone of these, shifts as much as one second of arc in a century, whether along a curved or a straight line it is of course at present impossible to tell. The slightly swifter movements ascribed to a score of minor objects harmonise very imperfectly (so far as can be judged) with preconceived notions of orderly circulation. They, on the contrary, rather suggest the advance of some slow process of systemic disintegration. M. Wolf's impression of centrifugal tendencies accordingly derives some confirmation from Dr. Elkin's chart, in which divergent movements seem to prevail, the region round Alcyone resembling, on a cursory view, a confused area of radiation for a flight of meteors, much more than the central seat of attraction of a revolving throng of suns.

There are indications, however, that adjacent stars in the cluster do not pursue independent courses; community of drift is perceptible in several quarters. At least four batches of stars seem to travel at nearly the same rate along approximately parallel lines, and may perhaps eventually set up as independent systems. Thus, inquiries into the condition of the Pleiades afford as yet little countenance to the view that their dynamical relations are of a permanent kind. On this point differences of radial motion, such as the Potsdam spectrograph is capable of detecting, ought to prove highly instructive. We may learn from them, for instance, whether a uniform method of circulation prevails in the cluster. For if so, its constituent stars should be found approaching (comparatively to Alcyone) on the one side, receding on the other. The reality, too, of local groupings of apparent

[1] *Astr. Nach.* No. 2074. [2] *Trans. Yale College Obs.* vol. i. p. 101.

movement might be similarly tested, if only the spectra of the objects concerned were bright enough for the purposes of photographic line-measurement. The Greenwich results showing Alcyone (η Tauri) to be approaching the earth at the rate of about thirty-six miles a second, need confirmation all the more that since the journey of the solar system is directed obliquely *from* the Pleiades, it could have no share in producing a velocity necessarily shared by the entire physical group.

A spectrum of the Sirian type characterises them. A simultaneous spectrographic impression obtained by Professor Pickering from close upon forty of these stars, January 26, 1886, demonstrated the nearly identical quality of their light, and furnished 'strong confirmation of their common origin.'[1] Only in two cases a stronger 'K line' recorded itself than such light ordinarily includes, and the divergence was, in one of the two, both accentuated and explained by diversity of motion. The star in question (*s* Pleiadum) has been already signalised as an incipient fugitive from the group to which it never truly appertained.

The stars of the Pleiades, while shining with so poignant a lustre as to make the sky-background they are relieved upon show to the eye as blacker than elsewhere, are in reality wrapt and entangled in an immense cosmical cloud. Some indications to this effect caught by optical means have been autographically amplified to so surprising an extent that the discovery of the nebulous condition of the Pleiades ranks among the most important achievements of celestial photography.

The 'Merope nebula' was compared by the late M. Tempel to a stain of breath upon a mirror. Discovered by him at Venice, October 19, 1859, it envelopes and stretches back in cometary shape from the star to which it is attached, covering a space of about 35′ by 20′.[2] But this large size only makes its perception more difficult, by impairing the effect of contrast with the surrounding sky. With high magnifying powers (which imply narrow fields of view), it is on this

[1] *Memoirs Amer. Acad.* vol. xi. p. 215.

[2] *Astr. Nach.* No. 1290; *Monthly Notices*, vol. xl. p. 622.

account completely invisible, and a haze so slight as to permit the observation of stars of thirteenth or fourteenth magnitude suffices to obliterate it. Thus, it has been only exceptionally seen, and has often been suspected to be variable. Tempel's contrary opinion, however, has of late been fully justified.

The idea, too, was entertained both by Goldschmidt [1] and Wolf that the filmy veil flung round Merope was but a fragment of a larger whole, and, as time went on, glimpses were snatched of misty shreds and patches in connection with other members of the group. Alcyone appeared to Searle at Harvard College, November 21, 1875, as surrounded by whitish light; [2] the attainment by the effusion about Merope of Electra and even Celæno was evident to Schiaparelli in 1875,[3] and to Maxwell Hall in 1880; [4] while a remarkable view afforded to Mr. Common by his three-foot reflector, February 3, 1880,[5] of three feebly luminous blotches between Merope and Alcyone, prompted his comment that 'there is a great deal yet to be settled as to the extent and number of the nebulæ in this cluster.'

Its significance, however, became apparent only when photography was brought to bear upon the subject. The first nebula discovered by the new method was a small spiral appendage to the star Maia, which printed itself on plates exposed by the MM. Henry, each during three hours, in December 1885.[6] Only the accumulating faculty of the 'chemical retina' could have revealed the presence of an object so excessively faint in a telescopic sense; but what is known to exist is, by that alone, rendered more than half visible, and the Maia nebula was accordingly discerned, February 5, 1886, with the Pulkowa thirty-inch refractor, then newly erected, and later with smaller instruments.[7]

Besides the Maia vortex, the Paris photographs depicted a series of nebulous bars on either side of Merope (partly

[1] *Les Mondes*, t. iii. p. 529. [2] *Harvard Annals*, vol. xiii. p. 74.

[3] *Astr. Nach.* No. 2045.

[4] *Monthly Notices*, vol. xli. p. 315. [5] *Ibid.* vol. xl. p. 376.

[6] Similarly recorded a month earlier at Harvard College, it was taken for a flaw in the negative.

[7] *Astr. Nach.* Nos. 2719, 2726, 2730.

seen by Mr. Common), and a curious streak extending like a finger-post from Electra towards Alcyone. But all these were mere samples of what lay behind. Impressions of the Pleiades secured by Mr. Roberts with his twenty-inch reflector in October and December 1886, showed the whole western side of the group to be involved in one vast nebulous formation.[1] 'Streamers and fleecy masses' of cosmical fog seem, in these astonishing pictures, almost to fill the spaces between the stars, as clouds choke a mountain valley. The chief points of its concentration are the four stars Alcyone, Merope, Maia, and Electra; but it includes as well Celæno and Taygeta, and is traceable southward from Asterope over an arc of 1° 10′. Tempel's nebula appears in its proper elliptical form in addition to the barred structure stamped upon the Paris plates; the little curved train of Maia is visible, though immersed in a far wider luminosity, reaching nearly to Asterope; Electra projects towards Alcyone a dim shaft, to which a thin streamer, 'resembling a detached nebulous straight line,' runs parallel further south.[2] These photographs, in fine, as Mr. Wesley wrote, 'not only prove beyond a doubt the existence of the much-disputed Merope nebula, but they also combine and harmonise in a very satisfactory manner the apparently irreconcilable drawings.'[3]

The matter was not, however, allowed to rest here. Early in 1888 the MM. Henry succeeded in giving to several plates exposures of four hours, with results identical in each case, and very curious. Their nature can be estimated from the frontispiece, which reproduces the final chart of the Pleiades prepared by the MM. Henry. The greater part of the constellation is shown in it as veiled in nebulous matter of most unequal densities. In some places it lies in heavy folds and wreaths, in others it barely qualifies the darkness of the sky-ground. The details of its distribution come out with remarkable clearness, and are evidently to a large extent prescribed by the relative situations of the stars. Their lines of junction are frequently marked by nebulous rays, establishing between them, no

[1] *Monthly Notices*, vol. xlvii. p. 24.

[2] *Ibid.* p. 90.

[3] *Jour. Liv. Astr. Soc.* vol. v. p. 150.

doubt, relations of great physical importance; and masses of nebula, in numerous instances, seem as if *pulled out of shape* and drawn into festoons by the attractions of neighbouring stars. But the strangest exemplification of this filamentous tendency is in a fine, thread-like process, 3″ or 4″ wide, but 35′ to 40′ long, issuing in an easterly direction from the edge of the nebula about Maia, and stringing together seven stars met in its advance, 'like beads on a rosary.'[1] The largest of these is apparently the occasion of a slight deviation from its otherwise rectilinear course. A second similar, but shorter streak runs, likewise east and west, through the midst of the formation.

Whether these luminous highways are due to material condensations, or merely indicate tracks of electrical excitement, they are equally significant upon one point. The connection by their means of rows of stars virtually demonstrates their real alignment, and thus considerably strengthens the presumption that the linear arrangement prevalent in clusters is no optical illusion, but depends upon intrinsic conditions, the outcome of universal laws.

Investigations of the Pleiades have led to many surprises; possibly the supply of them is not even yet exhausted. The main result so far has been to exhibit the group as combining with such singular completeness the properties of a great nebula with those of a cluster, that we are inevitably led to regard the gap between these two kinds of aggregation as less wide than had heretofore been supposed.

Stars of all orders are there gathered together into (it might be said) a miniature sidereal system. The largest are of such 'surpassing glory' as to dim by comparison the splendour of Sirius and Vega; the least are probably as inferior to them as the moons of Mars are to Jupiter. The 'act of order' in this 'peopled kingdom' is not easy to divine, but the mutual relations of its denizens are at once perceived to be highly intricate. Within the wide framework of the association room is found for subordinate groupings of various characters and degrees of closeness,

[1] Mouchez, *Comptes Rendus*, t. cvi. p. 912.

from stars far apart, but 'drifting' in company, to pairs as unmistakably united by contiguity as two nuts within the same shell. Thus, the polity governing the entire system of the Pleiades would seem to be of the federative kind. Nor can we be yet sure that its bonds, while evidently so loose as to give unshackled play to local liberties, are nevertheless sufficiently strong to restrain the slow workings of disruptive tendencies.

CHAPTER XVI.

STAR CLUSTERS.[1]

About five hundred clusters are at present tolerably well known to astronomers, and a large number besides, their character rendered ambiguous by distance, are probably included among both 'resolvable' and 'unresolved' nebulæ. Such aggregations may be broadly divided into 'irregular' and 'globular' clusters. Although, as might have been expected, the line of demarcation between the two classes is by no means sharply drawn, each has its own marked peculiarities.

Irregular clusters are framed on no very obvious plan; they are not centrally condensed, they are of all shapes, and their leading stars rarely occupy critical positions. The stars in them are collected together, to a superficial glance, much after the fashion of a flock of birds. Alcyone, it is true, seems of primary dignity among the Pleiades, and the Pleiades may be regarded as typical of irregular clusters; yet the dominance, even here, of a central star may be more apparent than real.

The arrangement of stars in clusters is, nevertheless, far from being unmethodical, even though the method discernible in it be not of the sort that might have been anticipated. It seems, indeed, inconsistent with movements in closed curves, and suggests rather the description of hyperbolic orbits. Obviously, however, its true nature must be greatly obscured to our perception by the annulment, through perspective, of the third dimension of space, whereby independent groupings, *flattened down* side by side, are rendered scarcely,

[1] See *Nature*, vols. xxxviii. p. 365, xxxix. p. 13.

if at all, distinguishable. That they should under these circumstances be to any extent traceable is more surprising than that they should sometimes be inextricably entangled with sprinkled stars belonging to the fore- or background. The nebulous linking together of a septuple set in the Pleiades assures us, however, that they are traceable, and that star-alineations are not illusory.

Nearly all observers have been impressed with the streaming and reticulated structure characterising many stellar assemblages. Thus where the feet of the Twins dip into the Milky Way, an object is encountered so 'marvellously striking' with a large telescope, that 'no one could see it for the first time,' Mr. Lassell declared, 'without an exclamation.' A field 19′ in diameter 'is perfectly full of brilliant stars, unusually equal in magnitude and distribution over the whole area. Nothing but a sight of the object itself can convey an idea of its exquisite beauty.'[1] Admiral Smyth described it as 'a gorgeous field of stars from the ninth to the sixteenth magnitudes, but with the centre of the mass less rich than the rest. From the small stars being inclined to form curves of three or four, and often with a large one at the root of the curve, it somewhat reminds one of the bursting of a sky-rocket.'[2] A beautiful photograph of this cluster,[3] taken by the MM. Henry, March 10, 1886, exhibits not less than two thousand of its components disposed in a kind of starfish pattern, the branches often connected by drooping chains, and composed in detail of sinuous lines, or the 'fantastically crossing arcs' of stars noticed by Father Secchi.[4]

The 'wonderful looped and curved lines' of conformation visible in a cluster in Auriga (M 37) attracted the attention both of D'Arrest and Lord Rosse;[5] about one hundred connected stars in Ophiuchus (M 23) 'run in lines and arches;'[6] a collection in Sagittarius (N. G. C., 6416) makes a 'zigzag'

[1] *Monthly Notices*, vol. xiv. p. 76. [2] *Cycle*, p. 168 (Chambers's ed.).

[3] M 35 = N. G. C. 2168. Nebulæ and clusters are throughout this volume distinguished, when among the 103 catalogued by him, by Messier's well-known numbers, otherwise by Dreyer's in the *New General Catalogue*.

[4] *Atti dell' Accad. Pont.* t. vii. p. 72.

[5] *Trans. R. Irish Acad.* vol. ii. p. 51. [6] *Phil. Trans.* vol. cxxiii. p. 460.

figure. The constituents of a large group near the Poop of Argo (N. G. C., 2567) struck the elder Herschel by an arrangement 'chiefly in rows,' illustrative, to his mind, of the mechanical complexities of such systems. Each row, he observed, while possessing its own centre of attraction, will at the same time attract all the others; nay, 'there must be somewhere in all the rows together the seat of a preponderating clustering power which will act upon all the stars in the neighbourhood.'[1] Speculations, indeed, upon the dynamical relations of 'stars in rows' are still premature, nor are they likely, for some time to come, to be accounted as 'of the order of the day.' But the continual recurrence in the heavens of this mode of stellar aggregation cannot fail to suggest the development of plans of systemic dissolution and recomposition on a grand scale, and involving the play of, to us, unknown forces.

The more attentively clusters are studied, the more intricate their construction appears. That which challenged Herschel's notice is not singular in intimating a league of several co-ordinate groups. There is rarely evidence in the conformation of irregular clusters of their being governed from a single focus of attraction; there are frequent indications of the simultaneous ascendency of several. A cluster in Sagittarius (N. G. C., 6451) is distinctly bifid. It was remarked by Sir John Herschel at Feldhausen as 'divided by a broad, vacant, straight band';[2] and his figure shows the separation as absolutely complete, the sections,

Like cliffs which had been rent asunder,

facing each other with a chasm between.

A splendid cluster in Sobieski's Shield (M 11) seems to be essentially trifid. Sir John Herschel, indeed, succeeded, by the use of high powers, in breaking it up 'into five or six distinct groups with rifts or cracks between them.'[3] But Father Secchi perceived in it a three-lobed central vacuity;[4] and the photograph reproduced in fig. 34 exhibits two great

[1] *Phil. Trans.* vol. civ. p. 269.
[2] *Cape Observations*, p. 116.
[3] *Phil. Trans.* vol. cxxiii. p. 462.
[4] *Atti dell' Accad. Pont.* t. vii. p. 75.

wings of stars in its neighbourhood to about the fourteenth magnitude, with a connecting mass between. Some partial clearings, however, accentuating the structure of the star-clouds, have been to a great extent obliterated in the process of reproduction. The cluster itself, compared by Admiral Smyth to a 'flight of wild ducks,' forms a sort of nucleus to the entire. The original of the picture was taken by Mr. E. E.

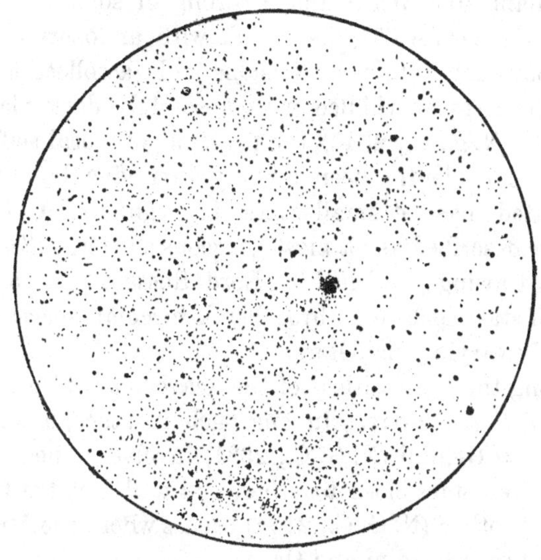

FIG. 34.—Star Cluster in Sobieski's Shield (M 11) photographed by Mr. E. E. Barnard at the Lick Observatory.

Barnard, of the Lick Observatory, with a portrait-lens of just six inches aperture and thirty-one focus, the exposure lasting two hours and three-quarters. The diameter of the field is 5·7 degrees.

This 'glorious object' (as Sir John Herschel called it) can just be made out with the naked eye on a perfectly clear night. Halley mentioned it in 1716 as 'of itself but a small obscure spot, but with a star that shines through it which makes it the more luminous.'[1] Some years later, Derham found it to be 'not a nebulose, but a cluster of stars, some-

[1] *Phil. Trans.* vol. xxix. p. 392.

what like that which is in the Milky Way.'[1] A catalogue of two hundred of the components (several of which apppear to be variable) prepared in 1870 by M. Helmert of the Hamburg Observatory,[2] provides material for the future investigation of relative changes.

The presence in a cluster in Monoceros (N. G. C., 2269) of 'a double seat of preponderating attractions' was observed by Sir William Herschel;[3] and a throng of some two hundred stars in Cancer (M 67), discernible with an opera-glass, falls no less obviously into two divisions.[4] In a collection seen at Parsonstown to be riddled with absolutely dark 'lanes and openings[5] (N. G. C., 2548), the principle of 'local self-government' has evidently been already carried a long way. A 'reticulated mass of small stars' in Cygnus (N. G. C., 6819) was there described as 'a most gorgeous cluster, *full of holes*;' and the drawing published by Lord Rosse depicts a winding ribbon of stars inclosing three blank circular spaces of symmetrically varying diameters.

Among the 'curiosities' of the heavens are to be reckoned *clusters within clusters*. In one such instance, a large loose collection in Gemini (N. G. C., 2331) involves a neat group of 'six or seven stars close together, and well isolated from the rest.'[6] Another (N. G. C., 2194) occurs where the Milky Way passes between Gemini and Orion.

Star-groupings of singularly definite forms are often met with. A triangular swarm (N. G. C., 7826) presents itself in Cetus; a rectangular area in Vulpecula (N. G. C., 6802) is densely strewn with fine star-dust. Clusters shaped like half-open fans are tolerably numerous. One situated in Gemini, if removed to a sufficient distance, would appear, according to Sir John Herschel, 'as a fan-shaped nebula with a bright point like a star at the vertex.' Another specimen of an 'acutangular' cluster 2′ in length (N. G. C.,

[1] *Phil. Trans.* vol. xxxviii. p. 72.
[2] *Publicationen der Hamburger Sternwarte*, No. 1, 1874.
[3] *Phil. Trans.* vol. civ. p. 268.
[4] Smyth, *Cycle*, p. 241; L. Fenet, *L'Astronomie*, t. vi. p. 145.
[5] *Trans. R. Dublin Soc.* vol. ii. p. 66. [6] *Ibid.* p. 56.

7510), is bounded by 'two principal lines of stars drawing to one.'[1]

Red or double stars frequently appear in some sort to dominate stellar assemblages. The chances of optical juxtaposition must indeed produce compound objects most freely where stars are most crowded, yet when they are marked out (as occasionally happens) both by superiority of lustre and by distinction of place, some significance may be attached to their presence. Thus, each of the oblique arms of a 'cruciform' group in Auriga (M 38), photographed at Paris, January 28, 1887, carries a pair of conjoined stars brighter than the rest.[2] A 'superb cluster' in Monoceros (N. G. C., 2548), standing out from a background of sky 'singularly dotted over with infinitely minute points,' has a double star in its most compressed part.[3] The central star in Præsepe is double, and there are many examples of more restricted gatherings round a composite luminary.

Groups apparently ruled by a conspicuous coloured object are met with in Auriga (N. G. C., 1857) and Cygnus (N. G. C., 7086). In Cygnus, too, is an oval annulus, 4′ across (N. G. C., 7128), of stars centrally surrounding a ruddy one of the ninth magnitude. A similar elliptical group, with a double substituted for the red star, constitutes a quasi-nucleus for one of the two great adjacent clusters in Perseus (N. G. C., 869). This superb object, like the still richer group (N. G. C., 884) it immediately precedes, was regarded by Herschel as merely a protuberance of the Milky Way, and its galactic affinities are undoubtedly very close. The two together form a telescopic pageant such as, in the wildest flights of imagination, Hipparchus could little have dreamed would one day be unrolled before the eyes of men out of the 'cloudy spot' in the sword-handle of Perseus which he (it is said) was the first to detect. Although the outlines of the two clusters can be brought within the same field of view, they are held to be really disconnected;[4] it is scarcely probable, however, that they were originally so. The second, and more considerable

[1] *Phil. Trans.* vol. cxxiii. pp. 476, 503.
[2] Smyth, *Cycle*, p. 140.
[3] J. Herschel, *Phil. Trans.* vol. cxxiii. p. 386.
[4] Smyth, *Cycle*, p. 60.

(known as 'χ Persei'), was micrometrically investigated by Vogel in 1867–70, photographically by O. Lohse in 1884,[1] with the result, from the comparison of 172 stars, of demonstrating their fixity during an interval certainly too short for the development into visibility of such tardy movements as were alone likely to be in progress. A rapid spectroscopic survey executed by Vogel with the Berlin nine-inch refractor, March 30, 1876,[2] disclosed no peculiarity in the light of any of the collected stars, although some of them have been called red, 'pale garnet,' and even 'ruby.' Their brilliancy suggests that this magnificent assemblage may be less exorbitantly distant from the earth than most other objects of its class. A fine photograph of the 'double cluster,' in which the 'festoon-like groupings' composing it are conspicuous, was taken by Mr. Roberts at Maghull January 13, 1890.[3]

The famous tinted cluster about κ Crucis can only be seen from southern latitudes. And it must be confessed that, with moderate telescopic apertures, it fails to realise the effect of colour implied by Sir John Herschel's comparison of it to a 'gorgeous piece of fancy jewellery.' A few reddish stars catch the eye at once; but the blues, greens, and yellows belonging to their companions are pale tints, more than half drowned in white light. Some of these stars are suspected of considerable mobility. During his visit to the Cape, Herschel determined the places of 110, all included in an area of about $\frac{1}{48}$ of a square degree,[4] and the process was repeated and extended to 130 components by Mr. H. C. Russell of Sydney in 1872.[5] The upshot was to bring out discrepancies which, if due to real movements, would be of extreme interest. But Herschel's measurements were necessarily too hasty to be minutely reliable, and changes depending upon their authority need to be confirmed by continuance before they can be unreservedly accepted. The same qualification applies to M. Cruls's supposed discovery of orbital revolution in three double stars within the precincts of the cluster.[6]

In the constellation Cancer may be seen any fine night in

[1] *Astr. Nach.* No. 2650.
[2] *Der Sternhaufen χ Persei*, p. 31.
[3] *Monthly Notices*, vol. l. p. 315.
[4] *Cape Observations*, p. 17.
[5] *Monthly Notices*, vol. xxxiii. p. 66.
[6] *Comptes Rendus*, t. lxxxix. p. 435.

winter, a blot of dim light placed midway between two fourth-magnitude stars. The stars were called by the ancients the Asses, *Aselli*, the interposed cloudlet representing to their fancy a 'Manger,' *Præsepe*. Since its disappearance was reckoned a sure presage of rain,[1] a good deal of popular attention was paid to it, and its stellar constitution was one of the earliest telescopic discoveries; but only preliminary steps have been taken towards its exact investigation. Of its components, thirty are measurable on Rutherfurd's photographs, and 363 were mapped over an area of three square degrees by C. Wolf some sixteen years later, eighty-two among them being carefully determined as 'fundamental stars' for the rest, by the same methods used for the Pleiades.[2] As yet, however, no materials are ripe for comparison. The date (1870) of Professor Asaph Hall's catalogue of 151 stars is too recent for the purpose; the results of Winnecke's observations of Præsepe with the Bonn heliometer have not been made public. Hence we are uninformed as to the nature of whatever changes may be progressing in this system.

The particles of a drop of water are not in more obvious mutual dependence than the constituent stars of globular clusters; 'the most magnificent objects,' in the elder Herschel's opinion, 'that can be seen in the heavens.' Were there only one such collection, the probability of its separate organisation might be reckoned 'infinitely infinite,' and one hundred and eleven of them were enumerated by Sir John Herschel in 1864. It does not, however, follow that the systems thus constituted are of a permanent or stable character; their configuration, in fact, points to an opposite conclusion. There may, of course, be an indefinite number of arrangements by which the dynamical equilibrium of a 'ball of stars' could be secured; there is only one which the present resources of analysis enable us distinctly to conceive. This was adverted to, many years since, by Sir John Herschel. Equal revolving masses, uniformly distributed throughout a spherical space, would, he

[1] Aratus, *Diosemeia*, vv. 160–180, 265; Theophrastus, *De Signis Pluviarum*, ed. Heinsius, p. 419.

[2] *Comptes Rendus*, t. xcv. p. 333.

showed, be acted upon by a force varying *directly* as the distance from the centre. The reason of this is easily seen; for the further out a component of such a system is located, the more matter there will be inside, and the less outside its orbit. The strength of the central pull thus reaches a maximum at the surface of the sphere, the velocity by which it is balanced growing in the same proportion. Ellipses described under these conditions would all, accordingly, have an identical period; whatever their eccentricities, in whatever planes they lay, in whatever direction they were traversed, each would remain invariable; and the harmony of a system in which no perturbations could possibly arise would remain unbroken for ever, provided only that the size of the circulating bodies, and the range of their immediate and intense attractions, were insignificant compared with the spatial intervals separating them.[1]

But this state of nice adjustment is a mere theoretical possibility. There is no likelihood that it has anywhere an actual existence; and the stipulations, upon compliance with which its realisation strictly depends, are certainly disregarded in all the stellar groups with which we have any close acquaintance. The components of these are neither equal, nor equably distributed. Central compression, over and above the merely apparent effect of the gradually increasing depth of the star-strata presented to the eye, is the rule in globular clusters. Three distinct stages of condensation were noted by Herschel in the effulgent southern object called '47 Toucani;' real crowding perhaps intensifies the 'blaze' in the middle of the superb group between η and ζ Herculis (M 13; see Plate II.); in other cases, the presence of what might be termed a nuclear mass of stars is evident. Here, then, the 'law of inverse squares' must enter into competition with the 'direct' law of attraction, producing results of extraordinary intricacy, and giving rise to problems in celestial mechanics with which no calculus yet invented can pretend to grapple.

Sir John Herschel allowed the extreme difficulty of even

[1] *Outlines of Astronomy*, 9th ed. p. 636.

imagining the 'conditions of conservation of such a system as that of ω Centauri, or 47 Toucani, &c., without admitting repulsive forces on the one hand, or an interposed medium on the other, to keep the stars asunder.'[1] Compacted into a whole, they might, he thought, instead of revolving individually, be supposed to rotate in their corporate capacity as a single body. But the establishment in such aggregations of a 'statical equilibrium' by means of an 'interposed medium' is assuredly chimerical. The hypothesis of their rotation as one mass is countenanced by no circumstance connected with them, and is decisively negatived by their irregularities of figure. The sharp contours of bodies whirling on an axis are nowhere to be found among these objects. Their streaming edges betray a totally different mode of organisation.

Globular clusters commonly present a radiated appearance in their exterior parts. They seem to throw abroad feelers into space. The great cluster in Hercules is not singular in the display of 'hairy-looking, curvilinear' branches. That in Canes Venatici (M 3) has 'rays running out on every side' from a central mass, in which 'several small dark holes' were disclosed by Lord Rosse's powerful reflectors;[2] showing pretty plainly that the spiral tendency, visible in the outer regions, penetrates in reality to the very heart of the system. From a well-known cluster in Aquarius (M 2), 'streams of stars branch out, taking the direction of tangents.'[3] That in Ophiuchus (M 12) is provided with long straggling tentacles, of a 'slightly spiral arrangement,' according to the late Lord Rosse. And a remarkable assemblage in Coma Berenices (M 53) was described by Herschel and Baily as 'a fine compressed cluster with curved appendages like the short claws of a crab running out from the main body.'[4] The peculiarity in question is the more significant that it is shared by many undoubted nebulæ.

We find it difficult to conceive the existence of 'streams of stars' that are not *flowing*; and accordingly the persistent

[1] *Cape Observations*, p. 139.
[2] *Trans. R. Dub. Soc.* vol. ii. p. 132.
[3] *Ibid.* p. 162.
[4] *Phil. Trans.* vol. cxxiii. p. 458.

radial alignment of the components of clusters inevitably suggests the advance of change, whether in the direction of concentration or of diffusion. Either the tide of movement is setting inward, and the 'clustering power' (to use a favourite phrase of Sir William Herschel's) is still exerting itself to collect stars from surrounding space; or else a centrifugal impulse predominates, by which full-grown orbs are driven from the nursery of suns in which they were reared, to seek their separate fortunes and enter on an independent career elsewhere. The somewhat hazardous conjecture that the process of their development may be attended by an increase of tangential velocity, appears to receive some countenance from the general superiority in brightness of outlying stars over those more centrally situated in globular clusters. But, in truth, the question as to whether separatist or aggregationist tendencies prevail in them is of too recondite a nature for profitable discussion at present. All that can be said is that, after the lapse of some centuries, photographic measurements may help towards deciding it. Even then, however, the answer may long appear dubious. For it is not impossible (though far from probable) that processions of stars, compelled by the attractive power of the cluster to deviate from their march past it, may here be arriving, there departing, after having effected an hyperbolic sweep round its denser portion. Comets, doubtless, from time to time dash through the system; it cannot be affirmed offhand that stars may not do likewise.

An object visually resembling a blurred star below the fourth magnitude was named by Bayer ω Centauri. It never rises in these latitudes, but Herschel's great reflector revealed it to him at the Cape as a 'noble globular cluster, beyond all comparison the richest and largest object of the kind in the heavens.'[1] The stars contained in it are literally innumerable; they are all excessively minute, and approximately equal, though irregularly distributed into little knots and groups.

The loveliness of the cluster 47 Toucani near the Lesser

[1] *Cape Observations*, p. 21.

Magellanic Cloud was, to Herschel's view, set off by a diversity of colour between an interior mass of rose-tinted stars and marginal strata of purely white ones.[1] This feature, however, has met with no later recognition; and to the present writer in 1888, the object appeared of the same silvery sheen throughout. Its diameter, inclusive of stragglers giving it the usual radiated aspect, measures about 20′; apart from them, it covers an area roughly equal to one quarter of the lunar disc, and is so obvious to unaided sight, that for several nights after his arrival in Peru it was mistaken by Humboldt for a comet.[2] About midway between the centre and circumference of 47 Toucani, a pair of stars much brighter than the rest are placed, perhaps by casual projection, although the blankness of the surrounding sky diminishes the probability of this being the case.

The gradations of lustre among the constituents of such assemblages commonly range over three or four magnitudes. Nor by any means unmethodically. As a general, if not an invariable rule, the smaller stars are gathered together in the middle, while the bright ones surround and overlay them on every side. Thus, of a magnificent cluster in Sagittarius (M 22), known since 1665, the central portion accumulates the light of multitudes of excessively minute, and is freely sprinkled over with larger stars. The effect, which probably corresponds with the reality, is as if a globe of stars of fifteenth were enclosed in a shell of stars of eleventh magnitude, some of these being naturally projected upon the central aggregation. Sir John Herschel remarked of a cluster in the southern constellation of the Altar (N. G. C., 6752): 'The stars are of two magnitudes; the larger run out in lines like crooked radii, the smaller are massed together in and around the middle.'[3] A similar structure was noted by Webb[4] in the Canes Venatici and Coma Berenices clusters already mentioned (M 3, M 53), as well as in an imposing collection in Libra (M 5), discovered by Kirch in 1702, the more condensed part of which (compared

[1] *Cape Observations*, p. 18.

[2] *Cosmos* (Otté's trans.), vol. iii. p. 1[illegible]2

[3] *Cape Observations*, p. 119.

[4] *Student*, vol. i. p. 460.

by Sir John Herschel to a snow-ball) seems as if 'projected on a loose, irregular ground of stars.' [1]

Irregularities of distribution in clusters assume at times a highly enigmatical form. At Parsonstown, in 1850, [2] three 'dark lanes,' meeting at a point considerably removed from the centre, were perceived to interrupt the brilliancy of the globe of stars in Hercules (M 13). They were afterwards recognised by Buffham and Webb, and recorded themselves with emphasis in a photograph by Mr. Roberts, of which Plate II. is a reproduction. On the original negative, one of the 'lanes' can clearly be seen to lead to a small but perfectly symmetrical oval ring of stars, surrounding one centrally situated; while further out, partial clearings recur in the same shape as before, markedly enough to demonstrate the dominating influence of some kind of law in their production. Globular clusters in Ophiuchus (M 12), in Pegasus (M 15), [3] and in Canes Venatici (M 3), appear to be similarly *tunnelled*. Preconceived ideas as to the mechanism of celestial systems are utterly confounded by phenomena not easily reconcilable with the prosecution of any orderly scheme of circulatory movement. Even if absolutely denuded of stars, the extensive vacancies indicated by the visibility of dusky rifts must be in part obliterated by the interposed light of the surrounding star-layers. They can hence become perceptible only when most pronounced; and are likely to exist, less fully developed, in numberless cases where they defy detection.

Differences of distance are alone adequate to account for the variety of *texture* observable in globular clusters. That in Aquarius, for instance, likened by Sir John Herschel to 'a heap of golden sand,' might very well be the somewhat coarse-grained Hercules group withdrawn as far again into space. At a still further stage of remoteness, the appearance would presumably be reached of a stellar throng in the Dolphin (N. G. C., 6934), which, with low powers, might pass for a planetary nebula, but under stronger optical compulsion assumes the granulated aspect of a true cluster. And many

[1] *Phil. Trans.* vol. cxxiii. p. 359. [2] *Ibid.* vol. cli. p. 732.
[3] Webb, *Cel. Objects*, p. 372.

PLATE 2.

CLUSTER IN HERCULES (M 13).

Photographed by Isaac Roberts, F.R.A.S., May 22nd, 1887. Exposure one hour.

such, their genuine nature rendered impenetrable by excessive distance, are doubtless reduced to the featureless semblance of 'irresolvable' nebulæ.

But there are real, as well as apparent, diversities in these objects. Although smaller and more compact clusters must, on the whole, be more remote than large, loosely-formed ones, yet 'this argument,' Sir William Herschel remarked, 'does not extend so far as to exclude a real difference which there may be in different clusters, not only in the size, but also in the number and arrangement of the stars.' There may be globular clusters with components of the actual magnitude of Sirius; others, optically indistinguishable from them, may be aggregated out of self-luminous bodies no larger than Mars, or even than Ceres, or Pallas. Our total ignorance of the real location in space of such objects, leaves us without the means of judging. Nor is ignorance likely to be replaced by knowledge for an indefinite time to come. We can, however, suppose it to be so replaced in a particular instance, and trace the consequences. Let our example be the great cluster in Hercules.

'This is but a little patch,' Halley wrote in 1716, 'but it shows itself to the naked eye when the sky is serene, and the moon absent.'[1] Messier termed it 'nebuleuse sans étoiles';[2] yet a 'twinkling' indicative of its stellar character may be caught with a telescope four inches in aperture; and a powerful instrument resolves it to the core. Within the precincts of Halley's 'little patch,' Sir William Herschel estimated fourteen thousand stars to be 'cribb'd, cabined, and confined'!

The apparent diameter of this object, including most of the 'scattered stars in streaky masses and lines,'[3] which form a sort of 'glory' round it, is 8′; that of its truly spherical portion may be put at 5′. Now, a globe subtending an angle of 5′ must have (because the sine of that angle is to radius nearly as 1 : 687) a real diameter $\frac{1}{687}$ of its distance from the eye, which if we assume to be such as would correspond to a parallax of $\frac{1}{20}$ of a second, we find that the cluster, outliers

[1] *Phil. Trans.* vol. xxix. p. 392. [2] *Conn. des Temps*, 1784, p. 233.
[3] *Phil. Trans.* vol. cxxiii. p. 458.

apart, measures 558,000 millions of miles across. Light, in other words, occupies about thirty-six days in traversing it, but sixty-five years in journeying thence hither. Its components may be regarded, on an average, as of the twelfth magnitude; for although the divergent stars rank much higher in the scale of brightness, the central ones, there is reason to believe, are notably fainter. The sum total of their light, if concentrated into one stellar point, would at any rate very little (if at all) exceed that of a third-magnitude star. And one star of the third is equivalent to just four thousand stars of the twelfth magnitude. Hence we arrive at the conclusion that the stars in the Hercules cluster number much more nearly four, than fourteen thousand.

If, then, four thousand stars be supposed uniformly distributed through a sphere 558,000 million miles in diameter, an interval of 28,300 million, or more than ten times the distance of Neptune from the sun, separates each from its nearest neighbour.[1] Upon a spectator in an intermediate situation, six stars (besides crowds of gradated inferiority) would then blaze with about 3,300 times the lustre that Sirius displays to us. Yet since a million and a half of stars of even this amazing brilliancy would be needed to supply the light we receive from the sun, the general illumination of the cluster can only amount to a soft twilight, excluding, it is true, the possibility of real night on any globe placed near its centre.

At its surmised distance, our sun would appear as a star of 7·5 magnitude; it would shine, that is to say, about sixty-three times as brightly as an average one of the grouped objects. Each of these, accordingly, emits $\frac{1}{63}$ of the solar light; and if of the same luminosity relative to mass, as the sun, it exercises just $\frac{1}{500}$ of the solar attractive power. The mass of the entire system of four thousand such bodies is thus equal to that of eight suns. This, however, may be regarded as a minimum estimate. The probabilities are in favour of the cluster being vastly more remote than we have here assumed it to be; hence composed of larger or brighter,

[1] See Mr. J. E. Gore's similar calculation from different data, in *Jour. Liv. Astr. Soc.* vol. v. p. 169.

and presumably more massive, individual bodies than results from our calculation.

The relations of clusters and nebulæ are evidently very close; but they are only just beginning to be effectually studied. The conjecture is, however, already fully justified that the two classes of object form an unbroken series—that clusters exist in every stage of development from nebulæ, and that the advancing condensation of many nebulæ will eventually transform them into veritable clusters. Suggestions to this effect derived from analogies of form[1] are corroborated by numerous recent observations of the actual co-existence with grouped stars of nebulous masses. The Pleiades is not the only example of a *hybrid* system. A beautiful, bright cluster of the same general character in Sagittarius (N. G. C., 6530) is obviously connected with a great nebula (M 8), in the meshes of which it seems as if entangled. The photographic investigation of this mixed display may be expected to yield results of high interest.

That potent method will also, it is hoped, be applied without delay to a combination of stars and nebulæ without a known parallel in the heavens. Mr. E. E. Barnard discovered at Lick in 1889 a cluster in Monoceros (N. G. C., 2244) to be completely surrounded by a vast nebulous ring 40′ in diameter.[2] The interior appeared perfectly free from luminous haze, the stars shining on an absolutely black sky.

Two rich clusters have long been known to include each a nebula of the planetary kind. One in Argo (N. G. C., 2818) has a central vacuity conspicuously occupied by a nebulous disc 40″ across; the other (M 46), in the same constellation, displays within its borders a fine annular nebula.[3] It is difficult, if not impossible, to believe either projected casually into such a remarkable position. Mr. Roberts's photograph, too, of the Hercules cluster at least suggests the presence of

[1] Lockyer, *Proc. R. Society*, vol. xliv. p. 29.

[2] *Astr. Nach.* No. 2918; Swift, *Sidereal Mess.* vol. ix. p. 47; Backhouse, *Observatory*, May 1890, p. 179.

[3] The whole neighbourhood of this cluster was perceived to be nebulous at Harvard College in 1870 (*Annals*, vol. xiii. p. 76).

a thin, nebulous residuum,[1] which may, with a longer exposure than sixty minutes, be rendered unmistakable. Nebulæ which thus linger on in clusters can scarcely be made manifest otherwise than with difficulty. For if, as we may be permitted to suppose, the stars replace their original more brilliant knots, what survives will usually be of the last degree of faintness.

The spectra of all clusters are, in the main, continuous, but give signs of possessing individual peculiarities, a systematic scrutiny of which might profitably occupy some one of the great telescopes now in existence. Dr. Huggins was struck, in 1866, with the absence of red rays from the analysed light of the great cluster in Hercules, and perceived in it irregularities due either to bright or dusky bands.[2] They were construed in the latter sense by Vogel in 1871,[3] since when the inquiry has been unaccountably neglected.

[1] *Monthly Notices*, vol. xlviii. p. 30.
[2] *Phil. Trans.* vol. clvi. p. 389. [3] *Astr. Nach.* No. 1864.

CHAPTER XVII.

THE FORMS OF NEBULÆ.

THE fantastic variety of nebular forms was long a subject of wonder, scarcely tempered by a speculative effort. Inchoate worlds, disclosed with astonishing profusion by Herschel's telescopes, seemed like mere 'sports of nature' in the sidereal spaces. Nebulæ were to be found in the semblance of rings, fans, brushes, spindles; they abounded in planetary, cometary, elliptical, branching varieties; nebulous shields, embossed with stars, or tasselled like the ægis of Athene, displayed themselves, as well as nebulous discs, rays, filaments, triangles, parallelograms, twin and triple spheres. One nebula, thought to resemble the face of an owl, was named accordingly; another suggested a crab; a third a swan; a fourth (the great Orion formation) became known as the Fish Mouth nebula, from its supposed likeness to the gaping jaws of a marine monster. Fancy ranged at large through this wide realm, attempting to familiarise itself with the strange objects contained in it by finding for them terrestrial similitudes.

Within the last few years, however—indeed, it may be said, since the completion of the Rosse reflector in 1845—nebular inquiries have entered upon a new phase. A 'glimmering of reason' has begun to hover over what long appeared a scene of hopeless bewilderment. With improved telescopic means—above all, with the aid of photography—*structure* has become increasingly manifest among all classes of nebulæ; structure, not of a finished kind, but indicating with great probability the advance of formative processes on an enormous scale, both as regards space and time. Masses that seemed all but amorphous when imperfectly seen, show to a keener scrutiny nodes and nuclei of condensation; curving lines of light, telling of the presence of movement and force, furrow them; they are perceived to be rifted as if by a colossal thunderbolt, or riddled as if by a portentous cannonade. Simple milky effusions prove to be far less common than had been supposed, and excessive complexity of constitution is already a recognisable characteristic of most nebulæ.

It is one which adds greatly to the interest of their study. For as the curious details of their organisation are laid bare by the intricate inequalities of their light, the prospect grows hopeful of gaining some insight into the nature of the systems formed by, or in preparation from them. Optical discoveries, while gradually acquiring physical significance, are helping to lay the foundation of a 'nebular theory' emanating from augmented knowledge, and the discreetly adventurous thoughts which it may be supposed to countenance.

Meanwhile, some mode of nebular classification has to be adopted for the guidance of our ideas; and since their rapid modification through fresh detections allows no arrangement to be at present more than provisional, it will be best to depart as little as possible from that already in use. We may, then, for descriptive purposes, divide nebulæ into the following eight classes, which, nevertheless, frequently overlap so widely as to be barely distinguishable: 1. Nebulous stars. 2. Planetary nebulæ. 3. Annular nebulæ. 4. Cometary nebulæ. 5. Spiral nebulæ. 6. Double nebulæ. 7. Elliptical nebulæ. 8. Irregular nebulæ.

In the course of one of his 'reviews of the heavens,' Sir William Herschel discovered a star in Taurus 'perfectly in the centre' of a 'faintly luminous atmosphere' about 3′ in diameter.[1] The consideration of this object (N. G. C., 1514) and of some others like it led him in 1791 to the memorable conclusion that there exists in space 'a shining fluid of a nature totally unknown to us.' Nothing, indeed, could be clearer than that 'the nebulosity about the star was not of a starry nature,' and there is just as little doubt that it is no atmosphere in the ordinary sense of the word. This is demonstrated by its extent alone. For the 'glow' round

[1] *Phil. Trans.* vol. lxxxi. pp. 71, 82.

Herschel's pattern nebulous star fills a sphere at least thirty times as wide, or 27,000 times as capacious, as that enclosed by the orbit of Neptune; and the Rosse telescope disclosed irregularities of illumination within it,[1] suggestive of unfolding activities, directed, perhaps, towards the production of a planetary scheme far vaster and more elaborate than our own. Thirteen nebulous stars were enumerated by Herschel, and many have since been added to them. A fine specimen in Eridanus was picked up by Swift in 1859;[2] a small star in Canes Venatici came out strongly 'burred' on one of Mr. Roberts's plates in 1889;[3] and among Sir John Herschel's southern discoveries was a close, sharply-defined double star surrounded by a bright luminous 'atmosphere' 2′ in extent[4] (N. G. C., 5367). The glow formerly apparent about 55 Andromedæ seems to have vanished. The star, which is of 5½ magnitude, was marked 'nebulosa' by Flamsteed and Piazzi;[5] Sir John Herschel regarded it as an especially fine example of that peculiarity, and his description was corroborated by Dr. Huggins in 1864;[6] but Lord Rosse examined the object eleven times without finding a trace of nebulosity;[7] and d'Arrest, Schjellerup, as well as Burnham with the eighteen-inch Chicago achromatic, were equally unsuccessful.[8] Another suspicious case of the same kind is met with in the still brighter star 8 Canum Venaticorum, perceived four times by Sir John Herschel, but by no one else, before or since, to be encompassed with a 'considerable atmosphere.'[9] Photography might usefully be employed to clear up the doubt as regards both stars. Illusory glows cannot impose upon the sensitive plate.

A 'nebulous star' proper forms the centre of an ill-defined aureola; but nebulous adjuncts to stars exist in every variety of branches and chevelures, wisps and whorls. In-

[1] *Trans. R. Dub. Soc.* vol. ii. p. 40. [2] *Sidereal Messenger*, vol. iv. p. 39.

[3] *Monthly Notices*, vol. xlix. p. 363. [4] *Cape Observations*, pp. 23, 107.

[5] According to Schjellerup (*Astr. Nach.* No. 1613), Piazzi merely copied from Flamsteed.

[6] *Phil. Trans.* vol. cliv. p. 442. [7] *Trans. R. Dub. Soc.* vol. ii. p. 50.

[8] *Memoirs R. Astr. Soc.* vol. xlvii. p. 220.

[9] *Phil. Trans.* vol. cxxiii. p. 427.

deed, the sequence is so continuous between bright stars with filmy appendages, and pronounced nebulæ involving minute stars, that it is often difficult to say whether the stellar or the nebular character predominates. Thus, 'planetary' nebulæ have often stellar nuclei; they can be discriminated, however, from nebulous stars, first, by their mainly gaseous spectra, next, by the comparatively definite termination of their discs.

It is no wonder, then, if among Herschel's nebulous stars one was found not strictly entitled to bear that name. This nondescript object (N. G. C., 2392) is situated in Gemini, and that it struck him as something unusual may be inferred from his designating it 'one of the most remarkable phenomena I have ever seen.'[1] With the Parsonstown reflector it presented a 'most astonishing' appearance. Herschel's 'equally diffused nebulosity' was replaced by several bright and dark rings, varying in breadth, and perhaps spiral in their arrangement.[2] The genuine nature of the combination might even yet be in doubt but for the dictum of the spectroscope. D'Arrest[3] found its light to be concentrated in the unidentified green ray of wave-length 5005—the central star or nucleus asserting its superior condensation by the display of a faint continuous radiance. It is then essentially a nebula, and generally passes for one of the annular, or perforated kind. A truly nebulous star of a reddish colour makes with it (not, we may surmise, fortuitously) an open pair at 105″.[4] The diameter of the system of rings is about 53″.

A somewhat analogous object, 30″ across, occurs in Cygnus (N. G. C., 6826). 'It is of a middle species,' Herschel wrote in 1802, 'between the planetary nebulæ and the nebulous stars, and is a beautiful phenomenon.'[5] To its greenish-blue disc and lucid centre corresponds a mixed spectrum of bright lines and uniformly dispersed light.

Planetary nebulæ were first distinctively adverted to by

[1] *Phil. Trans.* vol. lxxxi. p. 81.

[2] *Trans. R. Dub. Soc.* vol. ii. p. 59; see also H. C. Key, *Monthly Notices*, vol. xxviii. p. 154.

[3] *Astr. Nach.* No. 1885; *Abhandlungen*, Leipzig, Bd. iii. p. 321.

[4] Lassell, *Memoirs R. Astr. Soc.* vol. xxxvi. pp. 42, 61.

[5] *Phil. Trans.* vol. xcii. p. 522.

Sir William Herschel. Their classification caused him a good deal of perplexity. 'We can hardly suppose them,' he remarked at starting, 'to be nebulæ; their light is so uniform as well as vivid, the diameters so small and well-defined, as to make it almost improbable they should belong to that species of bodies.' After, however, he had weighed and found wanting the hypotheses of their being actual planets belonging to distant suns, or distended stars, or comets near aphelion, he at last decided—rightly, as usual—in favour of their nebulous nature.[1]

Fifty of them were known when Pickering began 'sweeping in 1881, for 'stars with remarkable spectra;'[2] and within a few years, upwards of twenty more were identified, through the quality of their light alone, by him and Dr. Copeland. These are, however, for the most part totally devoid of the visible disc which was the original badge of their class; they are either very small, or very remote planetaries; and are often distinguished as 'stellar nebulæ.'

A true 'planetary' aspect has not, indeed, in any case survived the scrutiny of modern observers. What had seemed equably illuminated discs are broken up by the powerful telescopes now in use, into brighter and darker portions, distributed in evident relation to some unknown conflict of forces. Some of these discs include strongly-marked nuclei; others a sprinkling of minute stars; condensation towards a spherical surface gives to many the aspect of a ring-shaped enclosure; few (if any) are clean at the edges.

The effects of progress in seeing may be exemplified by the history of a noted planetary nebula (M 97) discovered by Méchain, near β Ursæ Majoris, the second 'pointer,' February 16, 1781. To both Méchain and Messier it appeared a barely discernible spot of faint light;[3] but Sir John Herschel described it as 'a most extraordinary object—a large, uniform nebulous disc, quite round, very bright, not sharply defined, but yet very suddenly fading away to darkness.'[4] At Parsons-

[1] *Phil. Trans.* vol. lxxv. p. 265.

[2] *Observatory*, vols. iv. p. 81, v. p. 294; *Monthly Notices*, vol. xlv. p. 91.

[3] *Conn. des Temps*, 1784, p. 265.

[4] *Phil. Trans.* vol. cxxiii. p. 402.

town, in 1848, two stars[1] were perceived in the interior, each surrounded by a dark space encroached upon by nebulous whorls; and the object received the name of the 'Owl nebula,' from the appearance of two great oculi thus presented. Its dimensions were found to be 163″ by 147″; but the outlines of the disc seemed 'ragged,' its torn edges serving, in Professor Alexander's opinion, as 'marks of disruption and dispersion outward.'[2]

The 'Owl,' like every other planetary nebula yet examined, gives a spectrum of bright lines. In the main, then, it is a globular mass of hydrogen and other gases of such inconceivable size that literally thousands of solar systems could be accommodated within its bulk. If we suppose it to be sixty-five light years distant from the earth—and the probability that it is much more remote comes nigh to certainty—its diameter must exceed that of the orbit of Neptune upwards of one hundred times! Here, indeed, there is room and verge enough for the unfolding, in a dim futurity, of vast creative purposes.

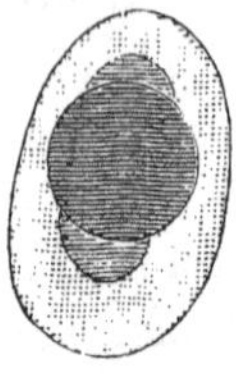

Fig. 35. Plan Sketch of a Nebula (Vogel).

This object might be called a *perforated* planetary nebula; others seem *multiplex*. Two or three superposed discs are traceable in them, recalling the complicated series of envelopes, of which the heads of many comets are constructed. A small oval planetary, for example, in Taurus Poniatowski (N. G. C., 6572), was resolved by Vogel with the great Vienna refractor, in 1883,[3] into three strata of nebulosity disposed as in fig. 35, representing, no doubt, successive spherical and ellipsoidal envelopes of diminishing luminous power.

An object (N. G. C., 6210) with an 'intense blue centre fading off to some distance all round,' and hazy at the edges,[4] was perceived by Vogel as triple. A faint oval *husk* (so to

[1] The fainter star disappeared in April 1850, perhaps through the tarnishing of the mirror, *Phil. Trans.* vols. cxl. p. 513, cli. p. 721; *Trans. R. Dub. Soc.* vol. ii. p. 93.

[2] *Astr. Jour.* vol. ii. p. 141.

[3] *Potsdam Publicationen*, No. 14, p. 34.

[4] *Trans. R. Dub. Soc.* vol. ii. p. 150.

speak) seemed to enclose a vivid *kernel*, and that again to include a stellar nucleus. This nebula is situated in the constellation Hercules; and one of a similar character in Eridanus (N. G. C., 1535) struck Mr. Lassell as made up of a faint circumferential disc surrounding a brighter one which included a speck of condensed (but it was thought) non-stellar light. The whole effect was 'extraordinary and beautiful.' [1]

Perhaps the most interesting of all the planetary nebulæ is one lying quite close to the pole of the ecliptic, near the star ω Draconis (N. G. C., 6543), also known as H iv. 37, or the thirty-seventh of Herschel's fourth class. Its longer diameter (for it is slightly elliptical) measures about 30″; it is of a blue colour, and shows a white star of eleventh magnitude giving a perfectly continuous spectrum exactly in the middle of a disc from which a purely gaseous one is derived. Dr. Huggins's first experiment in the analysis of nebular light was in fact made upon the planetary in Draco,[2] which has in various ways been used as a test object. Attempts to determine its parallax were vainly made by d'Arrest, Brünnow, and Bredichin.[3] For proper motion, too, it was tried by d'Arrest in 1872, with a similarly negative result. During the eighty-two years elapsed from a careful observation by Lalande in 1790, the nebula had remained to all appearance completely stationary. But this fixedness was really to some extent communicative as regards its *minimum* distance from the earth. D'Arrest showed that unless this exceed a light-journey of forty-seven years, the nebula must have become sensibly displaced in the course of eighty-two years by the simple perspective effect of the sun's advance at the rate of five miles a second.[4] And since there is little doubt that this estimate of the solar velocity ought to be trebled, the least admissible distance of the

[1] *Memoirs R. Astr. Soc.* vol. xxxvi. p. 40.

[2] *Popular Hist. of Astr.* p. 437, 2nd ed.

[3] Brünnow obtained for H iv. 37 a nominal parallax, but Bredichin, from nearly double the number of observations, derived a *negative* one implying the nebula to be more remote than the tenth magnitude star with which it was compared. *Astr. Nach.* No. 2916 (Oudemans); *Eng. Mec.* vol. xliii. p. 504 (H. Sadler).

[4] *Astr. Nach.* No. 1885.

nebula should also be trebled. Its light then can only reach our eyes after an interval of 140 years,[1] *if then*; for it may spend a great deal longer on the road. The real size of the globe it emanates from must be vast in proportion to such extreme remoteness. We find, accordingly, that at least forty-four diameters of the orbit of Neptune would be needed to measure it from side to side.

Recent observations, however, raise a question as to whether the figure of the Draco planetary is even approximately globular. With the great Lick refractor, Professor Holden perceived the disc to be made up of two superposed rings, one lying (it was inferred) behind the other in space, and generated either previously or subsequently to it in time (see fig. 36). Their arrangement, in short, seemed to be that of a true helix, and the object was hence ranked as the first member of a class denominated 'helical nebulæ.'[2] The great merit of this curious discovery consists, not in the establishment of a new and barren distinction, but in the rationalising of one already made. Helical nebulæ may then be regarded as spiral nebulæ under a novel aspect.

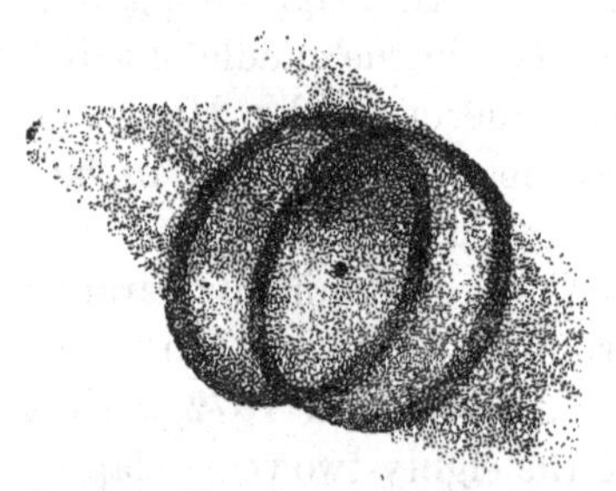

FIG. 36.—Planetary Nebula resolved into Helical Form (Holden).

A blue, or greenish tinge is more or less characteristic of all gaseous nebulæ; it is especially conspicuous in a planetary (N. G. C., 3918) discovered by Sir John Herschel in the Centaur, and described as 'very like Uranus, only half as large again,' its 'colour a beautiful rich blue, between Prussian and verditer green.'[3] The light of this object is about equal to that of a seventh-magnitude star.

A considerable number of planetary nebulæ, as we have already partly seen, manifest both annular and spiral tendencies. In some, a marginal brightening gives, with suffi-

[1] Corresponding to a parallax of 0″·023.

[2] *Monthly Notices*, vol. xlviii. p. 388.

[3] *Cape Observations*, p. 100.

cient telescopic power, the effect of a ring; in others, curving lines of light betray incipient spirality of conformation. Since they partake of the nature of all the three species, their classification is to a great extent arbitrary. Five of Herschel's planetaries assumed in fact at Parsonstown a ring-shape.[1] One of these (N. G. C., 2438), the nebula involved in the cluster M 46,[2] was observed not alone to be pierced with a nearly central cavity, but to contain two, perhaps three stars, towards one of which the exterior nebulous ring wound spirally inward.[3] A 'hole' too disclosed itself in a planetary in Gloria Frederici (N. G. C., 7662), observed by Lassell as bi-annular, 'a ring within a ring.' To Father Secchi, it had, with high powers, the effect of a 'magnificent horseshoe of scintillating points,'[4] the glitter of which was also evident to Vogel.[5] Yet the object is of a purely gaseous constitution, emitting the ordinary nebular trio of bright lines, to which, by a peculiarity shared with only two other known nebulæ, it adds the blue ray characterising the spectrum of stars like γ Argûs.[6] The 'bi-annular' planetary is either hazy or 'fringed' at the edges, of a bluish colour, and measures 32″ by 28″.[7]

A 'sky-blue likeness of Saturn' replaced, in Mr. Lassell's reflector, a round, faintly lucent object (N. G. C., 7009) discovered by Herschel in 1782 near the star ν Aquarii. With higher powers, the disc became a ring 26″ by 16″, hazy within and without; and the whole interior assumed, under Vogel's examination, a curious screw-shaped structure.[8] Professor Holden remarked an unexpected point of likeness between this nebula and the one last mentioned (N. G. C., 7662), in the possession by both of an interior oval ring, singularly warped and twisted out of the central plane; the peculiarity associating them also with the 'helical' planetary in Draco. The ansæ, or handle-

[1] *Phil. Trans.* vol. cxl. p. 507.
[2] See *ante*, p. 249.
[3] *Phil. Trans.* vol. cxl. p. 513.
[4] *Astr. Nach.* No. 1018; *Les Étoiles*, t. ii. p. 14.
[5] *Potsdam Publicationen*, No. 14, p. 37.
[6] Huggins, *Phil. Trans.* vol. cliv. p. 440; *Harvard Annals*, vol. xiii. p. 72. See *ante*, p. 78.
[7] Lassell, *Memoirs R. Astr. Soc.* vol. xxxvi p. 51.
[8] *Potsdam Publ.* No. 14, p. 37.

like appendages, producing in the object near ν Aquarii its resemblance to Saturn with half-opened rings, are a unique feature among nebulæ. First represented by Lord Rosse, they were resolved by Professor Holden [1] into distinct luminous masses, just traceably connected with the main body (see fig. 37), and not improbably satellite-stars in course of formation. The attendance of small stars upon nebulæ both planetary and annular, is too close and too constant to be accidental, and long ago attracted the attention of Sir John Herschel.[2] One such group (N. G. C., 6818) struck him as exactly like a planet and pair of moons; and stars in slow transit across a nebula of which they are the dependents may often appear projected upon it. But not the slightest evidence of movement in these ancillary objects has yet been detected.

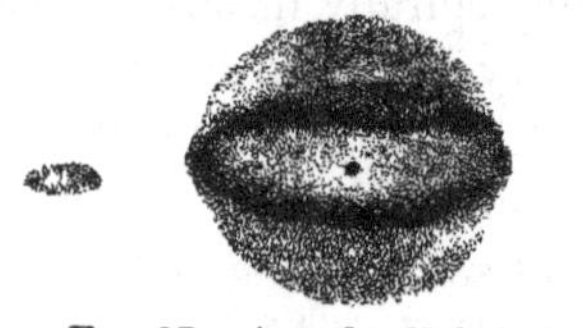

Fig. 37.—Annular Nebula in Aquarius (Holden).

The typical annular nebula (M 57) was first noticed by Darquier of Toulouse in 1779, between β and γ Lyræ. It consists of an oval bright ring, 80″ by 60″, the interior of which is filled with a dim nebulous haze like 'gauze stretched over a hoop.'[3] Harding, already in 1797, perceived irregularities in the illumination of the ring;[4] and vivid patches are especially conspicuous at either extremity of the minor axis. Minima of light, on the contrary, terminate the major axis; the nebula, as photographed at Herény, September 1, 1886, taking somewhat the shape of a pair of parentheses set a little apart, thus ⊂⊃, but with spiral links between.[5] With the Rosse reflector, filaments of nebulosity were seen streaming outward from the edges;[6] and to Father Secchi the ring appeared glittering with stars 'like finely powdered silver.'[7] Tempel too remarked upon the great number of stars visible without and within the nebula; and Professor Hall, at Wash-

[1] *Monthly Notices*, vol. xlviii. p. 391. [2] *Phil. Trans.* vol. cxxiii. p. 500.
[3] Sir J. Herschel's *Outlines of Astr.* p. 644, 9th ed.
[4] Holden, *Monthly Notices*, vol. xxxvi. p. 64.
[5] Von Gothard, *Astr. Nach.* No. 2749. [6] *Phil. Trans.* vol. cxxxiv. p. 322.
[7] Webb, *Cel. Objects*, p. 347.

ington in 1877, perceived it to be 'surrounded by a ring of faint stars,' of twelfth to fourteenth magnitudes.[1]

Views of this object afforded to Professor Holden by the Lick telescope in 1888[2] included a bewildering multitude of sharp stellar points, arranged, with evident reference to the nebulous oval in their midst, into elliptical chaplets clinging to its inner and outer edges, neither of which was bounded by a smooth curve. The perspective relations of these appearances form a very curious problem. To us they present themselves as a flat picture, but of course deceptively; we find it, however, difficult to realise the disposition of parts in the solidity of space, to which they truly correspond. The ellipticity of the ring, for instance, can scarcely be due to foreshortening; since if it were, the marking of the least diameter by maxima, of the greatest by minima, of luminosity, should be a pure coincidence, which it evidently is not.

A central star, discerned in this nebula by von Hahn at Remplin, towards the close of the last century, was missed by him in 1800,[3] and has often since evaded the scrutiny of better-provided observers. It appeared on the Herény and Liverpool photographs,[4] but left no trace on plates exposed at Paris. Satisfactory proof, however, of its suspected variability is still wanting.

A nebula in Cygnus (N. G. C., 6894) might be called a reduced copy of that in Lyra. It measures 47″ by 41″, the interior vacuity, which is partially filled with faint light, 20″. A conspicuous star is included within it.[5] An object of the same kind in Scorpio (N. G. C., 6337) was described by Sir John Herschel as 'a beautiful delicate ring of a faint ghost-like appearance, about 40″ in diameter.'[6] Two stars, or nebulous nodes, are placed in it diametrically opposite to each other, and the whole aspect of the nebula suggests extreme remoteness.[7]

[1] *Astr. Nach.* No. 2186.

[2] *Monthly Notices*, vol. xlviii. p. 383.

[3] *Astr. Jahrbuch*, 1802, p. 10.

[4] Von Gothard, *Astr. Nach.* No. 2749; Spitaler, *ibid.* No. 2800.

[5] Lord Rosse, *Trans. R. Irish Acad.* vol. ii. p. 156.

[6] *Cape Observations*, p. 114.

[7] Lassell, *Memoirs R. Astr. Soc.* vol. xxxvi. p. 47.

If, on the other hand, vicinity can be inferred from great apparent extent, Mr. Barnard's nebulous ring in Monoceros must be comparatively near the earth. Its outer diameter, as mentioned in the last chapter, was estimated at 40′, its inner at 20′; and the brightest among several 'knots' apparent in it had been detected by Mr. Swift as a separate object (N. G. C., 2237) so long ago as 1865. Variations both of its shape and brightness are, indeed, surmised by him to be in progress.[1] Strangest of all, there are indications that the formation really exists in duplicate. What seemed to be a section of another vast ring was perceived by Mr. Barnard close to, and perhaps in nebulous connection with, the first.[2] Their further investigation, especially by photographic means, may lead to very curious results.

Four ring-nebulæ, two in the northern, two in the southern hemisphere, were known to the Herschels; and, as we have seen, many so-called 'planetaries' show annular, as annular nebulæ show spiral proclivities. Rings, in some instances, visibly curve inward towards a nucleus, giving rise to the variety which we have designated 'cometary' nebulæ. Thus, a ninth-magnitude star with nebulosity attached (N. G. C., 1999) was found at Parsonstown to resemble 'a comet coiled into a ring,'[3] and was photographed, precisely under the same aspect, by Mr. Common in 1883.[4] The triple star, ι Orionis, less than a degree distant, is enveloped in a nebula of analogous form, which belongs also, in a measure, to the appendage of the star Maia in the Pleiades. Sir John Herschel's 'falcated' nebulæ are of the same kind. One such in Argo, 10′ in extent (N. G. C., 3199), appeared to him of a 'semi-lunar shape,' diffuse outside, but with a sharp inner edge.[5] Another (N. G. C., 346) occurs far to the south in Hydrus. 'A complete telescopic comet,' a perfect miniature of Halley's,[6] was encountered in the constellation Eridanus, (N. G. C., 1325), and star-like condensations, with brush or fan-like appurtenances, are not unfrequent on his lists.

The discovery of spiral nebulæ was beyond question the

[1] *Sidereal Messenger*, Jan. 1890.
[2] *Astr. Nach.* No. 2918.
[3] *Trans. R. Dub. Soc.* vol. ii. p. 50.
[4] *Observatory*, vol. xii. p. 84.
[5] *Cape Observations*, pp. 20, 94.
[6] *Ibid.* p. 61.

most important result of the construction of the great Parsonstown reflector. Its significance is continually enhanced as the wide prevalence of convoluted forms among this whole class of sidereal objects is rendered more fully apparent by the increasing advance of exploration.

The typical spiral nebula in Canes Venatici (M 51) presents, with a great telescope, a truly amazing appearance. The two nuclei separately catalogued by Sir John Herschel are then seen to be connected by an exterior faint sweep from an

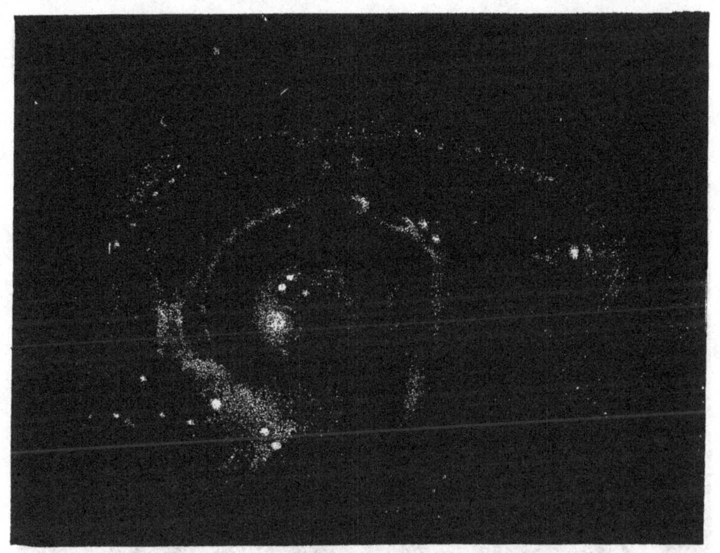

FIG. 38.—Spiral Nebula in Canes Venatici.
(*From a photograph taken at Herény.*)

inner system of wreathing nebulous bands. These, too, show nodosities and angularities [1] (well shown in fig. 38, from a photograph by M. von Gothard) in obvious mutual relations, as if the knots, instead of simply forming upon the spires, had determined, or at least deflected their course. A still clearer knowledge of their arrangement has been gained through a remarkable photograph taken by Mr. Roberts with four hours' exposure, April 28, 1889.[2] The nebula displays itself in it, no longer coiled like a watch-spring, but as com-

[1] Vogel, *Publicationen*, No. 14, p. 32. [2] *Monthly Notices*, vol. xlix. p. 38.

posed of a pair of curving arms issuing from opposite extremities of an oval central body. One of these loses itself in a vague effusion, as a comet's tail dies out into darkness; the other attains the secondary nucleus, and there terminates. The spiral character of this great vortex is perhaps rendered exceptionally conspicuous by its being more favourably placed than most others for our inspection. We seem to get nearly a bird's-eye view of it, and are thus enabled to take in the design of its construction at a glance. Its spectrum is continuous.

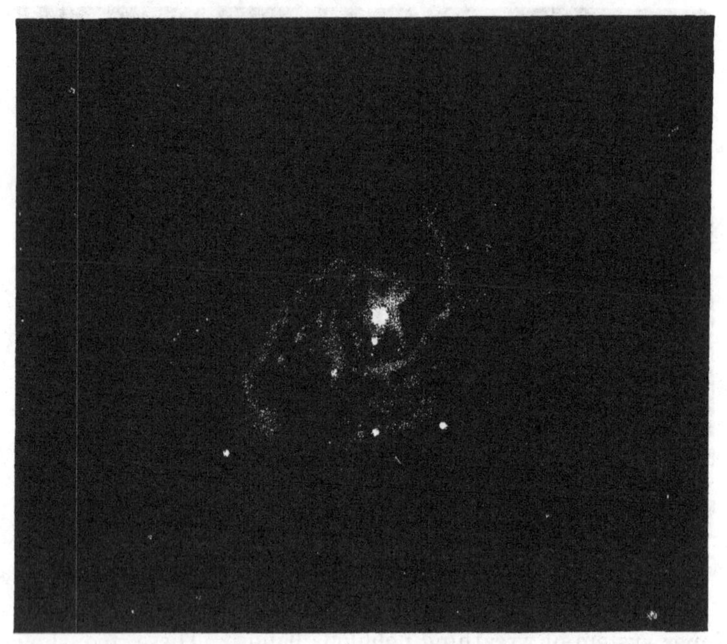

FIG. 39.—Spiral Nebula in Virgo. (*From a photograph taken at Herény.*)

An object in Virgo 3′ across (M 99) is a left-handed spiral; its branches turn the opposite way from those of M 51. Their tendency to form knots and angles is strikingly shown in fig. 39, copied from a photograph obtained in two hours by M. von Gothard, April 12, 1888.[1] A diffuse nebulous mass in Triangulum (M 33), just discernible with the naked eye, was shown by the Rosse reflector to be a 'large spiral, full of knots.'[2]

[1] Vogel, *Astr. Nach.* No. 2854. [2] *Phil. Trans.* vol. cli. p. 711.

Its light is of a continuous nature,[1] although it may not prove, on further examination, to be genuinely *stellar*.

Curved furrows of light, such as it is agreed to call 'spiral,' have been traced in many planetary and annular nebulæ; with still greater optical power than is now at the disposal of astronomers, they might possibly be brought to view in all. The kind of structure which they indicate seems indeed to characterise, in some degree, every form of cosmical agglomeration, and depends no doubt upon laws of their ordained development foreign to our terrestrial experience.

The tendency of winding nebulous bands to become *knotted* often proceeds so far that the knots all but completely absorb the bands. Multiple groups of nebulæ then appear ranged along curved lines, the intermediate faint luminosity becoming perceptible only with large telescopic apertures. Such is the curious double nebula in Perseus (M 76), noticed by the present Lord Rosse to constitute, with subordinate nodules and streamers, a system modelled on a 'reaping-hook' pattern.[2] A gaseous spectrum is derived from it. Similar combinations are met with in the southern constellation Mensa (N. G. C., 2046) where five nebulæ are disposed along an oval line, and in a 'falcated' nebula with three knots situated in Cepheus (N. G. C., 7008).

All the diversities of double stars, it was pointed out by Sir John Herschel,[3] 'have their counterparts in nebulæ; besides which, the varieties of form and gradation of light in the latter afford room for combinations peculiar to this class of objects.' Its members are surprisingly numerous, one in sixteen of the 5079 nebulæ catalogued by Herschel in 1864 being in unmistakable connection with other adjacent objects. Double and triple nebulæ are usually spherical, centrally condensed, and with traces of mutual coherence. Their separation is thus still visibly incomplete; orbital revolutions can scarcely be assumed as probable; nor is there any sign of their being in progress.

[1] *Harvard Annals*, vol. xiv. p. 288. The nebula is No. 242 of the *Harvard Photometric Catalogue*.

[2] *Trans. R. Dub. Soc.* vol. ii. p. 21.

[3] *Outlines of Astronomy*, p. 647.

A nebula in Ursa Major (N. G. C., 3690) divided by Swift in 1885,[1] makes probably the closest pair known; and a curious reproduction, with greatly widened spatial intervals, of star systems like that of γ Andromedæ occurs in a triple nebula in Virgo consisting of a bright round nebula, attended, at a distance of 5′, by an extremely faint one which is itself double[2] (N. G. C., 5813–14). Another compound object of a striking character was noticed by Sir John Herschel[3] in Canes Venatici (N. G. C., 4631), where an enormously long ray of nebulosity has a round, dimly luminous companion, a tenth-magnitude star placed between serving perhaps as a centre of attraction for both.

A very close double nebula in Gemini (N. G. C., 2371–12) has also an intervening star symmetrically located in the line joining their centres.[4] Cirrus-like streaks of nebulosity partially encircle the two objects. Duplicity is, in other cases, still less clearly defined. Thus, a pair of nebulæ near γ Leonis (N. G. C., 3226–27) are together enclosed in a faint luminous envelope, the effect recalling that of the celebrated 'Dumb-bell' nebula in Vulpecula (M. 27),[5] which is only perceived to be essentially single when the 'neck' uniting two conspicuous hazy masses is brought into view with a powerful telescope. Sir John Herschel first observed the elliptical outline of the entire to be rounded out by faint luminosity, and thus saw it in its true aspect as a large, diversified oval disc, measuring about 5′ by 8′. It might indeed be called a magnified planetary nebula not devoid of annular inclinations. The possibility that, by the progress of the central contraction and marginal spreading indicated by its present hourglass shape, the chief part of its mass may, in the course of ages, become diffused into a ring, is strongly suggested by the analogy of the bright spots at either end of the minor axis of the ring-nebula in Lyra. A planetary in the southern hemisphere (N. G. C., 1365) appears, in fact, to

[1] *Sid. Mess.* vol. iv. p. 39.

[2] D'Arrest, *Astr. Nach.* No. 1369.

[3] *Phil. Trans.* vol. cxxiii. p. 431.

[4] Lassell, *Memoirs R. Astr. Soc.* vol. xxiii. p. 62; Lord Rosse, *Phil. Trans.* vol. cxl. p. 512.

[5] D'Arrest, *Abhandlungen*, Leipzig, 1857, p. 325.

have already reached a more advanced stage on the same road, and several 'miniatures' of the 'Dumb-bell' are included among that class of objects. One especially in Cygnus (N. G. C., 6905), depicted by Vogel with the Vienna 27-inch, easily gives an impression of actual duplicity,[1] and showed at Parsonstown as a 'beautiful little spiral.' It has a central star, and four 'satellites.' Its spectrum, like that of the Dumb-bell nebula, is (so far as is yet known) absolutely monochromatic. The leading nebular ray at 5005 concentrates the whole of its light.

In a photograph of the Dumb-bell nebula, taken by Mr. Roberts with an exposure of three hours, October 3, 1888, the shaping-in towards the middle, so marked to the eye, is almost obliterated through the prolonged accumulation of chemical effects; but it intimates pretty clearly the approximate completeness of the oval bright border of the disc, as well as its superposition upon a fainter, more elliptical one, visible as a kind of effusion at the extremities of its longest diameter. Vogel's drawing too[2] suggests, though after a different fashion, the presence of two ellipses, one partially concealed behind the other; and there hence seems reason to think that this singular formation partakes, in more ways than one, of the compound character evident in many planetary nebulæ.

[1] *Potsdam Publicationen*, No. 14, p. 36. [2] *Ibid.* p. 35.

CHAPTER XVIII.

THE GREAT NEBULÆ.

THE elliptical and irregular classes of nebulæ are illustrated by such splendid examples that we have thought it well to devote a chapter to their separate consideration. One member especially of each towers above the rest, like Ajax among the Argive host, its rival alone excepted, and the two are so different that it is not easy to award the palm of superiority to either. Needless to say that we allude to the objects in Andromeda and Orion, the types respectively of the elliptical and irregular plans of nebular construction.

The former (M 31) is the only real nebula which can readily be detected with the unaided eye, and it is the only one, accordingly, which was discovered in pre-telescopic times. Al Sûfi was familiar with the 'little cloud' near the most northern of the three stars in the girdle of Andromeda; [1] and its place was marked on a star-map brought from Holland to Paris by De Thou, and believed to date from the tenth century.[2] Simon Marius, who was the first to turn a telescope upon it, December 15, 1612, called it 'stellam quandam admirandæ figuræ,' and compared its dull and pallid rays to those of a candle shining by night through a semi-transparent piece of horn. Yet this strange phenomenon was only rescued from neglect by Boulliaud, whose attention was directed to it by the passage of the comet of 1664 across that part of the sky. So surprising did the disregard of it by Hipparchus, Tycho, and Bayer then appear to him that he concluded it to vary in light, an hypothesis which, however, derives no support from recent observations.

[1] Schjellerup, *Description des Étoiles*, p. 120.
[2] Le Gentil, *Mémoires de l'Acad.* 1759, p. 459.

With powerful light-concentration this 'most magnificent object' (to borrow Sir John Herschel's phrase) assumes vast proportions. They were extended by G. P. Bond, using the fifteen-inch refractor of Harvard College, to cover an area of $4° \times 2\frac{1}{2}°$, and he probably did not reach their absolute limits. Two adjacent nebulæ, one (M 32) descried by Le Gentil in 1749, the other (N. G. C., 205) by Caroline Herschel in 1783, undoubtedly fall within their compass.'[1] The light of this nebula is 'of the most perfectly milky, absolutely irresolvable kind.'[2] It does not collect into 'floccules,' and produces none of the scintillating effect giving to many gaseous nebulæ a delusive appearance of resolvability. From the circumference towards the centre, however, it gradually brightens, then abruptly condenses to a small nucleus of indistinct outline under high magnifying powers, and possibly (like the nuclei of many comets) *granulated*, but assuredly not stellar.

This progressive brightening inward shows, nevertheless, interruptions. On September 14, 1847, Bond discovered two long dark rifts running nearly parallel to one another, and to the axis of the nebula.[3] Their detection was a consequence of the widened area of the luminosity perceived by him, the inner rift having been taken, until then, for its boundary in that direction. The outlines of Bond's drawing are given in the accompanying diagram by Mr. Wesley (fig. 40), in which the 'rifts' are marked A and B. C represents Le Gentil's, D Miss Herschel's attendant nebula, E an exceptionally lucent region crowded (it has since been found) with hosts of minute stars.[4]

These enigmatical appearances at last assumed an intelligible form in a photograph taken by Mr. Roberts, October 1, 1888.[5] The view given by this magnificent picture of the Andromeda nebula as a symmetrical, though still inchoate structure, ploughed up by tremendous, yet not undisciplined

[1] Bond, *Memoirs Amer. Acad.* vol. iii. p. 83.
[2] J. Herschel, *Memoirs R. Astr. Soc.* vol. ii. p. 496.
[3] *Memoirs Amer. Acad.* vol. iii. p. 80.
[4] Ranyard, *Knowledge*, vol. xii. p. 76.
[5] *Monthly Notices*, vol. xlix. pp. 65, 120.

forces, working harmoniously towards the fulfilment of some majestic design of the Master Builder of the universe, is of a nature to modify profoundly our notions as to how such designs obtain their definitive embodiment. Plate III. reproduces an impression secured by the same artist with four hours' exposure, December 29, 1888. Although not covering so large an area as the visual representation in fig. 40, it is

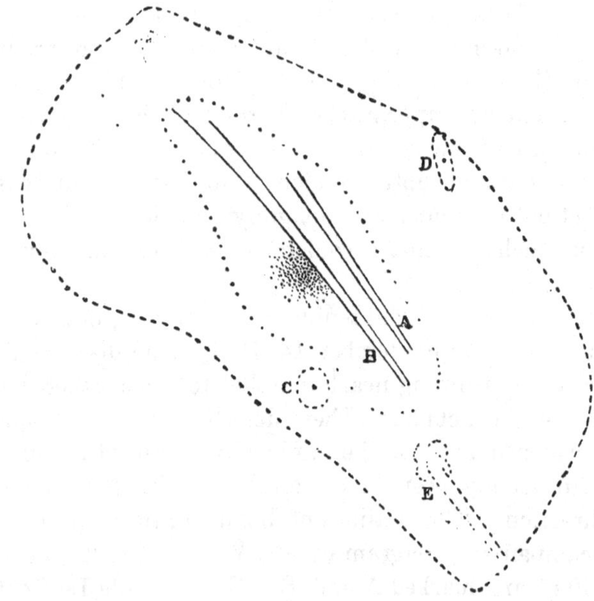

FIG. 40.—Diagram of the Great Nebula in Andromeda, from Bond's drawing in 1847.

of a totally different order of merit as regards clearness and consistency of details, the significance of which can be at once perceived by reference (in fig. 41) to Mr. Wesley's careful tracing of the lines of conformation brought out on the sensitive plate.

Bond's 'canals' are now seen prolonged and curved into two vast rings (AA and BB), which prove, on attentive consideration, to be dusky intervals separating the successive spires of a single great stream of nebulous matter, winding outward from near the primary to reach the secondary nucleus

PLATE 3.

GREAT NEBULA IN ANDROMEDA.

Photographed by Isaac Roberts, F.R.A.S., December 29th, 1888. Exposure four hours.

(M 32). The similarity of the relations between the two nuclei here and in the 'whirlpool' nebula in Canes Venatici scarcely needs to be emphasised by a remark. Thousands of stars are scattered over and around the Andromeda nebula; the situation of which in a prolongation of the Milky Way perhaps

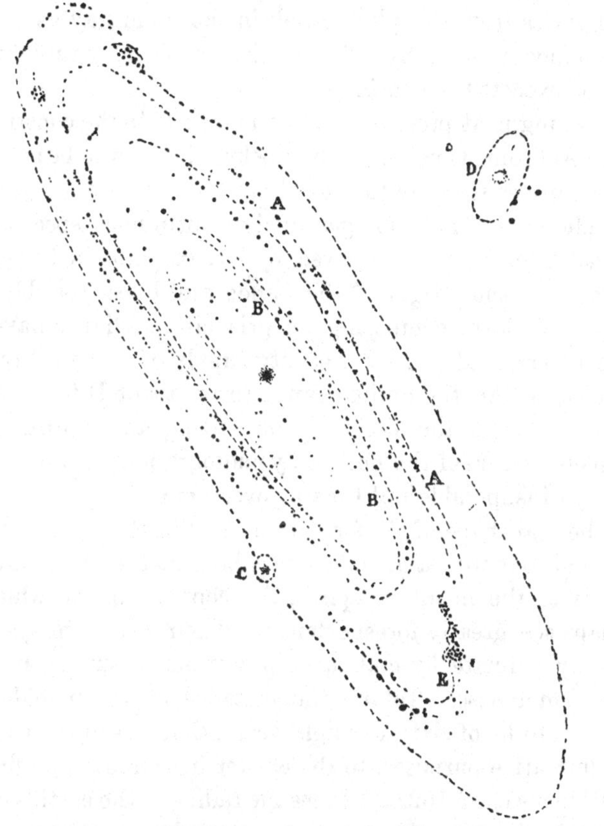

Fig. 41.—Diagram from Roberts's photograph of the Andromeda Nebula.

sufficiently explains their profusion. Many of them may be entirely disconnected from it; but not all. Shoals of sharp stellar points assume, in the photographs, the annular or spiral arrangement of the nebula, and 'lie along the edges of the dark rifts following all their sinuosities.'[1]

[1] Ranyard, *Knowledge*, Feb. 1 and Aug. 1, 1889.

A palpable discrepancy between the drawings and photographs of this stupendous object was adverted to by Mr. Roberts.[1] The small oval nebula marked with the letter D in fig. 40 lies, it is easy to see, much more obliquely towards the principal centre of the formation than the corresponding object in Plate III. And since a delineation of it by Trouvelot in 1876 varies from the photograph in the same way as Bond's, the change, if real, must have taken place very rapidly, and may be expected to continue.

Nothing is at present certainly known as to the constitution of the Andromeda nebula. Mr. Lockyer infers it to be meteoric, and compares its state to that of 'a comet within a month of perihelion.'[2] That no genuinely continuous spectrum is derived from it[3] was perceived by Dr. Huggins in 1864; but the nature and origin of the bands and lines, possibly both bright and dark, composing its prismatic radiance have yet to be ascertained. Only a scanty supply of chemical rays is included within the limited range (from about D to F) of the latter; so that the difficulty of getting an approximately complete picture of the nebula by photographic means seemed wellnigh insuperable until it was overcome.

The real shape of the formation must be that of a disc, oval or round, but with a globular mass both at the origin and extremity of the nebulous spires, the convolutions of which we perhaps see greatly foreshortened. Their extent in space we can only estimate by making a precarious assumption as to their remoteness. Taking the distance of the nebula, for instance, to be of sixty-five light years, that already attributed, for illustrative purposes, to the cluster in Hercules, we find its radius to measure 162,000 times the radius of the earth's orbit; so that the frontier of this glimmering realm, as determined by Bond, is much more than half as remote from its centre as the nearest fixed star (*a* Centauri) is from ourselves! In travelling from end to end of it, light spends nearly six years; and if it lie obliquely towards us, then our view of the further margin may be of an earlier date than our view of the hither margin, by a couple of years or more. Extensive changes of luminosity

[1] *Observatory*, vol. xii. p. 51. [2] *Ibid.* p. 98. [3] See *ante*, p. 81.

within the nebula might then manifest themselves to us successively, although they had really occurred simultaneously; while, on the other hand, the coincidence in time to our senses, of widespread variations, would argue a position of the nebula nearly square to the line of sight. That it is still (so to speak) in the *plastic* stage, there can be little doubt; and the outbreak of the 'new star' of 1885 has shown that the action of the powers engaged in moulding it to its appointed shape may occasionally be attended by a catastrophic liberation of energy.

FIG. 42.—Photograph of the Milky Way in Sagittarius.

But what is 'its appointed shape'? What is this marvellous system destined to become, as it emerges from the nebulous condition? In framing a reply to such questions we have only analogy to guide us; yet the presumption is very strong of a general similarity both of modes of action and of their results in every part of the sidereal world. Look, for example, at fig. 42. It represents, from an admirable photograph by Mr. Barnard, a great collection of Milky Way stars in Sagittarius. The area covered by it somewhat exceeds that of the Greater Magellanic Cloud—forty-two square degrees—

and the stars thronging it range to below the thirteenth magnitude. The fundamental method of their arrangement can at once be apprehended. They are evidently condensed into an oval nuclear mass fenced round (as was pointed out by Professor Holden) by a partially vacant ring. When the glass positive from which the figure has been printed is held so far from the eye that it begins to assume a nebulous appearance, not one alone, but several half-obliterated rings come into view constituting a series of spirally-connected clearings, the result, apparently, of some prevalent law by which the distribution of the clustered stars is governed. Thus seen, the photographed group presents so striking a resemblance to the self-depicted Andromeda nebula, that it is difficult to reject the evidence it affords of close relationship, or to deny the possibility that in the cluster we see the exemplar of what the nebula, at some future epoch, will have become.

It may be added that the apparently connected stellar groups M 24 and M 17 may be, in some sort, the equivalents of the attendants upon the Andromeda nebula; while the cascades of stars falling to the left of the main cluster are nebulously represented in the object portrayed (from a photograph by Mr. Roberts) in Plate IV.

This is an elliptical nebula in Ursa Major (M 81) which reproduces, on a small scale, most of the features of the Andromeda nebula. Lord Rosse estimated its greatest extent at about 16′, but this limit is considerably transcended in the photograph,[1] which brings out, in addition, the essentially spiral structure of the object. The presence of a two-fold system of whorls, like those forming the nebula in Canes Venatici, is, in fact, strongly suggested by the impression on the sensitive plate, of two sweeping trains of nebulous matter, issuing tangentially from opposite extremities of the formation, as if in continuance of two distinct original effusions from the nucleus.

A companion nebula (M 82), half a degree distant from M 81, was described by Lord Rosse as 'a most extraordinary object, at

[1] Roberts, *Monthly Notices*, vol. xlix. p. 362.

PLATE 4.

ELLIPTICAL NEBULA IN URSA MAJOR (M 81).

Photographed by Isaac Roberts, F.R.A.S., March 31st, 1889. Exposure $3\frac{1}{2}$ hours.

least 10′ in length, and crossed by several dark bands.'[1] It is bi-nuclear, of a double lance-head shape, and is doubtless an elliptical spiral viewed edgewise. Although it appears with its primary on Mr. Roberts's plate, its relegation to near the edge of the field has placed it at a disadvantage as regards definition, and so the instructive particulars about it which might otherwise have been gathered, remain unrecorded. A nebulous star completing the group[2] is perhaps physically connected with it. The continuous spectrum derived from M 81, 82, as from other elliptical nebulæ, proves to be no less deficient in red light than the spectrum of the Andromeda nebula.[3] We may then safely attribute an almost identical constitution to all the three objects.

Sir John Herschel noticed an approach to an annular conformation in many of this class of nebulæ, and made the curious remark, that 'as the condensation increases towards the middle, the ellipticity of the strata diminishes.'[4] This looks as if their ovalness were not a purely visual effect, and other circumstances tend to confirm the supposition. Thus, in cases where either the foci of a nebulous ellipse (N. G. C., 6595), or the extremities of its major axis (N. G. C., 6648), are occupied by a pair of stars, it may be considered certain that the appearance of elongation corresponds to the reality.

The longitudinal clefts often visible in ray-shaped nebulæ imply an emphatic assertion to the same effect. For otherwise, why should they run lengthwise, rather than in any other direction? If the bodies they characterise were, in fact, circular discs, none of their physical features could have any relation to the appearance they happen to assume by projection. Such a relation is, however, extremely conspicuous in an elliptical nebula in the Centaur (N. G. C., 5128), divided along its entire length by 'a perfectly definite, straight cut, 40″ broad.' A delicate nebulous streak runs between, and parallel to the halves, which are sharply bounded on the sides facing each other, but hazy on those averted. 'The internal edges of this very problematic object,' Sir John Herschel remarked,

[1] *Trans. R. Dub. Soc.* vol. ii. p. 79.

[2] See *ante*, p. 253.

[3] Huggins, *Phil. Trans.* vol. clvi p. 388.

[4] *Cape Observations*, p 22.

'have a gleaming light like the moonlight touching the outline in a transparency.'[1] It is by no means a solitary example. A long narrow nebula in Leo (N. G. C., 3628) is 'split into two parallel rays.'[2] A black chasm, into which the nucleus protrudes, separates a lucid from a faint streak in Coma Berenices[3] (N. G. C., 4565); and a nebula in Andromeda (N. G. C., 891), with 'a chink in the middle, and two stars,' supposed by Sir John Herschel to be 'a thin flat ring, of enormous dimensions, seen very obliquely,' would rather appear, from the Parsonstown observations, to belong to the numerous category of *cloven rays*. That the distinctive peculiarity of these depends upon some general constructive principle cannot readily be doubted; but we should vainly attempt to speculate upon its nature.

No descriptive formula is wide enough to include all the capricious forms of 'irregular' nebulæ. In regarding these singular structures, we seem to see surges and spray-flakes of a nebulous ocean bewitched into sudden immobility; or a rack of tempest-driven clouds hanging in the sky, momentarily awaiting the transforming violence of a fresh onset. Sometimes, continents of pale light are separated by narrow straits of comparative darkness; elsewhere, obscure spaces are hemmed in by luminous inlets and channels. The 'great looped nebula' (30 Doradûs), one of the inmates of the greater Magellanic Cloud, resembles a strip of cellular tissue. It serves not to conceal, but to ornament in a wide, openwork pattern the sky behind it, and was described by Sir John Herschel as 'an assemblage of loops,' the 'complicated windings' of which constitute it 'one of the most extraordinary objects which the heavens present.'[4]

The 'net-work nebula' (N. G. C., 6995) is of a somewhat similar nature. 'A most wonderful phenomenon!' Sir John Herschel exclaims. 'A very large space' (about 25′ by 12′) 'full of nebulæ and stars mixed. A network or tracery of nebulæ follows the lines of a similar network of stars.'[5] The 'dark hollows left by the interwoven nebulæ' were perceived

[1] *Cape Observations*, pp. 20, 105.
[2] *Trans. R. Dub. Soc.* vol. ii. p. 95. [3] *Ibid.* p. 118.
[4] *Cape Observations*, p. 12. [5] *Phil. Trans.* vol. cxxiii. pp. 469, 503.

by Mason and Smith at Yale College, in 1839, to be almost bare of stars; while a map of those involved sufficed to show with tolerable accuracy the windings of the nebula.[1] These observers ascertained, too (what Sir William Herschel had suspected),[2] its union with a 'bifurcate, milky ray' (N. G. C., 6992) in the neighbourhood into an immense complex formation, two or three degrees long, the intricacies of which might probably be studied with greater advantage by photographic than by visual means.

But all other irregular nebulæ sink into insignificance compared with that shown by an opera-glass as a silvery patch round one of the stars in Orion's sword. This extraordinary object (M 42) has been under effective observation for 234 years, and during the last seventy has been monographed, mapped, measured, figured, and photographed with a diligence worthy of its pre-eminence. Hence, future changes in it, should they take place, must be slight indeed, to escape detection.

The multiple star θ Orionis [3] might be called the foundation stone of the entire structure. All the lines of its architecture are laid down with reference to it, and the intimate physical association of the stars with the gaseous stuff surrounding them has been spectrographically demonstrated by Dr. and Mrs. Huggins.[4] This gaseous stuff, although it pervades the trapezium, seems less luminous there than elsewhere. The space about the stars usually forms a sort of oasis of comparative darkness in the midst of a wilderness of piled-up flakes of light. Usually, not always. D'Arrest, it is true, invariably saw the stellar group on an almost black ground; but O. Struve several times, and especially in 1861, found the trapezium as densely nebulous as the contiguous tracts;[5] and the same appearance was noted by both Schröter and Lamont.[6]

The brightest part of the nebula, called, from its first delineator, the 'Huygenian region,' is shaped like a right-

[1] *Trans. Amer. Phil. Soc.* vol. vii. p. 178. [2] *Phil. Trans.* vol. lxxv. p. 31.
[3] See *ante*, p. 216. [4] See *ante*, p. 79.
[5] *Mémoires de l'Acad. de St. Pétersbourg*, t. v. No. iv. p. 115.
[6] Holden, *Washington Observations*, 1878, App. I. p. 100.

angled triangle. Over this space the light is collected into 'flocculent masses,' which, with the best seeing, prove to be throughout of a 'hairy' texture. The effect was compared by Sir John Herschel to the 'breaking-up of a mackerel-sky, when the clouds of which it consists begin to assume a cirrous appearance,' [1] and suggested to Mr. Lassell [2] 'large masses of cotton-wool packed one behind another, the edges pulled out so as to be very filmy.' The 'pea-green colour' of the whole object struck him forcibly, but is apparent to most observers as little more than a greenish tinge.

Emanations (or what seem such) from the Huygenian core stretch away in wide curves to form the outlying portions of the nebula. One great effluence runs out into a 'proboscis' attached to the upper jaw of the nondescript creature limned, in almost unearthly radiance, on the sky. Another, representing the lower jaw, bounds on the northern side the chasm of the distended fauces (the 'Sinus Magnus'), and is subsensibly connected—as if by a shoal leading to an island—with the 'nebula minima' (M 43), a rounded mass which 'appears as if just drawing together into a star.' [3] Behind, and between the two, misty effusions spread far afield; resolved by G. P. Bond into an intricate fabric of convoluted and branching filaments, the brighter region from which they spring displaying a similar, but more compact mode of aggregation. 'It is now impossible,' he wrote after a particularly fine view of the nebula, on February 26, 1861, 'to see it in any other aspect than as a maze of radiating, spiral-like wreaths of nebulosity, or filamentous tentacles, the centre of the vortex being about the trapezium.' [4] And to Mr. Safford, using the then recently constructed 18½-inch Clark equatoreal, it appeared as 'an assemblage of curved wisps of luminous matter, which, branching outward from a common origin in the bright masses in the vicinity of the trapezium, sweep towards a southerly direction, on either side of an axis passing through the apex of the Regio Huygeniana.' [5]

[1] *Memoirs R. Astr. Soc.* vol. ii. p. 491. [2] *Ibid.* vol. xxiii. p. 54.
[3] J. Herschel, *Memoirs R. Astr. Soc.* vol. ii. p. 495.
[4] *Harvard Annals*, vol v. p. 158. [5] *Ibid.* p. 169.

PLATE 5.

GREAT NEBULA IN ORION.

Photographed by Isaac Roberts, F.R.A.S., December 24th, 1888. *Exposure* 81 *minutes.*

The application of photography to this amazing object has not only supplied records of its actual condition indefinitely more authentic than any producible by the human hand, but has served to combine the peculiarities of its conformation into a strikingly suggestive whole. Plates V. and VI. exhibit two impressions obtained by Mr. Roberts, the first, and more pictorially perfect, with an exposure of 81 minutes,

FIG. 43.—Index-Diagram to Structures photographed in the Orion Nebula.

the second in 3½ hours, February 4, 1889. The unprecedented comprehensiveness of the latter has indeed been secured with some sacrifice of characterisation. The central tracts appear 'dark with excessive bright'—their particular features lost in the general glare. But the main outline comes out with unlooked-for definiteness, and approximate completeness. The result can best be judged of by reference to fig. 43, exhibiting an 'index diagram' prepared by Messrs. Ranyard and Wesley from several negatives (all by Mr. Roberts) of varied exposures. Two points are made perfectly clear by it. First, that the whole fabric of the nebula is concave towards an axis 'passing through the trapezium in a north-easterly and south-westerly direction.'[1] Next, that the effluences from the trapezium have a predominant tendency to assume ramified, or tree-like forms. Thus the seemingly eruptive jet marked with the letter *b* mimics the shape of a stone-pine, and bears a less equivocal resemblance to the 'stemmed' type of solar

Fig. 44.—Group of Synclinal Structures from a photograph of the Corona of 1871. Drawn by Mr. W. H. Wesley.

prominences, as well as to certain arboreal structures visible in some photographs of the solar corona. The latter analogy is rendered still more apparent by comparing the vast nebulous out-growths, *e* and *h*, with the group of coronal rays represented in fig. 44. The same kind of structures seem, in both cases, at once to spring upward and to curve inward, as if under the influence of a two-fold action—outward from a centre, and inward towards an axis. This organic similarity—first detected by Mr. Ranyard—between the Orion nebula and the luminous appendages of the sun,

[1] Ranyard, *Knowledge*, vol. xii. p. 147.

Plate 6

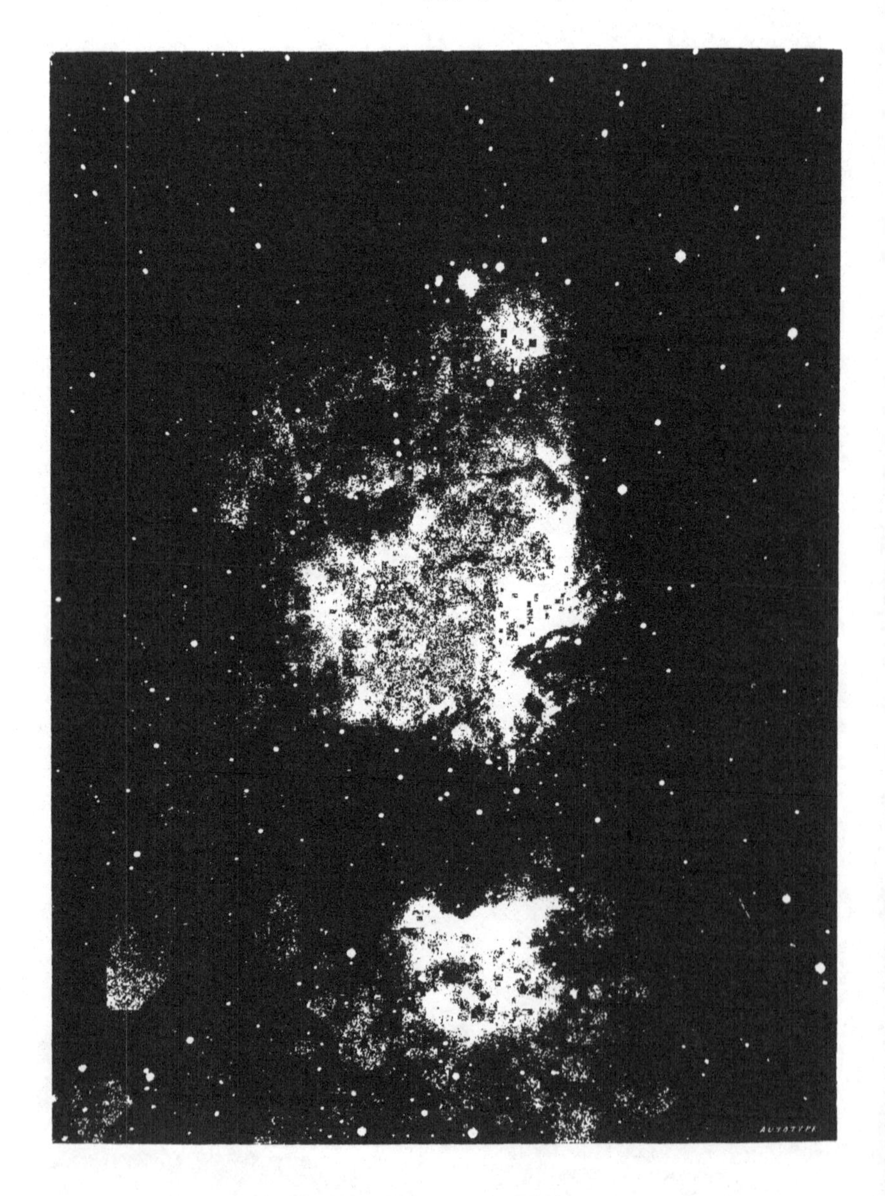

GREAT NEBULA IN ORION.

Photographed by Isaac Roberts, F.R.A.S., February 4th, 1889. Exposure 205 minutes.

is rendered more profound by the association of helium with hydrogen in the chemical composition of both.

The limits of the great 'sword-handle' nebula are continually being pushed further back, and there is no reason to believe the process nearly terminated. On Mr. Roberts's plates,[1] first the 'nebula minima' was joined on to the main body, then, with lengthened exposures, the cloudy mass to the north (N. G. C., 1977) shown near the bottom of Plate VI. was reduced to its true position of an offset, by which the forms of the parent-body are pretty closely imitated.[2] But even the combined object is far from representing the nebulous contents of this part of the sky. Over an area of 150 square degrees, in which it is nearly central, twelve new nebulæ were photographed at Harvard College in the spring of 1888, and indications were obtained that Sir William Herschel's surmise[3] of the union into one immense stratum of the 'great' nebula with others lying north and south of it might be verified in the immediate future.[4] A spiral form for the stratum is further strongly suggested by 'long-exposure' impressions secured by Mr. W. H. Pickering in California[5] and Peru.

It may be that we see in the Orion nebula a great undeveloped cluster on the model of the Pleiades. Nearly a thousand stars were catalogued by Bond in the portion of it (3·36 square degrees in area) examined by him, and his list seems to be practically exhaustive. Many of them should, indeed, in Professor Holden's opinion,[6] be accounted rather as nebular condensations than as true stars; but that they will eventually grow to be such, as they slowly absorb the nebulous material now enshrouding them, is a justifiable conjecture. The same process appears already far advanced in the Pleiades, with the result that the stellar now altogether predominates over the nebular element in the compound system. It can scarcely have been always so. The balance inclined perhaps, in the remote past, as decisively the other way as it now does in the Orion nebula.

[1] *Monthly Notices*, vol. xlix. p. 296.
[2] Ranyard, *loc. cit.* p. 148.
[3] *Phil. Trans.* vol. lxxix. p. 249.
[4] *Harvard Annals*, vol. xviii. p. 117
[5] *Sid. Mess.* vol. ix. p. 1.
[6] *Washington Observations* 1878 App. i. p. 221.

It is not easy, Sir John Herschel tells us, 'for language to convey a full impression of the beauty and sublimity of the spectacle offered by the Argo nebula, when viewed in a sweep, ushered in as it is by so glorious and innumerable a procession of stars, to which it forms a sort of climax.'[1] 'Situated in one of those rich and brilliant masses, a succession of which, curiously contrasted with dark adjacent spaces, constitute the Milky Way between Centaur and Argo,' its branches with their included vacuities cover more than a square degree, and are strewn by above twelve hundred stars. The peculiarity, giving the formation the name occasionally applied to it of the 'key-hole' nebula, is a large lemniscate-shaped opening in the central and brightest part, the blackness of which is qualified only by the veiling of one corner by a strip of thin nebulous haze. Four stars are placed precisely at the edges, none within the vacuity; and the famous variable η Argûs[2] lies close to its eastern border. Such was the brilliancy of the star in 1838 as almost to obliterate the 'key-hole,' now, and previously to the outburst, the special, individualising feature of what would otherwise seem a chaotic sea of luminous billows. The significance of the peculiarity is emphasised by its duplication. A second oval aperture, completely dark but for the faint sparkle of four minute stars, occurs in the southern, sparser part of the nebula. A similar *echo* of a characteristic form is met with, as our readers may remember, in the great Hercules cluster.[3]

A very peculiar gaseous nebula in Sobieski's Shield (M 17) has revealed itself piecemeal, and may still have disclosures in reserve. Messier, first of all, noticed there in 1764 a spindle-shaped, starless ray, about 6′ in length. Sir William Herschel added an arch, springing from its western extremity; and the combined object became known as the 'Horseshoe' or 'Omega' nebula, its form resembling that of the Greek capital letter Ω, with the left-hand base-line turned up obliquely.[4] Again, it suggested a 'Swan' to observers whose instruments were inadequate to show the complete arch, and Flammarion compared

[1] *Cape Observations*, p. 38. [2] See *ante*, p. 116. [3] See *ante*, p. 246.
[4] J. Herschel, *Phil. Trans.* vol. cxxiii. p. 461.

it to a smoke-drift, fantastically wreathed by the wind. Signs of incipient organisation are, nevertheless, traceable in it. At the 'elbow,' or angle between the base and the swan-neck, Sir John Herschel detected two bright *knots*, tending evidently to become insulated from the surrounding nebulosity; and perceived further, in the clear air of the Cape, a *second* very faint 'horse-shoe,' attached to the opposite end of Messier's streak from the first.[1] This new part of the nebula seems to have been independently re-discovered by Swift, of the Warner Observatory (N. Y.), July 4, 1883;[2] and both he and Herschel saw or suspected additional patches and convolutions. The edifice has yet to be 'crowned' with the assistance of the camera.

The nebula, as at present known, presents the outline rather of a two-arched bridge with a wide pier, than of a horseshoe. It appears to include two centres rotating, presumably, in opposite directions; since, of the inflected arms of light originating from them respectively, one follows the lines of a right-handed, the other of a left-handed spiral. It is possible, too, that their connection is not designed to be permanent. The 'pier' betweeen the 'arches' may eventually collapse. The conjecture at least is not unreasonable that, under the strain of counter-solicitations, the continuity of Messier's streak will at last give way, and the matter contained in it be gathered round the pair of vortices constituting the nuclei of a double spiral nebula.

The 'Trifid' nebula in Sagittarius (M 20) has a dark central space. The three great *lobes* of which it consists are divided by 'a sort of three-forked rift or vacant area, abruptly and uncouthly crooked, and quite void of nebulous light.'[3] These tracks of darkness are *thoroughfares*. They run right through from the brilliant centre to the dim circumference of the nebula. A conspicuous multiple star,[4] placed just at the sharp inner edge of one of the luminous detachments, probably exercises a predominant influence over them all; and corresponding ascendency may be claimed for a single

[1] *Cape Observations*, p. 10. [2] *Sidereal Messenger*, vol. iv. p. 38.
[3] Herschel, *Outlines*, p. 654. [4] See *ante*, p. 219.

star a little to the north, from which, as Herschel says, 'a fourth nebulous mass spreads like a fan or downy plume.' So far does it spread that Sir John Herschel[1] surmised a connection between it and another great nebula (M 8), which, since it is composed of 'nebulous folds and masses, surrounding and including a number of oval dark vacancies,' might be designated the 'Lagoon' nebula. Its intermixture with a fine cluster has been already mentioned.[2]

The spectrum of the trifid nebula consists mainly of the inevitable ray at 5005, with a trace of the hydrogen F;[3] and its light impresses the eye with a corresponding effect of greenish colour. Photographic investigations of this object can be pursued with advantage only in more southern latitudes than ours.

No logical distinction can be established between irregular nebulæ and those indefinite tracts of milky radiance termed by the elder Herschel 'diffused nebulosities.' The total area of fifty-two separate tracts perceived to be thus phosphorescent was estimated by him at 152 square degrees, and he added the judicious remark that 'the abundance of nebulous matter diffused through such an expansion of the heavens must exceed all imagination.'[4] From the few observations since made of these regions,[5] it would appear that their illumination is not absolutely uniform, cumuli of very slightly enhanced brightness hinting here and there at initiated condensations, and at dim beginnings of nicely-balanced trains of operations, set in motion by Sovereign Decree to evoke a cosmos from a chaos in those misty fields of space.

[1] *Loc. cit.* [2] See *ante*, p. 249. [3] Secchi, *Sulla grande Nebulosa di Orione*, p. 28. [4] *Phil. Trans.* vol. ci. p. 277.

[5] Dreyer, *V. J. S. Astr. Ges.* Jahrg. xxii. p. 62; Littrow, *Sterngruppen und Nebelmassen*, p. 29; Tempel, *Astr. Nach.* No. 2511.

CHAPTER XIX.

THE NATURE AND CHANGES OF NEBULÆ.

SPECULATIONS as to an identity of nature between nebulæ and comets are no novelty; they presented themselves, as they could hardly fail to do, to the mind of Sir William Herschel;[1] but some consistency was first given to them by the recent experimental researches of Mr. Lockyer. It is true that the results of light analysis are far from being decisive in their favour. The spectra of the two classes of bodies are fundamentally unlike. No gaseous nebula gives a trace of the carbon-bands which characterise nearly all comets; and no comet has yet furnished any direct evidence of the presence of hydrogen among its constituents. Moreover, nebulæ (apart from the stars contained in them) seem to emit no genuinely continuous light, while cometary nuclei glow in the ordinary manner of white-hot solid and liquid substances. Traces of a spectroscopic analogy can indeed be shown to exist;[2] but they are met with only in the secondary elements of each spectrum. The resemblance seems only incidental; the dissimilarity essential.

This does not, however, detract from the closeness of a *physical* analogy, the deep import of which cannot be too forcibly dwelt upon. Both comets and nebulæ consist of enormous volumes of gaseous material, controlled by nuclear condensations, whether of the same or of a different nature in the two genera we need not now stop to inquire. Both, there is the strongest reason to believe, shine through the effects of electrical excitement. In both, there are manifest signs of the working of repulsive, as well as of attractive agencies.

A telescopic comet is indistinguishable except by its motion from the ordinary, centrally brightening globular nebula, which

[1] *Phil. Trans.* vol. ci. p. 306. [2] See *ante* p. 76.

is itself indistinguishable from an exceedingly remote globular cluster. Superficial likenesses, however, do not count for much; one object may counterfeit another without bearing any true relationship to it. What it is really important to note is the *structural* resemblance of nebulæ to comets. The parts of a comet become differentiated exclusively under solar influence. Hence their symmetrical arrangement, as regards an axis passing through the sun, is modified only by the orbital dis-

FIG. 45.—Nebula in Leo (M 66) photographed by M. von Gothard.

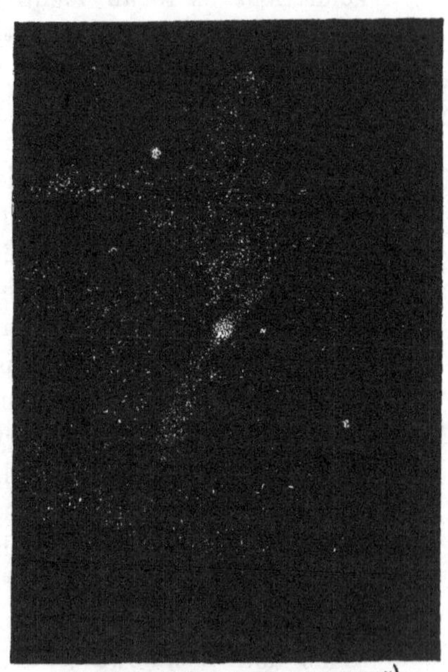

FIG. 46.—Nebula in Gemini (M 65) photographed by M. von Gothard.

placement of the body to which they belong, and from which they emanate. It is then extremely curious to find a corresponding kind of *polarity* impressed upon the features of certain nebulæ. One in Leo, for instance (M 66), photographed at Herény April 16, 1888 (fig. 45), might pass for the head of a very active comet. A series of incomplete envelopes, lying in the precisely opposite direction from the nucleus to that taken by a bright ray, seem the remnants of successive luminous effusions under the prescriptive guidance of some external force. Forms very similar showed themselves in the comets of 1861 and 1882, and in comet Sawerthal (1888).[1] A fine nebula in Gemini (M 65), also photographed by M. von Gothard, displays a different kind of axial symmetry. Four *wings*, not all equally well developed, are attached to a nucleus, in an arrangement like that of the sails of a windmill (fig. 46). Two-branched spirals no less plainly betray the action of polar forces, and confirmatory examples might be drawn from other classes of nebulæ.

From many of these objects, however, as from many comets, only a single stream of effusion is manifest; and we then get 'stars with tails,' which might well pass for miniatures of the bearded travellers through our constellations. These effusions are, in some cases, bounded by straight lines; they are fan-shaped; in others, they are curved, often so strongly that the 'brush' is bent into a 'coil.' Such differences may be plausibly associated with the varied conditions of axial rotation in the star-like nucleus. Where there is none, the issuing matter naturally proceeds straight outward; its curvature depends upon its being left continually further behind in the continually widening circles reached by it as it ascends from an advancing point of efflux.

It is easy to see that this process, if carried far enough, will result in the production of a 'spiral' nebula. And there are indications that many of these objects are in fact composed of matter expelled—perhaps by almost imperceptible degrees—from a slowly rotating nucleus, each branch of the spiral corresponding to a separate, though possibly simultaneous efflux. Their convolutions are, however, probably of a 'helical' nature,—they follow the lines rather of a corkscrew than of a watch-spring. The effect seems to be as if the proper or advancing movement of the rotating nucleus had been accelerated or retarded through influences from which they were exempt, while the formation of the branches was still progressing. It is not meant to suggest that this actually took place, but only to supply a hint towards a mental picture of the phenomena we have to deal with.

[1] C. H. F. Peters, *Astr. Nach.* No. 2550; *Washington Observations* for 1880, App. I.; *Sirius*, Bd. xxi. p. 165.

With very rare exceptions, nebulæ are seen by us, not in plan, but in perspective. 'The only thing,' Professor Holden says,[1] 'we really know about the form of a nebula, in general, is that it is projected into a certain shape. The problem is to find the true curves in space, knowing only the projected curves.' With this problem, all but insoluble as it appeared, he has found the means of dealing, in at any rate a hopeful manner.

By persevering trials with helices of wire projected in all imaginable varieties of position, a curve was at last found which might be called typical. In some one of its innumerable perspective shapes (a few specimens of which are given in fig. 47), it proved capable of representing with considerable accuracy the main outlines of nearly all the nebulæ classed as spiral, among which the peculiar 'Omega' nebula is included. A strong presumption is thus raised that the 'type-curve,' or something like it, is really conformed to in nature, and that this first attempt to 'determine the actual situation of the different branches of a nebula in space of three dimensions from the data afforded by the projection of these branches upon the background of the sky,' has met with a certain measure of success.

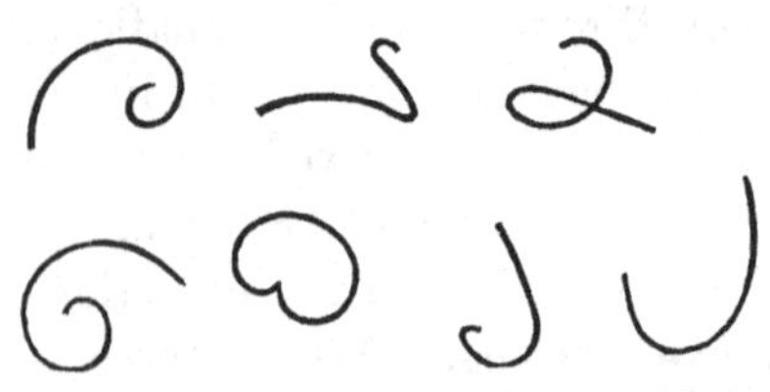

FIG. 47.—Type-Curve of Helical Nebulæ viewed in various positions (Holden).

Should this presumption be borne out by further experience, 'many interesting questions,' Professor Holden concludes,[2] 'may receive a solution. For example, what are the directions in space of the *axes* of these different nebulæ? Is there anything systematic in these directions? What is the law of the force by which particles of matter are expelled from (or attracted to?) the central nucleus? Have we here in the nebulæ different types of spirals somewhat analogous to the different types of comets' tails so ably discussed by Professor Bredichin? Some of the parts of these nebulæ must be approaching the earth, some receding from it. Can we by

[1] *Publications Astr. Soc. of the Pacific*, No. 3, p. 25. [2] *Loc. cit.* p. 31.

the spectroscope discriminate between such motions? A suggestion which holds out even the hope of successfully attacking such problems' has undoubtedly a very high value.

Nebulæ of the spiral and 'brush' kinds are not the only ones exhibiting cometary relationships. A characteristic feature of nebulous trains and wings, exemplified both in the Orion and Pleiades formations, is their vague diffusion on one side, but sharp termination on the other, just as in the wing-like appendages to the heads of comets. The continued ejection of matter *against* a counter-current of force, by which it is unceasingly driven backward, seems indicated in both cases.

Planetary nebulæ, too, imitate, in a fashion of their own the heads of comets under energetic solar action. The multiple discs correspond most strikingly to the multiple envelopes of comets, and intimate a similar origin through interior expansive or repulsive agencies. Only that, in the absence of the directive power of the sun, the waves of emanation spread equally in all directions, producing successive approximately globular, instead of parabolic surfaces. Under the combined influences of rotation and contraction, such shells might be expected eventually to subside into rings;[1] but it would be extremely rash to affirm that annular nebulæ did, in point of fact, come in this way to exist.

Cometary transformations may even help to explain those of 'new stars.' On January 23, 1835, Halley's comet showed as in no respect different from a star of the sixth magnitude; by the 25th it had distended to twice the linear dimensions of the planet Jupiter. And Pons's comet, observed September 23, 1883, as almost stellar, and eight times brighter than the day before, had resumed, forty-eight hours later, its habitual foggy aspect. The alternate appearances of Nova Cygni as star and nebula were perhaps due to causes not unrelated to those here concerned. In ascribing to them an electrical nature, we risk, it is true, laying ourselves open to the taunt *obscurus per obscurior*; yet we can only speak as we know, and trust to time and research for better counsel. Elec-

[1] Roche, *Mémoires de l'Acad. des Sciences*, Montpellier, t. viii. p. 244.

tricity is not the less a potent and universal reality that we are ignorant of its mode of working.

The nebulæ are the only class of heavenly bodies from which no signs of movement have, up to this, been derived. To all appearance, they are absolutely stationary. Accurate observations of them are, indeed, comparatively recent; they go back, nevertheless, far enough to justify the statement that not one among about four hundred well-determined nebulæ becomes progressively displaced by so much as one second of arc yearly. The nebulæ in the Pleiades ought perhaps to rank as exceptions to this general immobility, since we need no direct proof to assure us that they drift with the cluster of which they form an integral part. The drift, to be sure, is slow; but it is securely ascertained, and affords grounds for the sole estimate, that is not mere guess-work, of the distance of a nebulous system.[1] The apparent indifference of all other nebulæ to the perspective effects of the sun's swift advance through space, leaves little probability that any of them lie as near to us as the nearer stars; and there are indications besides that whatever individual movements they possess are of an extremely sluggish nature. For none yet experimented upon display, to a measurable extent, the spectroscopic line-displacements indicating motion in line of sight; and since these have no relation to remoteness or vicinity, the inference is obviously suggested that nebulæ are really slower moving bodies than stars. Decisive and sufficient information on this point cannot, however, be looked for until photography has been brought to bear upon it.

There is the same total want of evidence of orbital, as of translatory movements in nebulæ. The systems they in numerous cases presumably constitute remain rigidly fixed. The contrary has often been asserted; yet revolutions, alleged on the strength of inexact observations, and brought to a standstill by precise ones, must plainly be dismissed as illusory; 'unless,' as Dr. Dreyer says,[2] 'we are to believe that nebulæ in the good old days moved about as they liked, but have been on their good behaviour since 1861 and kept quiet.'

[1] See *ante*, p. 226.

[2] *Monthly Notices*, vol. xlvii. p. 418.

The existence, so frequently observed, of a nebulous connection between grouped nebulous objects intimates a state of things hardly reconcilable with mutual circulation. The relations of these, as yet, imperfectly separated individuals are perhaps in a state of transition, like those of multiple cometary masses, at times enclosed together, like double nebulæ, in a dimly luminous sheath.

The idea of the accompaniment of planetary nebulæ by *satellites* was suggested to Sir John Herschel by the frequent and close proximity to such objects of minute stars,[1] and he recommended their careful micrometrical measurement as a criterion of possible future changes. But none have yet been detected. D'Arrest found the attendant stars just in the positions Herschel had assigned to them,[2] and they have not visibly shifted since. The probability, indeed, of the stars genuine association with the nebulæ they seem to wait upon has gained in strength with the general increase of knowledge regarding stellar and nebular relations; nevertheless, circulatory movements, if in progress, are likely (considering the enormous remoteness and consequent spaciousness of these systems) to be so exceedingly slow that centuries are but as days in the reckoning of their periods.

Systemic alterations in nebulæ may then long evade recognition, and their geometrical *status quo* long be apparently preserved. Possibly, however, an exception to this rule may be found in the second companion to the great Andromeda nebula, should the pivoting movement suspected to have been executed by it between 1876 and 1888 be verified by continuance. Otherwise, the assertion of recent photographs regarding the direction of the longer axis of the attendant nebula must be considered to hold good for every anterior epoch. Changes arrested by the exclusion of the possibility of mistakes may pretty safely be set down as having originated in them.

Variability in light is a quality of nebulæ as surely as of stars, although one, in particular cases, by no means easy to establish. Nebulæ are peculiarly sensitive to atmospheric

[1] *Phil. Trans.* vol. cxxiii. p. 500. [2] *Leipzig Abhandlungen*, Bd. iii. p. 308.

influences. Their finer details, always hovering on the verge of visibility, are completely shrouded by the lightest mist. Hence, even to the same eye, and with the same instrument, the aspect of the same nebula often varies greatly from night to night; and since personality is nowhere stronger than in the perception of the delicate luminous gradations delineating to our sight the forms of nebulæ, a difference of observers adds a further incalculable element of uncertainty. Rumours of change then easily arise, but are with difficulty substantiated.

The presumption they start from is, however, fully warranted by facts. The occurrence of luminous fluctuations in nebulæ has been proved by the total disappearance of three, all, strange to say, situated in the same constellation. On October 11, 1852, Mr. Hind discovered, near the group of the Hyades in Taurus, a small round nebula (N. G. C., 1555) with slight central condensation. It was then very faint, but brightened steadily until 1856, when d'Arrest ranked it as belonging to the first, although verging towards the second class of brilliancy.[1] His amazement then was extreme to find on October 3, 1861, its place apparently vacant! Some glimmer of its light was indeed made out for a year or two longer with the Pulkowa fifteen-inch refractor; but that too faded, and the object has now, for above a quarter of a century, been invisible in the most powerful telescopes. The apparition, so far as can be judged, was a strictly temporary one. Hind's notice, there is reason to believe, did not lag far behind its first perceptibility with moderate instruments.

A curious feature of the occurrence was the sympathetic, or at any rate simultaneous, decay in light of a small star—since known as 'T Tauri'—placed almost in contact with the nebula. The star, however, recovered in 1868 about the same time that a second new nebula (N.G.C., 1554) came into view. First discerned by O. Struve, it was observed by d'Arrest, who was fully convinced of its novelty; and his opinion was borne out by its subsequent total disappearance.[2]

[1] *Astr. Nach.* Nos. 1366, 1689; Auwers, *ibid.* No. 1391.

[2] Dreyer, *Memoirs R. Astr. Soc.* vol. xlix. p. 214.

The third 'temporary' nebula within astronomical acquaintance was seen by Chacornac at Paris, attached to an eleventh magnitude star near ζ Tauri, October 19, 1855 (N. G. C., 1988). A conspicuous gain in lustre allowed it to be perceived, after three months, as covering a nearly rectangular area, $3\frac{1}{2}'$ by $2\frac{1}{2}'$, with thin parallel striæ, like those of a cirro-stratus cloud.[1] On November 20, 1862, it was nevertheless utterly gone, the star remaining unaffected alike by its visibility and by its disappearance.

The light-changes of nebulæ do not offer the same diversity as those of stars. Only two kinds of variability—those producing respectively ephemeral appearances and capricious brightenings and fadings—are represented among them. No periodical nebulæ have yet been shown to exist. The influences, of whatever nature, bringing about the rhythmical pulsations of stellar light would seem to be absent from the nebular kingdom. A distinction, however, peculiar to themselves, can be established among variable nebulæ. Their fluctuations may be either general or partial. They may affect the whole of a moderately compact object, or certain sections of an extensive formation. Examples of both kinds, and of all degrees of authenticity, abound; but we will only mention a very few, in which the reality of change seems scarcely disputable.

One such is afforded by an elliptical nebula in Leo (N. G. C., 3666), 'very bright' when discovered by the elder Herschel in 1784, but noted by his son as abnormally faint for the first class. Subsequently observed alternations have made it all but certain that the discrepancy indicated genuine change.[2] A nebula in Cetus, too (N. G. C., 955), is evidently subject to similar vicissitudes. Schönfeld in 1861, and Vogel in 1865, failed to see it, although it was, at sundry other epochs, easily visible to the former observer, as well as to d'Arrest and Winnecke, and fully justified in 1887, in Dr. Dreyer's opinion,

[1] *Comptes Rendus*, t. lvi. p. 637.

[2] Winnecke, *Astr. Nach.* No. 2293; Dreyer, *Memoirs R. Astr. Soc.* vol. xlix. p. 218.

Herschel's classification of it as of the second order of brightness.[1]

Again, one of a group of nebulæ in Virgo, observed by Schmidt at Athens in 1862 (N.G.C., 5655), could not be found by d'Arrest in 1865, two minute stars appearing as its *locum tenentes*.[2] If, as seems probable, the identical object was recorded by Herschel, from an observation of December 28, 1785, as No. 498 of his second class, its re-emergence to view may at any time be looked for. The collection of objects with which it is associated were judged by d'Arrest (no doubt rightly) to be the brightest 'knots' of a wide-spreading nebulous structure (M 49). The variability of one of them approximates then to the local changes of irregular nebulæ, exemplified with most certainty in their chief, the great Orion nebula.

The occurrence of such in this well-watched object has been placed beyond reasonable doubt by the researches of Professors O. Struve[3] and Holden.[4] Even while the observations of the latter were in progress, fluctuations in the lustre of some of the component masses were distinctly recognisable, and one new nebulous patch was actually seen to develop out of a faint stellar point. No measurable change of place, however, could be detected by a most diligent comparison of the records of over a century, in either the misty floccules or the contained stars. Some of these, of the minuter sort, are certainly variable to the extent of one or two magnitudes. And their changes, although apparently lawless, are at times very rapid, as was evidenced by the striking inequalities of the coupled images of ten or twelve of them on a plate doubly exposed by Mr. Roberts, at an interval of four days, early in 1889.

The partial light fluctuations, strongly suspected in both the Omega and the Trifid nebulæ, have resulted chiefly in a modification of what we may call *coast-lines*, here advancing, there encroached upon by the sea of darkness which surrounds

[1] Winnecke, *Monthly Notices*, vol. xxxviii. p. 104; Dreyer, *loc. cit.* p. 213.

[2] D'Arrest, *Astr. Nach.* No. 1520.

[3] *V. J. S. Astr. Ges.* Jahrg. xix. p. 35; *Mélanges Math.* t. ii. p. 530; d'Arrest, *Astr. Nach.* No. 1366.

[4] *Washington Observations*, 1878, App. i. pp. 121, 225.

them. In the latter object, a singular apparent alteration in the relative places of the multiple star and the nebulous masses involving it, is perhaps due to this instability of outlines. Sir John Herschel in 1827 and 1833 described the star as located 'exactly in the central vacuity' of the nebula, and just at the point of convergence of the three rifts dividing it throughout.[1] But a drawing made by him at the Cape in August 1835 exhibits the star no longer as central, but as adhering to the eastern mass of nebulosity; and virtually the same state of things was noted by Mason and Smith in 1839,[2] and subsists so obviously to the present time as to render a mistake about it inconceivable. The implied change, however, must have taken place abruptly between 1833 and 1835, and then ceased; so that proper motion, either of the star or nebula, had certainly nothing to do with it.[3] There seems then no alternative but to admit that the frontier-lines between luminosity and obscurity were, at the epoch in question, very considerably 'rectified.'

The relative variability of the parts of nebulæ will henceforth, in all probability, be investigated by photographic means alone. Data of perfect definiteness on the subject can now easily be obtained by the use of a light-scale previously impressed upon the plates to be exposed; and their accumulation during some years will do much to abolish the doubts that still remain as to the extent and reality of the shiftings of the luminous level in these still enigmatical objects.

One point about them appears, however, tolerably clear. It is that the glow of those shining with discontinuous light is of an electrical nature.[4] For the gases of which they are mainly composed cannot be rendered incandescent by the transference of heat from diffused solid particles or bodies not themselves highly incandescent; and the absence from

[1] *Memoirs R. Astr. Soc.* vol. iii. p. 63; *Phil. Trans.* vol. cxxiii. p. 460; Holden, *Amer. Jour. of Science*, vol. xiv. p. 434 (1877).

[2] *Trans. Amer. Phil. Soc.* vol. vii. p. 175.

[3] Dreyer, *Monthly Notices*, vol. xlvii. p. 419.

[4] In his Bakerian Lecture, Mr. Lockyer stated this to be 'proved' as regards the glow of hydrogen in the Orion nebula (*Proc. R. Soc.* vol. xliv. p. 12), and his argument is valid for all gaseous nebulæ.

nebulæ of glowing solids is shown by the nature of the nebular spectrum, which probably includes no element of truly continuous light, apart from that of the 'stars,' apparently associated in some way with all nebulæ. The unmistakable analogy, again, between solar-coronal and cometary forms on the one side, and nebular forms on the other, indicates for all a kindred origin in the play of opposing forces, generated by certain foci of condensation, one of which is our sun, while the others may be vaguely but safely designated as 'nuclei.' Where there is only one such nucleus, the enveloping gases assume a simple globular or oval shape; where there are many, the result is exceedingly complex. Irregular nebulæ are thus, in our view, potential star-clusters; they consist of a stellar framework, draped with nebulous folds, spirals, and festoons, disposed along lines of force laid down by the rival or concurrent energies of the compact masses which it is permissible to regard as inchoate suns.

CHAPTER XX.

THE DISTANCES OF THE STARS.

THE most arduous among the problems of stellar astronomy was, singularly enough, the first to be attacked. It was attacked, indeed, before the possibility was even remotely discerned that stellar astronomy might come to be regarded as a substantive branch of science. In the hope, not of penetrating the inscrutable secrets of the remote sphere of the fixed stars, but of solving doubts about the motion of the earth, Copernicus, Tycho, and Galileo led the way in the long series of experiments on the apparent displacements of the stars resulting from our own annual travels round the sun. The interest of the question whether such displacements existed or not was for them of a wholly 'parochial' kind; it lay in the test they afforded as to the reality of the terrestrial revolutions. Should the stars be found to shift ever so little by the effect of perspective, then the heliocentric theory could no longer be gainsaid; if, on the contrary, they ignored sublunary circlings, the 'pill' (as Kepler termed it) to be swallowed by Copernicans was indeed a huge one. For the distances to which the fixed stars had, in that case, to be relegated, seemed in those times monstrous and incredible; and monstrous and incredible they would appear still, were we not forced by irrecusable evidence to believe in them.

From the beginning to the end (so far) of the history of these inquiries, it may be taken almost as an axiom that the largest ostensible parallaxes have been obtained by the worst means. With each successive improvement in methods and instruments, as the limits of possible error shrank, the displacements apparently measured dwindled, and the stars became less accessible to attempted determinations. During some three centuries, the ill-success in this matter of an astronomer was a measure of his skill and judgment. Results obtained with suspicious facility by inexpert observers utterly evaded the guarded scrutiny of such men as Tycho Brahe, Bradley, and Pond. Flamsteed, indeed, just at the close of the seventeenth century, detected in the pole-star annual variations which were certainly *not* illusory. Yet here too there was a *caveat*. Theory and fact did not correspond.

Let us consider for a moment what must be the visual effects upon very distant objects of the comprehensive and unceasing rounds of the planet upon which we are borne as spectators. Unmistakably, to begin with, we see them in different directions at different times of year. In January and July, in March and September, and so on, we are at opposite ends of base-lines 186 millions of miles in length. The stars then must be continually thrown, now a little to one side, now to the other, of the true, or 'mean' places which they would severally occupy if viewed from the immobile sun. In other words, each describes round its mean place in a period of a year a small apparent orbit, which is nothing else than the orbit of the earth projected in miniature on the sky. For stars situated in the ecliptic—that is, in the plane of the earth's travelling—this orbit contracts into a right line, along which the star merely swings to and fro; for stars near the pole of the ecliptic, the perspective orbit is virtually a circle; while intermediate latitudes afford all degrees of foreshortening. Every star—unless those few lying close to the pole of the ecliptic—has thus its epochs of maximum parallax, six months apart, when it seems to stand alternately at opposite extremities of the major axis of the parallactic ellipse, and it is then that measures of its apparent displacements can be most advantageously made. These opportune seasons occur when the earth's longitude falls short of, or exceeds by ninety degrees the longitude of the star. They are accordingly different for stars with different longitudes.

The precise *form* of displacement due to the earth's revolution round the sun is thus strictly calculable for each individual star; the *amount* alone cannot be predicted, but must be obtained by observation; and from this amount the distance of the star is deduced. For each parallactic orbit is a perfect model, both in shape and size, of the earth's orbit as it would be seen from the star, abridgment of compass (down to contraction into a virtual point) corresponding to a more and more profound immersion of the point of survey in the abysses of space.

The parallax of a star is then the difference between its positions as seen from either side, and from the centre of the earth's orbit. It is, in short, the angle subtended, at the distance of that particular star, by the mean interval between earth and sun. Now we can tell in a moment how far off a spectator must be to see a line ninety-three millions of miles in length diminished to the angular dimension of, let us say, one second. He must be 206,265 × 93 millions of miles distant. But no star has yet been found so *near* to us as this. That is to say, the shift of no known star amounts to as much as the width of a sixpence held up at Charing Cross to a spectator at Stanhope Gate or at Millbank.

We are now in a position to understand why it was that Flamsteed's observations of the apparent displacements of Polaris could not, when critically examined, be set down to the account of parallax. The star seemed indeed to describe, regularly each year, a little ellipse of exactly the right *shape*; and as to its *size*, there was no *à priori* reason why the pole-star should not have a parallax of upwards of twenty seconds. But there was one irreconcilable discrepancy. The displacements noted occurred at wrong times. Had they been of a parallactic nature, the position of the star in its minute fictitious orbit should have been invariably ninety degrees in advance of what it actually was. They were not then due to parallax; but obtained their proper explanation from Bradley's discovery of the aberration of light in 1729.

During the ensuing century and a quarter, the only valid results obtained in this direction consisted in demonstrations, renewed and enforced from time to time as more conclusive evidence presented itself, that with the instrumental means then available, stellar parallax was an inappreciable quantity. Bradley showed that it must fall short of half a second, and although his reasoning applied strictly to only a limited number of stars, it was rendered at once more general and more cogent by the investigations of Pond and Airy, of Struve and Bessel. It thus seemed that astronomers should content themselves with the knowledge that the stars were exorbitantly remote—so remote that light spent at least four or five years in travelling to us from the brightest of them, and might, for anything that appeared, spend indefinitely longer. The labours and refinements of two centuries had issued in fixing a *lower* limit for distance, an *upper* limit for parallax; in isolating the sun from his compeers by setting between him and them an unmeasured abyss of desert space; in widening to a startling extent the boundaries of the visible universe. Kepler's 'mighty bolus had to be swallowed in its entirety.

At last, in 1827, Savary of Paris brought forward a method (already referred to [1]) for fixing an *upper* limit to the distances, a *lower* to the parallaxes of binary stars moving in known orbits. The further off from us such orbits are, the greater of course their real size, and the longer the time taken by light to cross them. Hence, the deviations from their true places of the moving stars caused by the delay in our vision of one of them, serve in theory, since they increase with remoteness, to determine the distance from the earth of the pair. Or if no such deviations are apparent, it should at least be possible to fix an amount which they could not exceed without becoming so. Savary, accordingly, professed to demonstrate in this way that ξ Ursæ Majoris, the couple most favourably situated for the purposes of the inquiry, must have a parallax exceeding $\frac{1}{300}$ of a second [2]—must, that is, be at a less distance than would be traversed by light in 109 years. But the information, however credible in itself, was not fully authenticated. Villarceau showed in 1878 [3] that the method was inapplicable except to stars of known relative masses; and these as yet scarcely exist. The most that can be hoped for from Savary's light-inequality is that eventually, in some rare cases, it may serve to determine indirectly parallaxes too small for direct measurement. Among the few pairs besides ξ Ursæ in

[1] See *ante*, p. 202.

[2] *Conn. des Temps*, 1830, p. 169.

[3] *Ibid.* 1878, p. 68.

which it might be perceptible are ζ Herculis and η Coronæ Borealis.[1]

By-ways, however, are of secondary importance since the straight road to the end in view has been made practicable. The engineer who carried it over the barrier long helplessly confronted was Fraunhofer. The improvements effected by him in the power and perfection of telescopes, as well as in the application to them of micrometrical apparatus, alone rendered possible the exquisitely refined measurements upon which the detection of stellar parallax absolutely depends.[1]

These measurements are for the purpose of determining variations in the angle between the star chosen as the subject of experiment and one or more 'comparison-stars' in its neighbourhood, treated as fixed points of reference because assumed to be indefinitely remote. From progressive annual changes of the intervals separating them from the star they serve to test, the amount of its parallax, hence of its distance from the earth, is learned. But if the comparison-stars are themselves affected by sensible parallactic displacements, then the result is, to a certain extent, if not wholly, vitiated. Should they chance, for instance, to be at the same distance from us as the compared object, then all the stars under observation will shift together, giving the effect of immobility, and implying the absence of measurable parallax, when in reality a large one may be present. Again, a parallax sought by the aid of a comparison-star itself possessing one-half the parallax of the star investigated, will come out only one-half of its true value. It may even happen that the comparison-star is, of the two, our nearer neighbour in space, when a 'negative parallax' (as it is called) will emerge, showing how great is the discrepancy in the wrong direction.

It is, indeed, the weakness of the 'differential method of parallaxes' that it gives relative, never absolute results. Not only does some degree of doubt always attach to them, but their deviations from the truth are always on the same side. They tend in all cases to diminish the parallax, and exagge-

[1] Birkenmajer, *Sitzungsberichte*, Wien, Bd. xciii. ii. p. 738.

[2] C. A. F. Peters, *Zeitschrift für populäre Mittheilungen*, Bd. iii. p. 96.

rate the concluded distance. Nevertheless, the advantages of the method are so overwhelming, it abolishes so many causes of error, and strikes at the root of so many illusions, that it has gained universal and exclusive preference. No one any longer thinks of attacking this delicate problem by the comparatively clumsy mode of determining absolute right ascensions and declinations at intervals of six months, and trying to distil, from a confused mass of pervading infinitesimal errors, the all but insensible evidence of those periodical changes which reflect across the gulf of sidereal space the movement of the earth in its orbit.

Nevertheless, the first genuinely measured stellar parallax was so far a casual result that it was arrived at in the ordinary course of observation. It presented itself, as it were, unsolicited. Alpha Centauri combines Struve's three criteria of vicinity. It is exceedingly bright; it has a large proper motion; and its components revolve swiftly in a wide orbit. No other star in the sky seemed beforehand so likely to come within the grasp of terrestrial determinations; and none has yet proved to be so little remote. Henderson's observations of it, made at the Cape in 1832-3, were discussed with a view to sifting out from them a parallactic element, after he had learned that the star's rapid onward movement afforded a presumption of relative nearness to the earth. Nor were his expectations belied. A parallax of about one second manifestly implied by annual changes of declination, was partially confirmed by observations of the same object in right ascension made by Henderson's assistant, Lieutenant Meadows.

His announcement to this effect, January 3, 1839,[1] was however, received with doubts, justifiable perhaps, considering the numerous precedents for illusion on the point, but not justified by the event. Bessel's similar (and slightly prior) communication regarding 61 Cygni inspired, on the other hand, general confidence. The Königsberg series, indeed, though by no means fortified with all the precautions now deemed necessary, seemed beautifully complete. The observations were of the differential kind, and favourable opportunities for

[1] *Memoirs R. Astr. Soc.* vols. xi. p. 61, xii. p. 329.

measurement were afforded at intervals of three months by the use of two comparison-stars so situated that the lines of maximum displacement with respect to them of 61 Cygni lay at right angles to each other. The close accordance in their results of operations thus wholly independent, and executed under widely different influences of season and temperature, and the harmonious flow of the curves into which the observations were projected by Mr. Main,[1] prompted the conviction that here at last was a stellar parallax the genuineness of which was beyond cavil. The extreme importance of its detection, pronounced by Sir John Herschel the greatest triumph ever achieved by practical astronomy, can be estimated from Bessel's declaration that, until it was actually compassed, he was unable to form an opinion as to whether the parallaxes of the nearest stars should be reckoned by tenths or by thousandths of a second ! [2]

The distance from the earth of 61 Cygni has been more frequently investigated than that of any other star, and, some trifling discrepancies notwithstanding, may be considered as satisfactorily ascertained. Bessel's parallax of about a third of a second turned out to be slightly too small, but was augmented to 0″·42 by Auwers's rediscussion in 1868 of the same data.[3] The two most reliable determinations are, perhaps, Sir Robert Ball's in 1879, giving a value of 0″·465, and Professor Pritchard's, giving 0″·432 from photographic comparisons in 1887. On the whole, the parallax of this star can hardly differ much from 0″·45.

Although the third largest known, it yet implies an actual distance so inconceivably vast that light spends seven years and three months in flying over the forty millions of millions of miles serving to measure it. Thus we see the coupled stars, not where they *are*, but where they *were* seven and a quarter years ago; that is (since their proper motion is about 5″·14 yearly) just thirty-eight seconds of arc behind their true places. The effulgent points terrestrially determined are then mere simulacra of the real stars; they pursue, without ever overtaking them; they would continue to shine and to

[1] *Ibid.* vol. xii. p. 42.
[2] *Astr. Nach.* No. 385.
[3] *Abhandlungen Kön. Akad.* Berlin, 1868, p. 114.

travel for seven years and upwards after their originals had been blotted out of the visible creation. Our views of all moving objects are of course to some extent affected by this curious kind of light-aberration; but in the sidereal heavens it attains proportions that are not only large, but, for the most part, incalculably large. Our survey of the background of the sky may lag centuries, even millenniums, behind our simultaneous survey of its foreground; and the disturbed synchronous relations between the varied luminous contents of the sphere are, so far as our perception goes, unsusceptible of reconstruction.

Transported to the place of 61 Cygni, our sun would appear more than eight times as bright. It would represent, not a star below the fifth, but one above the third magnitude, such as δ Cassiopeiæ, one of the lucid five constituting the 'W' by which that constellation strikes the eye. The Swan binary, indeed, ranks at present (for it may have seen better days) among the least luminous of the stellar host.

Fraunhofer's construction of the instrument with which Bessel observed it marked the turning-point from failure to success in parallactic inquiries. The heliometer, as a star-measuring machine, not through superimposed contrivance, but by original design,[1] is specially adapted to facilitate them, and may almost be said to have, in this department, superseded the ordinary equatoreal and micrometer. It has two chief points of superiority. In the first place, much wider pairs of stars can be grasped with it, its compass being, by the mobility of the semi-lenses, extended far beyond the limits of a single field of view. The range of selection for comparison-stars is thus greatly enlarged, and the chances of a systemic connection with the central star fatal to the purpose of the designed operation are reduced to a minimum. In the next place, the stars under observation can be visually equalised by placing a wire-gauze screen of any desirable opacity over the segment of the object-glass forming the image of the brighter one, whereby tendencies to personal error, difficult to be otherwise got rid of, are perfectly eliminated. These

[1] See *ante*, p. 15.

are not the only features of the heliometer tending to promote critical precision, but they are the most important.

Bessel's success with 61 Cygni gave the impulse to numerous undertakings of the same kind. Their result depended mainly on the skill or luck of the observers in picking out from innumerable, indefinitely remote stars the few near enough to be sensibly displaced by being viewed from opposite sides of the earth's orbit. Two circumstances mainly determined their choice.

That distance is a factor of stellar brightness is so obvious a truth that it may almost be reckoned a truism. Admitting the widest possible range of variety in actual light-power, it is still certain that the most lustrous objects are likely to be found in closest proximity to the earth. That is, on a wide average. Individual exceptions abound; but the more numerous the stars considered, the more approximately does the theoretical inverse ratio between distance and the square root of total light correspond to actual fact.[1]

Several conspicuous stars were, on this ground, fixed upon for investigation by C. A. F. Peters in 1842-3. His instrument, the Pulkowa vertical circle, was perfect of its kind, and was used with consummate skill. The results, moreover, so far as they went, were *absolute*, being irrespective of comparisons with other stars; but they did not go far. The deduced parallaxes were so small, and their 'probable errors' so relatively large, that it was difficult to place much confidence in them. Vega came out with a parallax of one-tenth of a second, Arcturus with 0″·13, each uncertain to the extent of more than half their respective values; the possible displacements of Capella were completely masked by observational discrepancies; *a* Cygni remained rigid, the measures only serving to show four hundred chances to one against its possessing a parallax exceeding 0″·1.[2] Only the result for Polaris was, owing to the exceptionally advantageous position

[1] Struve, *Mens. Microm.* p. clxii.

[2] *Zeitschrift für pop. Mitth.* Bd. iii. p. 105; *Mémoires*, St. Pétersbourg, t. vii. p. 140, 1853; *Astr. Nach.* No. 1147.

of the star, thoroughly trustworthy, and has since been ratified.

A second criterion of nearness was found in the appearance of rapid motion. This varies in the same proportion as distance, but in the reverse sense. At twice the distance, an identical velocity is only half as effective in producing angular displacement; at three times the distance, its *seeming* amount is reduced to one-third, and so on. Thus, apparent swiftness, no less than apparent lustre, depends in part upon vicinity, and the largest proper motions must belong, on the whole, to the nearest stars.

And, on the whole, parallax-hunters, taking rapidity of advance for their guide, have prospered the best. A seventh-magnitude star in Ursa Major, flitting annually over $4\frac{3}{4}$ seconds of angular space, was found by Winnecke in 1858[1] to have a parallax ($=0''\cdot5$) inferior only, among those as yet determined, to that of α Centauri. This insignificant object, numbered 21,185 in Lalande's great catalogue, is separated from the earth by a light-journey of $6\frac{1}{2}$ years, and to that extent our observations of it are retarded. So that it is in reality always $31''$ in advance of the place we are compelled to assign to it. For a body claiming the rank of a sun, it is either very small or very obscure. In its position, our ruling orb would show as fifty-seven times more luminous.

An 8·5 magnitude star in the same constellation (Lalande, 21,258), also distinguished for apparent velocity, disclosed to Auwers's measurements with the Königsberg heliometer in 1860–2 a parallax of $0''\cdot26$,[2] corresponding to a light-journey of $12\frac{1}{2}$ years, and a permanent displacement on the sphere, due to its proper motion in that interval, of $55''$. The real brilliancy of the star is only $\frac{1}{63}$ that of the sun. Krüger's simultaneous and identical determination of its distance with the Bonn heliometer showed, as he remarked, the peculiar aptitude for such researches of that description of instrument.[3] A still smaller star in Draco (Oeltzen, 17,415) gave an even more emphatic warrant to confidence in swiftness, rather than

[1] *Astr. Nach.* No. 1147. [2] *Monthly Notices*, vol. xxiii. p. 74.
[3] *Acta Societat. Scient. Fennicæ*, t. vii. p. 373.

in conspicuousness, as a certificate of proximity. Krüger, induced by its yearly movement of 1″·27 to subject it to experiment, obtained a parallax of one quarter of a second;[1] while the fine binary system, 70 Ophiuchi, with an annual motion of 1″, proved to be removed from the earth by twenty years of light-travel (parallax 0″·16).[2] And all these results seemed, from the smallness of their 'probable errors,' to be exceedingly trustworthy.

The probable error of any result, however, represents only what we may call the *uncaused* inaccuracies of the observations upon which it is founded. It sums up, according to the doctrine of probabilities, the effect of their deviations, in either direction, from the mean. But it takes no heed of 'systematic' errors due to causes working steadily in one sense, but, so to speak, underground. These are the real sources of mischief from which fallacious parallaxes have abundantly sprung in times past, and which cannot, in the present and future, be too carefully guarded against. Especially formidable are certain slight idiosyncrasies of perception, by which measures of distance become modified with the varying positions of the line of direction between the objects measured, relative either to the vertical or to the line joining the observer's eyes. And since this subtle spring of error rises and falls harmoniously in a period of a year (because dependent upon the uranographical situation of the stars under scrutiny), it would be capable, if not adverted to, not only of completely vitiating observations apparently accordant, but even of simulating parallactic changes that had no real existence. Instrumental errors, too, connected with changes of temperature, or the deforming power of gravity (as conditioned by the shifted positions of the telescope at different seasons), take the same cyclical course; and there can be no doubt that to some such lurking deceptive influence the parallax of 0″·07 attributed to *a* Herculis by Captain Jacob in 1858,[3] owed its origin. Since

[1] *Acta Societat. Scient. Fennicæ*, t. vii. p. 383.

[2] Krüger, *Astr. Nach.* Nos. 1210–12.

[3] *Madras Observations*, 1848-52, Appendix; *Memoirs R. Astr. Soc.* vol. xxvii. p. 44.

the star chosen for comparison was no other than the well-known physical attendant of the object examined, the fact of illusion is patent.

The exigencies of this kind of work were first recognised, and its principles fully explained, in an elaborate paper published by Dr. Döllen of St. Petersburg in 1855;[1] and his principles were ably carried into effect by Dr. Brünnow in a series of investigations of stellar parallax at Dublin between 1868 and 1874. The example thus set of the thorough elimination of errors, at once personal and periodical, has since been generally followed. More effectually than by most other men, the famous 'Know thyself' of the old Greek philosophers has been taken to heart by astronomers. Their anxious and elaborate inquiries regard not merely microscopic inequalities of scale-divisions and screw-values, changes in refraction, corrections for aberration and proper motion, but the cunning tricks of their own nerves, the caprices of cerebration, all the varying conditions of perception in the organism at their individual command.

None of these precautions were neglected in the important work executed by Drs. Gill and Elkin at the Cape in 1881–3.[2] Fully alive to its subtle requirements, they gave to their determinations a precision which entitles them to rank among the most valuable of astronomical data. Dr. Gill's discussion, especially, of the parallax of *α* Centauri is a model of what such an inquiry should be. It leaves, one may say, no stone unturned beneath which a source of illusion might lie concealed.

Of the nine southern stars undertaken, three were investigated concurrently by both observers. Among these was *α* Centauri, and the parallax of $0''.75$, resulting from independent comparisons with no less than four pairs of adjacent stars, is probably more nearly accurate than any value of the sort yet registered. The fact is then assured that light, which flies from the sun hither in eight minutes, spends four years and four months on the journey from the nearest fixed star. The

[1] *Bulletin de l'Acad.* St. Pétersbourg, t. xiii. Suppl.

[2] *Memoirs R. Astr. Soc.* vol. xlviii.

corresponding distance is, in round numbers, twenty-five billion miles.

On Sirius, too, a double attack was made, of which the upshot was a parallax of 0″·38, or a distance of 8·6 light-years. These were the first measures of the bright dog-star made under perfectly suitable conditions. They are at present being repeated by Dr. Gill with different comparison-stars.

The third parallax simultaneously determined was that of ε Indi, a fifth-magnitude star with a proper motion of nearly five seconds a year. It proved to be 0″·22, representing a light journey of 14½ years, and showing that the total radiative power of the star is about half that of the sun, while its rush through space, taking only the directly measurable, *thwart-wise* part of its motion, is at the rate of sixty-three miles per second, or more than three times as swift as the earth's movement round the sun.

Of the six remaining stars, four, chosen for their exceptional mobility, did not fail in some degree to justify the presumption it afforded of relatively near neighbourhood. Thus, the chief star of the remarkable triple system, o_2 Eridani, yielded to Dr. Gill's investigation a parallax of 0″·166, implying a light-journey of 19·6 years, and a 'proper motion displacement' for the trio of 80″. The swiftest moving star, however, in the southern hemisphere is one of 7·5 magnitude in Piscis Australis, known as 'Lacaille 9352.' Its angular rate of progress only just falls short of 7″ a year; and the parallax of 0″·28 (distance 11½ light-years) obtained for it by Dr. Gill proves its linear velocity to be at least 73 miles per second. A still quicker traveller is ζ Toucani, of fourth magnitude and circumpolar at the Cape, which, with a proper motion of two seconds, is discovered to be advancing at the express rate of 101 miles a second. That is, if Dr. Elkin's small parallax of 0″·06, corresponding to a light-journey of fifty-four years, be correct. In proportion to its amount, it must be admitted that the margin of error qualifying it is somewhat unduly large. The last of the rapid stars measured was *e* Eridani, for which Dr. Elkin's operations brought out a parallax of 0″·14, representing a light-journey of twenty-two years and nine months. This star is of 4½ magnitude, has a proper motion of more than 3″, and is about thrice as luminous as our sun.

The two remaining stars on the Cape list, though both splendid objects, are virtually fixed in the sky. They have no perceptible onward motion; and neither of them, accordingly, showed any sign of possessing a sensible parallax. This is really, when we come to consider it, an astonishing result.

Second only to Sirius in the southern hemisphere, Canopus far outshines every star north of the celestial equator. As the chief luminary of the great constellation Argo, it seems to command, while standing slightly aloof from, the dazzling array of all stellar ranks spanning the heavens from the Greater Dog to the Cross. But since its bluish-white rays are almost undimmed by absorption, the probability is strong that their intensity largely exceeds the solar proportion of luminosity to mass.

Now Dr. Elkin's failure to detect any parallactic shifting in Canopus obliges us to suppose it at such a distance that its light needs *at least* sixty-five years to reach us; how much longer, it is impossible to tell. At this minimum remoteness, our sun would shrink to a 7·5 magnitude star; it would be one of the dense shoal of telescopic objects imperceptible to unaided sense, and individualised only by the industry of astronomers. But 2,500 stars of 7·5 magnitude give only the light of one Canopus, whence it follows that Canopus is certainly brighter, and may be very greatly brighter, than 2,500 suns like ours. There is only one way of escaping from this startling conclusion. It is possible, though far from probable, that both Dr. Elkin's reference-stars are physically connected with the brilliant object they lie near; in which case, his result was of course null and void. This doubt, however, will shortly be set at rest through Dr. Gill's remeasurement of Canopus with a fresh and much wider pair of stars.

The ninth star tested at the Cape in 1881–3 was β Centauri, one of the so-called southern Pointers, and taking rank among the lower grades of the first magnitude. Dr. Gill's *negative* result made it all but certain that the considerable *absolute* parallaxes previously derived from meridian observations of the star were illusory; but here again, a new series of measures by the same eminent investigator must prove decisive.

The time had now come when a change in the system upon which inquiries of this kind were prosecuted seemed feasible. Hitherto, observers had been content to select the most promising subjects for their experiments, without any regard to the co-ordination of results. The outcome was a collection of detached statements as to stellar distances, interesting, each by itself, in a high degree, yet incapable of being combined for the purpose of any general conclusions. So long ago as 1853, Dr. Peters had pointed out that what was needed for obtaining a fundamental acquaintance with the structure of the sidereal world was not so much the determination of exceptional parallaxes, as the steady compilation of data for some well-grounded inference relative to the distances of defined *star-classes*.[1] But it was not until thirty years later that it became possible to act on the suggestion.

Encouraged by the success of the work just accomplished, Dr. Gill proposed, January 11, 1884, a scheme for dealing with the problem of star-distances in its widest bearings. Two 'great cosmical questions,' he saw, pressed for answers, which might be obtained by the judicious distribution of some years' continuous labour. They are:

1. 'What are the average parallaxes of stars of the first, second, third, and fourth magnitudes respectively, compared with those of lesser magnitude?'

2. 'What connection subsists between the parallax of a star and the amount and direction of its proper motion, or can it be proved that there is no such connection or relation?'[2]

Some advance has already been made towards procuring at least partial replies. A plan was concerted by which Dr. Elkin, now in charge of the new Yale College heliometer, undertook the measurement of a considerable number of

[1] *Mémoires de St. Pétersbourg*, t. vii. p. 149.
[2] *Memoirs R. Astr. Soc.* vol. xlviii. p. 191.

representative northern stars, while Dr. Gill dealt with a corresponding southern list at the Cape, where a seven-inch heliometer, constructed for the purpose by Messrs. Repsold with the utmost perfection of instrumental art, was erected in the summer of 1887. In the previous determinations, the Dunecht heliometer, become, by purchase from Lord Crawford, the private property of Dr. Gill, had been employed; but its aperture of only four inches, restricting the choice of comparison-stars to brighter objects than were always to be found in the most advantageous situations, made it unsuitable for carrying out the comprehensive plan later decided upon.

It included an investigation of the parallaxes of all first-magnitude stars—ten in the northern, ten in the southern hemisphere. This done, the average distance of stars of the highest light-rank becomes known, no longer by inference or guessing, but by direct measurement. A scale-unit for the stellar universe will then, at last, be available. For, once we know the distance in billions of miles or light years corresponding to the first magnitude—the distance, that is, at which a 'mean star' would shine with about the lustre of Spica or Capella—the distances corresponding severally to the lower magnitudes follow as a matter of course. They are linked together by an invariable proportion. We have already explained what is meant by the 'light-ratio,'[1] but it may here be repeated that a star of any given magnitude is, by definition, one 2·512 times brighter than a star of the magnitude next below, and 2·512 times less bright than a star of the magnitude next above it. But, since light varies inversely as the square of the distance, any star removed to $\sqrt{2{\cdot}512}=1{\cdot}585$ times its actual distance, would show exactly one magnitude fainter than it did before. This number, then, 1·585, the square-root of the light-ratio, may be designated the 'distance-ratio.' It represents the difference of distance corresponding to a difference in light of one stellar magnitude. The *relative* mean distances of the various classes of stars are then known; to render them *absolute,* we only need to ascertain the real mean distance of any one of those classes.

[1] See *ante*, p. 20.

It is true that, within each class, enormous disparities exist. Small stars, comparatively near the earth, take their stand on the same level of apparent brightness with indefinitely large, but indefinitely remote bodies. What is invariable for each magnitude is the *proportion* between real brilliancy and the square of the distance. Symbolically expressed, $\frac{l}{d^2}$ is constant. That is to say, photometric uniformity results from a certain balance being struck between remoteness and light-power, by which the effect of equality is produced. The law, however, connecting average distance with apparent lustre is not invalidated even by the limitless variety included in the above expression. The extremes are vastly wide apart, but the mean remains, for all practical purposes, the same. It should, nevertheless, be always clearly borne in mind that the conclusions thus obtained are *general*, and should only be generally applied. Referred to particular cases, they are utterly fallacious, and more likely than not to mislead.

The interest attaching to Dr. Elkin's deduction, in 1888, of a mean parallax of 0″·089 for the first-magnitude stars of the northern hemisphere,[1] can now be appreciated. Even as half a result, it is worth attentive consideration.

It establishes provisionally, at a distance of 36½ light-years, the first halting-place for explorations of sidereal space. Thus inconceivably remote, on an average, are the brightest and, we must add, the nearest stars. Our sun, thus placed, would sink below ordinary naked-eye visibility; it would be of 6·5 magnitude. So that the conclusion of its insignificance among its fellow suns is rendered at least plausible.

On the scale measured at Yale College, the mean distance of stars of the second magnitude is 58 light-years (parallax 0″·056); stars of the third magnitude are at 92 light-years (parallax 0″·035), and so on; the invariable ratio of 1·585 regulating the increase of distance and decrease of magnitude for each descent of one step (provided only that

[1] *Sid. Mess.* Nov. 1888, p. 395. The result actually obtained was 0″·085, relative to the comparison-stars, for which the observations indicated a probable average parallax of 0″·04. This amount, added to 0″·085, gave 0″·089 as the *absolute* mean parallax of the ten stars measured.

light suffers no diminution of intensity in interstellar space). When we get down to the sixteenth magnitude, which is about the *minimum visibile* in the largest telescopes (the Lick refractor may perhaps go a magnitude lower still), we find the theoretical light-interval lengthened to 36,000 years; but there is no certainty that any such far-travelled rays do, in point of fact, reach us. The regular progression of distances may not extend so far. It must stop somewhere, if the stellar system be—as we have reason to think it is—of finite dimensions; at what particular magnitude, however, the break occurs, it would at present be futile to conjecture. All that can be said is that, distance becoming at length eliminated as a factor of magnitude, the differences of the faintest stars represent, chiefly or solely, real inequalities in shining. There may possibly, for instance, be no 'mean distance' corresponding to the sixteenth magnitude. The stars of that rank would not then, on the whole, be further off than those of the rank next above them, but would, on the whole, possess only $\frac{1}{2{\cdot}512}$ of their real light. This *must* be the case—so far as we can see—at some stage of the descent into the abysses around us.

Of the ten stars measured at Yale, Procyon, with a parallax of 0″·266, giving a light-journey of 12·3 years, was found to be the nearest to the earth. Altair, at a distance of 16·4 light-years (parallax 0″·199), came next; Aldebaran, at 28 light-years (parallax 0″·116), third; Regulus, with a parallax of 0″·093, might be termed a mean first-magnitude star at mean distance. Of the remainder, four gave no reliable indications of possessing *any* sensible parallax. That is to say, the concluded value—which was negative for Betelgeux and Deneb, positive for Arcturus and Vega—was in each case qualified by a probable error larger than itself. As to Vega, there is still some doubt, a considerable parallax relative to its well-known optical attendant having been found for it by several previous observers; but that *a* Orionis, *a* Cygni, and Arcturus, are plunged in depths of space unfathomable by any method yet brought into use, may be admitted without hesitation. The excessive remoteness of Arcturus, especially, enables us to recognise in it perhaps the most stupendous sun within our imperfect cognisance. For, although less lustrous than Canopus, it is probably (judging from the character of its spectrum) much more massive proportionately to its light, while the rate of its movement can only be termed portentous. Assuming the accuracy of Dr. Elkin's nominal parallax of 0″·018, its velocity across the line of sight alone must reach 372 miles a second, 380 being the utmost that our sun can generate in a body reaching its surface from infinite space.

So closely and so consequentially have advances in this arduous branch followed the growth of improvement in heliometers, that direct visual measurements for the purpose with any other instrument might almost seem waste of labour. But direct visual measurements are no longer the only practicable ones. By the energy of Professor Pritchard, the determination of stellar parallaxes by photography has been converted into a working reality of incalculable value and promise. His first experiment was with the classic 61 Cygni, of which 330 separate impressions, obtained in 1886–7, furnished the materials for 30,000 measurements, or 'bisections' of star-images.[1] For the immediate purpose, these extraordinary pains were largely superfluous; but they had the ulterior object, fully attained by their means, of establishing the credit of a novel and unfamiliar method. Not only is the resulting parallax of 0″·43 for 61 Cygni of the highest authority, but the most delicate of all astronomical inquiries was at once, with the full assent of experienced judges, admitted to be within the full competence of the photographic camera.

Similar determinations have since been made of Polaris, α, β, and μ Cassiopeiæ. The results will be found in the Table of Parallaxes appended to this volume. At the distance of Polaris, our sun would be discoverable with an opera-glass as a 7·5 magnitude star; in the place of μ Cassiopeiæ it would sink to 8·4 magnitude. The pole-star hence emits the light of 158, μ Cassiopeiæ of nineteen suns, while the large

[1] *Monthly Notices*, vol. xlvii. p. 87.

proper motion of the latter comes out, in linear velocity, 363 miles per second!

The advantages of photography for stellar parallax-work are manifold. Perhaps the chief of them is the nearly indefinite power of control afforded by it. Any star on the plates, situated at all near the prolongation of the major axis of the parallactic ellipse (in other words, with a tolerably large 'parallax-factor'), may be used as a point of reference. Comparisons can thus be multiplied almost at pleasure, and inferred displacements with regard to one star checked by recourse to another, duplicate plates being at hand for additional safety. By the proper use of such safeguards, delusive results can be all but certainly excluded. Moreover, *relative* parallax becomes virtually *absolute* when comparisons are made with a great number of stars, most of which are presumably too remote to complicate the result by perspective movements of their own.

The peculiar province of photography is with stars too faint to be conveniently dealt with by visual means: for the images of the brightest ones, over-exposed through the necessity of giving the small stars in their neighbourhood time to imprint themselves, become diffused into blurred discs beyond the power of accurate bisection. Thus, the investigation of stellar parallax relatively to magnitude falls naturally to workers with the heliometer, while its relation to proper motion can best be elucidated with the aid of the photographic camera.

The main object just now of inquirers in this branch is to obtain a wider basis for general conclusions regarding the distances of the stars. For this purpose it is more important to secure a considerable number of parallaxes reasonably well determined than a few reduced by scrupulous care within the narrowest possible bounds of error. Research in this sense is already well on its way. It has been pointed out by Mr. Monck [1] that the average parallax of stars of the second will prove a more reliable datum than that of stars of the first magnitude, both because they are more numerous, and because they are more nearly equal in brilliancy; and there

[1] *Sid. Mess.* Feb. 1889, p. 62.

is a prospect that this datum will before long be provided, at least in an approximate form, by the Oxford plates reinforced by Cape measures. Its actual ascertainment will constitute a step of the very highest importance.

For the present, we have only to ask ourselves, what are the indications derivable from the work so far as it has gone? They are decidedly, in the first place, against the existence of any large parallaxes. It is, of course, still amply possible that stars may be found much nearer to the solar system than *a* Centauri, but their discovery is growing every year less and less probable. Sir Robert Ball[1] examined some years ago about 450 objects in a manner which, though summary, would have sufficed to bring to view any parallax of a single second of arc. None was forthcoming. His list included a number of red and variable stars, Nova Cygni, Webb's planetary nebula, and the Wolf-Rayet gaseous stars in Cygnus; and it may be noted in passing that no star with a banded, or a bright-line spectrum has yet exhibited the slightest tendency towards parallactic displacement. Until, however, many more efforts have been made for its detection, much significance cannot be attached to its absence. The best chance of success would perhaps be with Mira Ceti, the well-determined proper motion of which is suggestive of proximity.

The cardinal truth emerging from these inquiries is that of the extreme isolation of the solar system. A skiff in the midst of a vast, otherwise unfurrowed ocean, is not more utterly alone. About the same proportion would be borne by an oasis one mile across to a desert twenty times as extensive as the Sahara, that our sun with his entire planetary household bears to the encompassing void of space. The enormity of its blank extent is strikingly illustrated by Father Secchi's remark that the period of a comet reaching at aphelion the middle point between our sun and the nearest fixed star, would be of one hundred million years;[2] and by recent measures, the nearest fixed star has been pushed further back into space by one quarter the distance assigned to it when he wrote. Yet the sun is no isolated body. To each individual of the un-

[1] *Dunsink Observations*, vol. v.

[2] *Les Étoiles*, t. ii. p. 146.

numbered stars strewing the firmament, down to the faintest speck of light just shimmering in the field of the Lick refractor, it stands in some kind of relationship. Together, they master its destiny, and control its movements. Independent only so far as its domestic affairs are concerned, it is bound, as a star to the other stars, by influences reaching efficaciously across the unimaginable void which separates it from them. The outcome of those influences in the translatory motion of the solar system, we shall consider in the next chapter.

CHAPTER XXI.

TRANSLATION OF THE SOLAR SYSTEM.

The study of the stars inevitably leads us to consider the advancing movement in the midst of them of the sun and its attendant train of planets. There can be no reasonable doubt—and the thought is an astounding one—that we are engaged on a voyage through space, without starting-point or goal that we can know of, but which may prove not wholly uneventful. Its progress may possibly bring about, as millenniums go by, changes powerfully influential upon human destinies; nay, an incident in its course may, at any time, by the inscrutable decree of Providence, terminate the terrestrial existence of our race, and consign the records of its civilisation, in dust and cinders, to the arid bosom of a dead planet. A curious sense of helplessness, tempered, however, by a higher trust, is produced as we thus vividly realise, perhaps for the first time, how completely we are at the mercy of unknown forces—how irresistibly our little 'lodge in the vast wilderness' of the universe is swept onward over an annual stretch of perhaps five hundred millions of miles, under the mysterious sway of bodies reduced by their almost infinite distances to evanescent dimensions.

But, as things are constituted, the translation of the sun's household is a necessity, albeit one of startling import to ourselves. The stellar system is maintained by the balance of forces, and motion is the correlative of force. As a star among stars, the sun can only maintain a separate existence by contributing its share to those harmonies of movement by which 'the heavens show forth the glory of God.' Destruction would be the eventual penalty of even a moment's immobility. A penalty, indeed, which might not be exacted until after the lapse of many millions of years. It may reasonably be assumed that *a* Centauri exercises upon the sun the strongest attraction of any individual star; but a collision would ensue very tardily upon abandonment to its influence. The sun (if undisturbed by competing pulls) would fall from a position of rest towards its next neighbour, less than the third of an inch in one month; the second month would see despatched nearly a full inch of the journey of twenty-five billions of miles; and although the acceleration would of course grow more rapid as the distance diminished, close upon fifteen million years should elapse before the fires of sun and star, probably become extinct during their gradual approach, could be rekindled by the catastrophe of their impact.

There is then an *à priori* certainty that the sun moves; and assurance on the point is rendered doubly sure by inferences from observed facts. For besides their annual parallax due to the earth's motion round the sun, the stars have a 'secular' or 'systematic' parallax depending upon and attesting the reality of the sun's motion round an unknown centre. Let us see how this systematic parallax can be investigated.

If the sun alone were in motion, and the stars at rest, the results in perspective displacements would be simple and unmistakable. Each star would appear to travel backward along a great circle of the sphere, passing through the two points towards and from which the sun's course was directed. So that there would be the semblance of a general retreat from the 'apex,' or solar *point de mire*, coupled with a thronging-in from all sides towards the opposite point, or 'anti-apex.' For each particular star, the amount of displacement should vary, inversely as its distance from ourselves in space, directly as the sine of its angular distance from the apex. Hence, if the annual parallax of even one such sensibly shifting star were determined, not only the rate, in miles per second, of the solar progression would at once follow, but the parallax of every other sensibly shifting star in the heavens could be deduced by a simple calculation from the relative quantity of its apparent movement.

But the stars are not at rest. They have movements of their own, greatly swifter, in many cases, than that of the sun. Perspective effects are thus to a great extent masked. Yet they subsist. It is mathematically certain that every star, whatever its own course or speed, reflects the sun's motion in the strict measure of its position with regard to it. What are called the 'proper motions' of the stars are then made up of two parts, one real, the other apparent. They include a common element, the separation of which from the heterogeneous admixtures disguising it, constitutes the problem to be solved.

With the instinctive appreciation of genius, Herschel went straight to the heart of the matter. What had to be done, he saw clearly, was to find out the direction which should be given to the sun's course, in order to make it account for as large a proportion as possible of the sum-total of stellar movements. 'Our aim must be,' he wrote in 1805, 'to reduce the proper motions of the stars to their lowest quantities.'[1] And again: 'The apex of the solar motion ought to be so fixed as to be equally favourable to every star.' But how is this to be done? Very simply, if we only consider, as Herschel did, a few of the brightest stars.

Take, for example, four stars with conspicuous movements, two in the northern, two in the southern hemisphere—namely, Vega, Capella, Sirius, and Fomalhaut. The great circles, of which each yearly describes a minute arc, traced backward on the sphere, very nearly intersect in a single point situated in the constellation Hercules.[2] Had we only the motions of those four stars to consider, we should accordingly infer without hesitation the 'sun's way' to lie thitherward. Nor should we be very far wrong. The most refined modern determination of the solar apex, founded upon the motions of several thousand stars, differs by only five degrees from the result of the extremely summary proceeding just indicated.

The graphical method, however, is evidently applicable only to a very restricted stock of data. When a crowd of stars

[1] *Phil. Trans.* vol. xcv. p. 248.

[2] This was remarked by Klinkerfues, *Göttingische Nachrichten*, 1873, p. 350.

have to be taken into account, the points of intersection of their respective circles of motion become spread over too wide an area for a 'mean apex' to be struck out fairly between them, even by the exercise of a judgment as discriminating as that which in 1783 led Herschel to set the present goal of solar travel in the vicinity of λ Herculis. The accumulated facts must then be dealt with by a method at once stricter and more comprehensive. A glance at the nature of the task in hand easily suggests to a mathematician what that method should be.

The proper motions of the stars give, as already hinted, the plainest evidence of individuality. The lines traced by them on the sky run in all possible directions. But a substratum of regularity underlies this seeming confusion. A mere inspection of the signs *plus* and *minus*, signifying respectively east and west, and north and south, attached in catalogues to the components in right ascension and declination of stellar movement, suffices to show a general prevalence of law through the unequivocal tendency of the signs to vary together in passing from any one to an adjacent region of the heavens.[1] At a *coup d'œil*, Argelander fixed the point from which this under-current of motion flowed, and so gave an improved apex for the course of the sun confirmed in general by all subsequent research.[2] It is then clear, in the first place, that no movement possibly assignable to the sun can explain all stellar displacements; a large residuum being real, and therefore by no ingenuity to be got rid of. While in the second place, the nearer the truth is approached as regards the direction and amount of the sun's motion, the smaller obviously this residuum will be. In other words, the most probable value of the solar motion will be that which renders the 'sum of the squares of the residuals' of stellar motion a minimum.

But why the sum of the squares, and not the simple arithmetical sum of the outstanding proper movements? It needs only common sense, aided by the most elementary geometry, to get a sufficient insight into the reason. Any one can see,

[1] Stone, *Monthly Notices*, vol. xxvii. p. 239.

[2] *Mémoires présentés à l'Acad.* St. Pétersbourg, t. iii. p. 569.

with the help of a pencil and a piece of paper, that if a line be divided into two segments, and squares be constructed on the segments, the sum of those squares will be the least possible when the line is equally divided, and will increase continually with the inequality of the segments. This simple fact gives the clue to the principle of 'least squares.' Its object is to elicit such a quantity as will make the outstanding errors of observation, or any other kind of residuals, as small as possible *all round*. Not merely small taken in the aggregate, but reduced impartially to a uniform level of insignificance. Under these circumstances, as we have seen from the consideration of our divided line, the sum of their squares will be a minimum; and it can be mathematically demonstrated that the most probable result in such investigations as are susceptible of this kind of treatment is arrived at when the condition of 'least squares' is fulfilled.

This mode of attack upon the problem of the sun's translation was first employed by Argelander in 1837. Assuming provisionally the correctness of Herschel's apex, he proceeded to compute for each of 390 stars with ascertained proper motions the lines along which those motions should proceed if due to systematic parallax alone. Their deviations from the prescribed directions gave him 'angles of error,' which, placed in the category of casual errors of observation, and treated by the method of least squares, indicated a corrected apex, such that by its adoption the sum of the squares of the differences between what was calculated and what was observed—that is, between the purely parallactic drift of the stars and their actual displacements—was reduced to the least possible amount. The solar movement was, in a word, so fixed as, in Herschel's phrase, 'to be equally favourable to every star.' According to this determination, the sun's way lies towards a point in right ascension 260° 51′, north declination 31° 17′, occupied almost exactly by a sixth-magnitude star numbered in Piazzi's catalogue 143 in the XVIIth hour.[1] Nor is there any sign that Argelander's confidence in its substantial accuracy was misplaced.

[1] *Mémoires présentés*, t. iii. p. 590.

An important modification of his method was introduced by Sir George Airy in 1859.[1] Abolishing the conception of a spherical surface of reference, he defined the linear movements in space of the sun and stars with regard to three directions at right angles to each other ('rectangular coordinates'). No assumption of any kind was then needed; the subject was treated with the utmost strictness and generality, and some possible causes of error were removed. Airy's had many points of theoretical superiority over Argelander's method. That, however, of introducing the consideration of the *quantity* of each star's movement was to a great extent counterbalanced by the necessity which it involved of adopting precarious suppositions as to the distances of the classes of stars employed. The apex for the solar movement resulting from the consideration of 113 stars was situated in R.A. 261° 29′, Dec. + 24° 44′; while Mr. Main's similar treatment of 1165 stars shifted it to R.A. 263° 44′, Dec. + 25°.[2]

In the latest, and most thorough investigation of this great subject, by M. Ludwig Struve,[3] Airy's method was adopted. The incitement to undertake a task rendered formidable by the very wealth of the materials at his disposal was afforded by Auwers's fresh reduction of Bradley's Greenwich observations. From a comparison of the star-places thus authoritatively determined for 1755, with those given in the St. Petersburg catalogue for 1855, a list of 2,814 proper motions was derived, of which 2,509 were available for M. Struve's purpose. Among the stars for various reasons excluded were the seven swiftest travellers, as unduly affecting the result through motions no doubt mainly original.

As the outcome of this exhaustive discussion, the apex of the solar motion was placed in R.A. 273° 21′, north declination 27° 19′, and a rate was assigned to it such that the space traversed in a century, viewed square from the average distance of a sixth-magnitude star, would subtend an angle of 4″·36. Admitting that stellar distance varies inversely as the square root of stellar brightness, hence that stars of the first

[1] *Memoirs R. Astr. Soc.* vol. xxviii. p. 143. [2] *Ibid.* vol. xxxii. p. 27.
[3] *Mémoires de St. Pétersbourg*, t. xxxv. No. 3, 1837.

are, on an average, only one-tenth as remote as stars of the sixth magnitude, we can, with the help of Dr. Elkin's mean parallax for the former class, translate this angular into linear velocity. It comes out $14\frac{1}{2}$ miles a second.

Well-nigh the whole of the stars visible to the naked eye in the northern hemisphere concurred in M. Ludwig Struve's determination. No principle of selection was employed; they were taken as they came; exclusion was only resorted to in the few cases where misleading peculiarities were obvious. But the surest proof of success was derived from the application of a test which had served only to impair confidence in previous results—the test, namely, of residual errors. The transport of the solar system must, unless it be shared by the observed stars, add, in the proportion of its velocity, to their apparent movements; the subtraction from them of the common drift it produces and is disclosed by, should then leave those apparent movements materially diminished. But this had hitherto been the case only to a very unsatisfactory extent. The withdrawal, however, of the parallactic element discriminated from them by M. Struve, at once reduced the sum of the squares of the 'corrected' proper motions nearly to one-half its original amount, thereby testifying emphatically to its own genuineness.

Combining his with fifteen prior results, M. Struve finally concluded for an apex in R.A. 267°, declination +31°. This position, as he said, cannot be far wrong. Apart from a source of uncertainty to be presently adverted to, the *aim* of the sun's motion may now be regarded as approximately known. Above a score of investigators have agreed in directing it towards a restricted area in the heavens marked by the outstretched left arm of the asterismal Hercules. Varying methods have been employed; hypotheses of sundry kinds have been adopted; different stars have been appealed to, with the same general result. The conviction of its truth becomes irresistible when we find that the entire stellar heavens pronounce in its favour. Indications of an identical slow steady flow of motion *away* from the constellation Hercules, and *towards* the constellation Columba, come from the southern, as well as from the northern hemisphere. Galloway's experiment in 1847, giving an apex in R.A. 260°, Dec. + 34° 23',[1] was striking, but could hardly be regarded as decisive, owing to the doubtful character of the data available to him. But M. de Ball's research thirty years later,[2] while of similar upshot, was not open to the same reproach, a small, though sufficient stock of authentically determined southern proper motions having in the meantime accumulated. And his conclusion was virtually re-affirmed by Mr. Plummer, with the help of 274 stars from Mr. Stone's 'Cape Catalogue for 1880.'[3]

But when, from the *direction*, we attempt to pass to the *amount* of solar motion, the case becomes widely different. Flagrant contradictions abound. Estimates of velocity range at large between five and 150 miles a second; the criteria of truth are at the mercy of individual judgment. The cause of these discrepancies lies in the uncertainty still prevailing as to the distances of the stars. Quite irrespectively of the remoteness of the objects whose perspective displacements serve as the index to it, the line of our advance through space can be searched out; but its rate can only become known when we know how far off the displaced objects are. The relative distances, however, assumed by most observers in this field being for the most part untrustworthy, the resulting angular velocities of the sun are no less so. And angular can of course be converted into linear velocity only through acquaintance with absolute distances.

It is nevertheless tolerably certain that the solar pace has nothing headlong about it. We are not whirled in the train of such a stellar projectile as 1830 Groombridge or ζ Toucani. Our condition, were it so, would be betrayed by unmistakable tokens. Everything, on the contrary, suggests the inference that our sun is among the more sedately moving stars. The ascription to it of a very high velocity entirely unwarranted by facts has had its spring in erroneous assumptions. From a

[1] *Phil. Trans.* vol. xxxvii. p. 79.
[2] *Wochenschrift für Astr.* Bd. xx. p. 169.
[3] *Memoirs R. Astr. Soc.* vol. xlvi p. 341.

statistical survey of the elements of motion in 1167 stars, Mr. Stone concluded their real to exceed their parallactic movements in the proportion of four to three;[1] and there is much reason to believe this estimate tolerably correct. A combination with M. L. Struve's of three previous values placed the linear velocity of the solar system at about sixteen miles a second. On the plan of inquiry heretofore considered, it is scarcely possible, for the present, to get nearer to truth on the point; but another is rapidly becoming available.

We have elsewhere explained the principle of spectroscopic determinations of motion.[2] Their peculiar value consists in their independence alike of distance and of visible displacement. Referring to movements visually imperceptible, they complete knowledge of stellar velocities by giving their otherwise unknown 'radial components.' Apart from this wonderful application of the spectroscope, the real lines of travelling of the stars could never have been ascertained, since we can immediately discern only that part of their motion lying *across* the line of sight, which, in individual cases, may be all or none. By the spectroscopic revelation, however, of motion *in* the line of sight, the missing element is supplied, precise and particular knowledge may be had for the asking, and the stars pursue their journeys under astronomical scrutiny, no longer as mere flitting bright specks on the surface of an imaginary sphere, but as suns in space, each with its own secret in reserve, and each contributing to swell and deepen the marvellous harmony of the whole.

The effects of recession and approach on the light emitted by moving objects being physical and real, they remain unimpaired by distance. Out at the verge of the sidereal system, or close at hand within our own atmosphere, they are the same for the same velocities, and can, with a sufficient light-supply, be detected with equal facility. Hence their special applicability to the problem of the sun's speed. To determine it with very approximate accuracy, it will only be necessary to compare the average radial celerity of a good number of stars lying in front of the sun's way with that of

[1] *Monthly Notices*, vol. xxvii. p. 239. [2] *Hist. of Astr.* pp. 245, 440.

others he is leaving behind. Movements of approach must, on the whole, predominate in the one direction, movements of recession in the opposite, half the mean difference, elicited by appropriate processes of computation, representing the rate of transport of our system. The method, however, cannot be successfully employed with such data as were, until lately, the only ones to be had. In a matter of such excessive delicacy as the measurement of minute shiftings of lines in stellar spectra, eye-observations are subject to too many and too grave disturbances for implicit reliance to be placed on them. They have done their work in showing the validity of the principle and giving preliminary indications of no small importance; and now the turn of the photographic camera has come. The quality of the materials furnished by it under Professor Vogel's able treatment leaves little or nothing to be desired; when their quantity has been sufficiently augmented, the rate, in miles per second, of our transport through space will become known with ease and certainty.

Some experimental attempts have indeed already been made to extract the desired information from visually determined line-of-sight movements; and M. Homann's especially, notwithstanding the defective nature of the information at his command, may serve as an index to the future value of the method. Forty-nine of the brightest stars, spectroscopically determined between 1878 and 1884, gave him a goal for the sun's way, no longer in the accustomed situation in Hercules, but transferred some four hours eastward to very near the place of 61 Cygni.[1] This new apex, it may be remarked in passing, lies in one of the densest parts of the Milky Way. The corresponding velocity of fifteen miles per second agrees so closely with M. Ludwig Struve's estimate as to encourage the persuasion that it needs little amendment.

Thus, both the direction and rate of transport of our system are not only included in the category of things *knowable*, but there is every prospect of their becoming known with more

[1] Homann, *Astr. Nach.* No. 2714; Schönfeld, *V. J. S. Astr. Ges.* Jahrg. xxi. p. 58.

and more satisfactory exactness in the immediate future. All that is needed is a closer and a wider application of means already in the hands of astronomers. Still our curiosity will not even then be satisfied. The value of the two items of information within our reach is indeed incalculable. They are a *sine quâ non* for the furtherance of inquiries into stellar mechanics ; are they to be a *ne plus ultra* as well ?

The sun, we are well assured, is not travelling along a straight line. The universality of gravitation makes rectilinear movement next to impossible, since no cosmical body can traverse space under the sole guidance of its own primitive velocity. It is true that, supposing primitive velocities altogether abolished (and we know of no reason why they should necessarily exist), any number of bodies might be united into a system endowed only with pendulum-like motions. The sun and stars might thus, by an abstract possibility, be totally devoid of advancing or circulatory movements, each swinging for ever to and fro through their common centre of gravity. But it is practically certain that this plan is not realised in the sidereal system.

The path of the sun is then a curve, but a curve most likely of such vast proportions as to remain for ages indistinguishable from a right line. Strictly speaking, however, its direction is continually changing ; the apex of to-day will not be the apex of to-morrow ; still less will it be the apex of a million years hence. Yet in a million years it may quite conceivably not have shifted from its present place in the sky by more than the width of the full moon ; and our best determinations still fall far short of the accuracy which would enable us to detect a change of half a dozen times that amount. Directly, that is to say ; indirectly, a much more insignificant alteration might disclose itself. It can easily be explained how.

Pond, who succeeded Maskelyne as Astronomer Royal in 1811, made the remark that the sun's motion must produce a kind of secular aberration of light, by which the stars are permanently displaced from their true positions.[1] The effect

[1] Liagre, *Bull. de l'Acad.* Bruxelles, t. viii. p. 168, 1859 ; O. Struve, *Mémoires*, St. Pétersbourg, t. v. p. 106, 6e Série.

of ordinary, or what we may call *annual* aberration, is to make them appear to describe little ellipses, the semi-axes of which depend upon the ratio of the velocity of light to the velocity of the earth in its orbit. But the sun's orbital movement being conducted, so far as experience yet goes, in one direction, the aberration due to it is in one direction too, and is hence constant, and for the present beyond the reach of observation. It is, however, constant only so long as the movement producing it remains sensibly so. As the latter changes, it will change too, and may in this way be brought within the domain of human cognisance. For upon the acceleration, retardation, or deflection of the sun's movement systematic displacements among the stars should ensue, the nature of which would at once betray their origin.

The total amount of this secular aberration may be roughly stated as one second of arc for every mile per second of the sun's velocity. Hence, stars 90° from the solar apex are pushed forward towards it by perhaps 15″, the effect upon other stars diminishing with the sines of their distances on the sphere from the same point. These aberrational can be distinguished from the parallactic displacements flowing from the same source, by their indifference to remoteness in space. Stars far and near, bright and faint, swift-moving and tardy, are equally affected by them. But while it is quite certain that visual disturbances of this kind take place, and to an extent possibly greatly in excess of that above assigned to them, their interest must for a long time remain purely theoretical. Indeed, it may well be that the modifications rendering them sensible and instructive will proceed with such exorbitant slowness that not even astronomical patience will avail to unmask them.

We do not know the plane of the sun's orbit—only the direction of one line in it. And that line, pointing towards the constellation Hercules, makes an angle of about 60° with the sun's equator. Thus, the solar movements of rotation and translation would seem to be unrelated one to the other; and the same remark applies to the planetary revolutions conducted, on the whole, along levels of space differing very little from that of the greater globe's axial movement. Our whole system is then driven obliquely upward by a power which, taking no apparent account of its domestic economy, owned doubtless an origin totally disconnected from that of gyrations given, through its influence, the helicoidal shape illustrated in fig. 48.

The sun's course, being inclined some twenty-eight degrees to the central plane of the Milky Way, is at present gradually removing it from that stupendous collection. This, however, does not necessarily imply real separateness. The movement of withdrawal actually progressing may be only temporary, in the sense that, after countless millions of years, it will be compensated by a return movement of approach. It is difficult to conceive that the combined attractions of the galactic myriads can, in the long run, be resisted. The most probable supposition as to the situation of the centre of force swaying our system, is that it lies somewhere in the cloudy zone which so enhances the mysterious beauty of our skies. If the orbit we are pursuing be approximately circular, then its centre must be divided by a quadrant of the sphere from the apex—it must lie somewhere on a great circle of which the apex and anti-apex are the poles. Now this great circle cuts the Milky Way at two opposite points in Perseus and Hydra, and there, accordingly, two alternative centres of the solar motion might be looked for. Argelander chose as the more promising position the spot marked by the great cluster in the sword handle of Perseus;[1] but the conjecture made no pretension to scientific authority, and the postulate upon which it was based, of the sun's

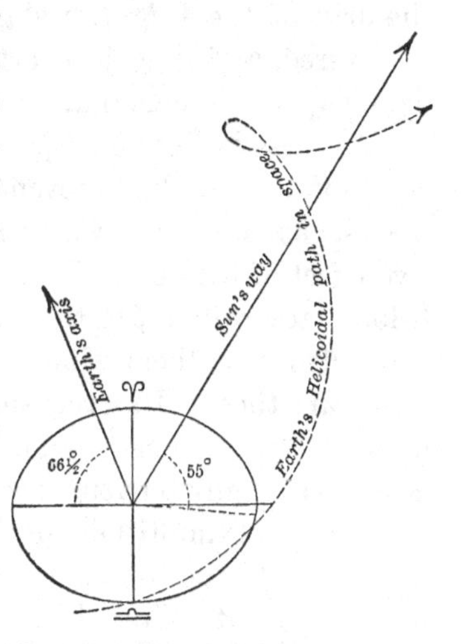

FIG. 48.—The Earth's Motion in Space.

[1] *Mémoires présentés*, St. Pétersbourg, t. iii. p. 302.

path being at all nearly circular, is in truth of a highly precarious nature.

We are even ignorant whether the translation of our system towards the constellation Hercules represents a primary or a secondary order of stellar revolution. It perhaps merely indicates the *interstitial* movement appertaining to the sun as a member of a restricted group of stars, the common transport of which proceeds undetected in a totally different direction. Hence the possibility, suggested by Herschel, of the presence of a higher kind of systematic parallax than that revealed in the drift of the brightest stars.[1] It has, however, yet to be discovered, and time is short for the investigations which its discovery would demand.

Progress is here only possible through careful and minute study of the residual movements of the stars—of the movements, that is to say, which remain after the general perspective effect of the sun's motion has been subtracted, and which belong, accordingly, to their individual selves. The questions connected with them which most immediately present themselves are these: Has the sun companions on its journey, or does it travel alone? and, Are real stellar displacements governed by any obvious law?

The great multitude of the stars are, to all appearance, indifferent to the transport of our system. They have clearly no share in it. Just because they stand aloof, and act as indicators of the way, its progress becomes sensible to us. For motion is not alone undiscoverable, it is even unimaginable without some fixed point of reference. We cannot, however, as yet pronounce with certainty against the existence of a particular dynamical bond connecting the sun with some few of the stars, which together with it form a company associated by subjection to identical influences, and engaged with it on the same journey through space. As to the criteria by which such associated stars, if present, can be discriminated from the rest, something will be said in the next chapter.

There, too, we will consider what answer should be given to our second query. A great deal depends upon it as regards

[1] *Phil. Trans.* vol. lxxiii. pp. 1276–7.

our conception of the sidereal universe. Nay, the result of inquiry upon the point has a vital bearing upon the subject we have just attempted, however inadequately, to deal with. For the assumption that the absolute movements of the stars have no preference for one direction over another forms the basis of all investigations hitherto conducted into the translatory advance of the solar system. The little fabric of laboriously acquired knowledge regarding it at once crumbles if that basis has to be remcved. In all investigations of the sun's movement, the movements of the stars have been regarded as casual irregularities; should they prove to be in any visible degree systematic, the mode of treatment adopted (and there is no other at present open to us) becomes invalid, and its results null and void. The point is then of singular interest; and the evidence bearing upon it deserves our utmost attention.

CHAPTER XXII.

THE PROPER MOTIONS OF THE STARS.

When the relative positions of the stars are compared at considerable intervals of time, they are in many cases found to have undergone small, but unmistakable changes of a seemingly capricious character. These are termed 'proper motions,' to distinguish them from merely nominal shiftings due to the slow variation of the points of reference which serve to define the places of all the heavenly bodies as seen projected on the inner surface of an imaginary concave sphere. Proper motions are by no means easy to get at. Only from the most delicate observations, and with stringent precautions for bringing those at distant dates under precisely similar conditions, can they be elicited with satisfactory accuracy. Otherwise, some trifling systematic discrepancies in the compared catalogues, or accidental errors of computation, might pass for genuine effects of movement, with disastrous influence upon sidereal investigations. Hence, proper motions cannot generally be regarded as established unless, in addition to the terminal observations showing a sufficiently marked change of place in the course of thirty, fifty, or one hundred years, at least one intermediate observation is at hand to prove that the suspected motion has proceeded uniformly in the same direction, and is accordingly not the creation of personal or instrumental inaccuracy.

Although not one of the millions of telescopic stars can, with any show of reason, be supposed at rest, less than five thousand of the stellar army are at present securely credited with measurable and progressive displacements. Many of these, including nearly all the *lucidæ* of the northern hemisphere, were observed by Bradley between 1750 and 1762;

while in the southern, Lacaille's simultaneous labours serve to authenticate the changes of some three-score of objects to which he devoted especial care. So that a large stock of data of the required high degree of accuracy already possess an antiquity of one and a third centuries; the multiplied observations of the last sixty years affording a further supply from which fresh and well-determined proper motions are continually being, as it were, harvested from the seed planted by an earlier generation.

The aspect of the heavens is, to the unaided sense, virtually unchanging. The constellations disclosed at the present time by the nightly withdrawal of the veil of twilight would be familiar, could they revive to survey them, to the watchers from the towers of Babylon. And most of the star-alignments given in our text-books might be as useful to students of celestial physiognomy a couple of thousand years hence as they are to-day. Every one of the indicated stars will indeed most probably, by that time, have shifted its place to the extent of many thousands of millions of miles. Yet so overwhelmingly vast is the sidereal scale that thousands of millions of miles measured upon it sink into insignificance.

Stars advancing in a century as much as 30″, or about $\frac{1}{60}$ of the width of the full moon, are counted rapid travellers; and the swiftest class, with secular motions of 100″ and upwards, now embraces close upon eighty members (see Appendix, Table IV.). Each of these, were it bright enough for casual perception, would in a couple of millenniums become very sensibly displaced even to an unskilled observer. But of the eighty quickest stars, less than one-half are visible to the naked eye, while only ten reach the fourth magnitude; hence their shiftings make very little difference in the general effect of the starry skies.

As might have been expected, the stars in most rapid apparent movement are among those nearest to the earth. Vicinity, in fact, and *angular* velocity vary together. Displacements on the sphere through identical rates of travel in space are large just in the proportion that the distances of the objects affected by them are small. Were there any approach to uniformity in the real velocities of the stars, we could then fairly estimate, from their seeming movements, their relative situations as regards ourselves. But there is no such approach to uniformity. Inexhaustible variety prevails here, as in every other branch of sidereal statistics. Stars with large proper motions are sometimes enormously, even immeasurably remote; and, if stars with large parallaxes and little or no movement have not been discovered, it is perhaps because they have not yet been looked for. Their occurrence, for a reason to be presently explained, would be of great interest, and is not unlikely to be certified by measurements on photographic plates.

But, however great the range of variety, it seems certain beforehand that, on the whole, the amount of visible motion in a given number of stars must decrease as their distance increases. And since their brightness falls off at the same time, although much more rapidly, there appears no escape from the conclusion that motion and magnitude must, on a wide average, vary together according to a definite ratio. From stars of the sixth photometric magnitude, for instance, we receive only one-hundredth part of the light sent to us by stars of the first magnitude; they must then, one with another, be ten times more remote. Otherwise, we should be driven to the unwarrantable assumption of a systematic difference of real lustre between apparently large and apparently small stars. But, if the average distance of sixth-magnitude stars be ten times, then their mean motion should be only one-tenth that of stars of the first magnitude. In point of fact, however, this is not so. The proper movements of classes of stars diminish indeed very notably with their brilliancy, but not in the computed proportion. The discrepancy deserves attentive study.

The average proper motion appertaining to the sixth magnitude, as determined directly by M. Ludwig Struve from 647 of Bradley's stars, is 8″ in one hundred years.[1] Ten times this quantity, or 80″, *ought* to be the average movement of stars of the first magnitude. But the mean derived

[1] *Mémoires de St. Pétersbourg*, t. xxxv. No. 3, p. 8.

from the actually observed shiftings of the twenty brightest stars in both hemispheres, is only 60″. And since more than half of these transcend the standard lustre of their nominal rank, an excess, rather than a deficiency, of motion might have been anticipated. The deficiency, it is true, may prove to be more apparent than real. Stellar magnitudes are, in general, not photometrically determined, but adopted from Argelander's great 'Durchmusterung.' Argelander's estimates, however, as M. Lindemann has lately shown,[1] were not governed by a uniform light-ratio. He unconsciously curtailed the intervals between ranks of stars above the sixth, and widened them for stars below the eighth magnitude. As the result, his sixth-magnitude stars being undervalued in point of light, are at a less mean distance, and possess a greater proper motion than if they were truly of the grade assigned to them. The same conclusion applies to objects of ranks intermediate between first and sixth; and we thus see that general statements about the proper movements of star classes need to be received with caution.

Yet the sluggishness of stars of the second magnitude seems to be a genuine fact. They should be, on the photometric scale, 6·3 times nearer to the earth than stars of the sixth magnitude. This would give for their mean secular motion $8'' \times 6{\cdot}3 = 50''{\cdot}4$. Twenty-two such stars, however, from Bradley's and the Pulkowa catalogues, show no more than 17″. And even this low figure more than doubles that representing the average movement of forty-two southern stars of 1·7 to 2·7 magnitudes, forming a descending sequence from the ten of first magnitude. Nor is this average improved by considering only the first twenty on the list, from β Crucis of 1·7 to κ Orionis of 2·2 magnitude. The swiftest of these (γ Crucis) travels only 20″ a century; taken all round, they move 8″, or with exactly the speed of stars presumably more than six times as remote!

The low apparent velocity of this class of stars is a very curious, and at present an inexplicable circumstance. It is accentuated by the close agreement in M. Struve's results for

[1] *Observatory*, vol. xii. p. 409.

stars of all three ranks from the second to the fourth. A glance at the accompanying Table from his Memoir will serve better than verbal explanation to make the matter intelligible. The object of its compilation was to exhibit the divergence between the proper motions actually determined and those computed from the basis of the mean secular displacement corresponding to the sixth magnitude. In the fourth column, however, we have substituted figures derived from strict photometric star-distances for others depending upon a scale of distances involving precarious, even if plausible assumptions.

Table of Secular Mean Proper Motions of all Bradley's Stars differing by not less than eight-tenths of a magnitude.

Magnitude	No. of Stars	Mean Motion	
		Observed	Computed
1	9	66″·5	80″·0
2	22	17″·2	50″·4
3	51	16″·5	31″·8
4	106	16″·2	20″·6
5	318	8″·3	12″·7
6	647	8″·0	8″·0
7	92	6″·8	5″·0
8	11	12″·5	3″·2

It will be observed that the velocity of each order brighter than the sixth falls short of its theoretical amount, while that of the fainter orders exceeds it. We hasten to add, however, that (as M. Struve points out) little or no dependence can be placed on the above mean rate of eighth-magnitude motion deduced from measurements of only eleven objects.

And now, what are we to think? How can we account for the indicated deficiency of proper motion in the brighter stars? Three possible explanations present themselves. It is conceivable that stars, say of the sixth and seventh, are really less effulgent bodies on an average than stars of the second and third magnitudes, and are consequently less remote than they should be on the more natural supposition of their equality. Their diminished distance would then at once render their extra celerity intelligible. Again, there *may* be a systematic increase of motion outward from the sun, producing in apparently small stars preponderating rates of displacement. Or thirdly, there may exist a special class of stars deficient in light-power, but travelling with exceptional speed, by the influence of which the balance of seeming swiftness is turned in favour of the less brilliant classes of stars.

There is some probability that the last alternative represents at least a partial truth; but facts might be arrayed against, as well as for it; and it is, at any rate, far too soon to adopt it definitively. Subsidiary questions of great interest are connected with it; as, Whether real velocity bears any relation to physical constitution? Are swift stars found equally in all the spectral classes, or have they, on the whole, reached a later stage of development than more inert luminaries? For satisfactory answers, however, we must wait yet some time.

The anomalous results of Professor Eastman's recent inquiries[1] into the comparative proper motions of classified stars might be satisfactorily explained on the basis of our third alternative. Distributing 550 of the swiftest stars according to their brightness into nine groups, he 'found an almost uniformly *increasing* proper motion as the stars grew fainter, until the ninth-magnitude stars were found to have a proper motion nearly *three times as great* as those of either the second, third, fourth, or fifth magnitudes.'[2] His figures are, nevertheless, very far from representing average movements. Only stars distinguished for apparent velocity were considered by him, and the smaller they were, the more entirely exceptional, obviously, their status among their equals in point of lustre.[3] Yet the disproportionate representation of faint objects on any list of rapidly moving stars is certainly a fact, and a very remarkable one.

It comes out with strong evidence in Table V. of the Appendix, printed by the kind permission of Professor Schönfeld. More than half of the seventy-six stars enumerated in it are invisible to the naked eye; the three swiftest are of 6·9,

[1] *Bull. Phil. Soc. of Washington*, vol. xi. p. 143.
[2] *Ibid.* p. 167.
[3] Se: r. Monck's remarks in *Nature*, vol. xli. p. 392.

7·5, and 8·5 magnitudes respectively; no less than fifteen range from the eighth to below the ninth, while only four stars of the first, *none* of the second, and two of the third magnitudes are included in the collection. The largest proper motion yet detected belongs to a seventh-magnitude star situated in the Great Bear, and numbered 1830 in Groombridge's Circumpolar Catalogue. Argelander discovered in 1842 its pace to be such as would carry it round the entire sphere in 185,000 years, or in 265 over as much of it as the sun's diameter covers. Its annual advance, in fact, amounts to 7″; and it has nearly equal competitors in two small southern stars observed by Gould during his stay at Cordoba. One is a 7·5 magnitude star in the Southern Fish (Lacaille 9352), the other, one of 8·5 in the constellation Sculptor. Next on the roll comes 61 Cygni with a proper motion of 5″·1; and *α* Centauri, with 3″·7, has tenth place. Twelve double stars are conspicuous for rapid movement, besides two wide pairs (bracketed together in the list), the components of which are severally associated by a community of swift progress. It is noticeable that three out of the four first-magnitude stars with proper motions exceeding one second of arc yearly—namely, Sirius, Procyon, and *α* Centauri—are binary combinations; nevertheless, the elder Struve's general inference as to the quicker translation of multiple than of simple objects, has scarcely been borne out by further experience. The 'flying stars' with which we have since made acquaintance are all unaccompanied.

The 'proper motions' of stars include, as was explained in the last chapter, an apparent, as well as a real element; they consist, in technical phraseology, of the *motus parallacticus*, optically transferred to the whole stellar multitude from the single real motion of the sun, and the *motus peculiaris*, belonging to each individual star. Means are being prepared of separating these two elements, at present largely blended together.

But the *motus peculiaris* itself is only a projection upon the sphere of a line of travel which may make any angle with the line of sight. Its conspicuousness then varies with direction, no less than with distance and actual velocity. A star may appear devoid of motion simply because the whole of it is 'end-on'; while the movements of others seem large because, lying square to the line of sight, they are completely effective for apparent displacement. Here, just where ordinary observation is baffled, the prismatic method comes to the rescue. The spectroscope 'takes up the running' for the telescope.

But eye-observations of the motion-displacement of spectral lines are, as we have said,[1] hampered by many difficulties and uncertainties. Although consistently pursued at Greenwich during the last twelve years, no results were, or could be, secured concordant enough to form the groundwork of any extended research. Contradictions abounded; even the *sense* of the movements registered often varied from night to night with the still or troubled state of the air; it could only be depended upon where accentuated by a rate of going somewhat above the average. Rapid recession was, however, fairly well ascertained in Aldebaran, β Andromedæ, Regulus, and Castor, the velocities assigned ranging from twenty-five to fifty-eight miles a second. On the other hand, Arcturus, *a* Cygni, Vega, *a* and δ Andromedæ, Pollux, and *a* Ursæ Majoris were found to be approaching the earth with an average speed of somewhere about forty miles a second.

Of these, *a* Cygni is, from a telescopic point of view, all but completely stationary, so that the whole of its motion appears to be directed towards our system, to which it will eventually, Professor Newcomb tells us,[2] become so near a neighbour as to outshine during several thousand years every star now visible in the sphere. The prodigious remoteness, however, attributed to this object by Dr. Elkin leaves open the possibility of its being animated by a considerable thwartwise movement rendered insensible by distance; and tends accordingly to invalidate any conclusions as to its present or future course.

Two bright stars in Orion, also virtually devoid of visible proper motion, prove to be rushing straight away from our system. The rate of recession of the brilliant Rigel, as deter-

[1] See *ante*, p. 328. [2] *Popular Astronomy*, p. 471 (ed. 1882).

mined by Professor Vogel in 1888,[1] is thirty-nine, of ε Orionis, the middle star of the Belt, thirty-five English miles a second.

A special interest and authority attach to the last-named results as being the earliest arrived at by photographic means. The prerogatives of the camera in this line of work are enormous. Not only do the worst mischiefs of atmospheric disturbance vanish with its employment, but the upshot of measurements executed upon one line can be checked or ratified by comparisons with other lines in the same spectrum, and on the same plate. Where motion is in question, all must be equally affected by it; hence perfect security against illusion is afforded. The full realisation of these advantages through Vogel's skilful use of the spectrographic apparatus erected by him at Potsdam in 1888,[2] thus constitutes an advance of incalculable moment in practical stellar astronomy.

The light of the stars subjected to these delicate determinations is collected by a refracting telescope of $11\frac{1}{2}$ inches aperture, and dispersed by transmission through a couple of large prisms. The blue ray of hydrogen (near G) is that primarily relied upon for measures which have their zero point fixed by the simple expedient of photographing with each star-spectrum the spectrum of hydrogen in its natural position as derived from a Geissler tube illuminated by electricity. Relative then to this fiducial line, the star-line is shifted towards the *blue*, when the motion is one of approach—an effect very conspicuously shown by Arcturus; towards the *red*, by such a movement of recession as Rigel exemplifies (fig. 49, from a *negative*).

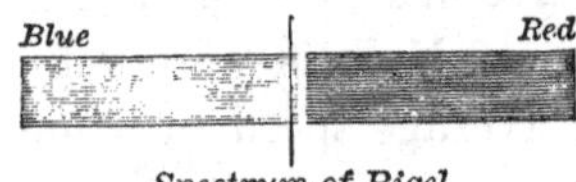

FIG. 49.—Displacement of Hydrogen line in the Spectrum of Rigel.

The more closely the data furnished from Potsdam are scrutinised, the more entirely satisfactory they appear. The margin of possible error qualifying them was at first estimated at about five English miles a second; but subsequent improve-

[1] *Astr. Nach.* No. 2839.

[2] *Monatsberichte*, Berlin, March 15, 1888, p. 397.

ments have reduced it to a much smaller quantity. A searching test of precision is afforded by the orbital movement of the earth as regards each star examined. Sometimes, according to the time of year, directed more or less obliquely towards, sometimes away from the star, it *ought*, since it is just as effective in shifting spectral lines as the movement of the star itself, to produce fluctuations in the gross amount of displacement, while the net result, this known element of terrestrial motion having been eliminated, *ought*, as representing the relative movement of star and sun, to remain constant. The test is triumphantly sustained by the Potsdam determinations. A long series of measures of Capella reflects with such approximate fidelity the changing effects of the earth's revolution, that the extreme differences of the results deduced for the star only slightly exceed four English miles a second.[1] It is scarcely possible that the concluded rate of withdrawal of sixteen miles a second can be wrong either way by as much as two miles. The much smaller approaching movements of *a* Persei and Procyon (seven miles a second in each case) were arrived at with still more satisfactory accuracy. The outcome of each of the two pairs of measures executed, agreed within less than half a mile! A marginal uncertainty of one mile and a half appears to attach to the rate of approach of sixteen miles a second attributed to the pole-star; and it diminishes to about one mile for Aldebaran's swift recession of thirty miles a second. At last then, firm ground has been reached in these critical investigations, the data supplied by which may now be fearlessly employed for the solution of many outstanding problems. That, for instance, of the variable movement of Sirius, found at Greenwich to alternate recession and approach with indications of cyclical regularity in a period of several years. The reality of the phenomenon (upon which the most recent experience of the Greenwich observers has thrown some doubt) will now before long be established or decisively disproved.

To the Table of Parallaxes included in the Appendix, a column has been added showing the linear velocities *across* the

[1] *Astr. Nach.* Nos. 2896-7.

visual ray of the fifty-two stars at known distances. They range from nearly four hundred to one and a half miles, and average sixty miles a second. And this for only one component of their motion! Nor is any reason imaginable why the other component—that in the line of sight—should be inferior n amount, even though Vogel's results, so far as they have gone, give a much lower mean rate. Later on, perhaps, the spectrographic method may bring to light radial velocities as extraordinary as certain tangential ones, by which the average of thwartwise movement is forced up to the above high value.

The first of these startling examples to become known was in 1830 Groombridge. The large proper motion and small parallax of this star compel the ascription to it of a speed, taking into account only that part of it lying square to our view, of at least two hundred miles a second—a speed uncontrollable, according to Professor Newcomb, by the combined attractive power of the entire sidereal universe. For his calculations show that the maximum velocity attainable by a body falling from infinity towards and through a system composed of 100,000,000 orbs, each five times as massive as our sun, and distributed over a disc-like space 30,000 light-years in extent, would be twenty-five miles a second.[1] But 1830 Groombridge possesses fully eight times this speed; and because velocity varies with the square root of the attracting mass, a world of stars of more than sixty-four fold the potency of that assumed as probable, would be required to set this object moving as it does unquestionably move!

Now the velocity producible by an attractive system is the limit of the velocity it can control—that is, bend into a closed curve. It is then certain that unless the stellar system possesses what we may call occult gravitational energies, the star in question cannot be one of its permanent members. Virtually in a straight line and without slackening, it will pursue its course right across the starry stratum it entered ages ago on its unknown errand, and will quit ages hence to be swallowed up in the dusky void beyond. There is, how-

[1] *Pop. Astronomy*, p. 499.

ever, an alternative supposition. The star *may* be acted upon by unknown compulsive influences.

This gains increased probability as the number of abnormally swift stars multiplies. Groombridge 1830 is no longer the only ‘runaway’ of our acquaintance. Linear stellar speed, apart from that share of it directed along the line of sight, exceeds in fact the computed maximum of twenty-five miles a second in nearly half the cases in which it has been ascertained, and the excess is here and there enormous. Thus, Arcturus ‘moves palpably through heaven’ at the rate of 375 miles a second, and the velocity of μ Cassiopeiæ is but slightly inferior, Groombridge 1830 standing only third in point of real celerity. Next to it comes ζ Toucani with its 101 miles, and four southern stars besides progress at above sixty miles per second.

‘Flying stars’ can then no longer be regarded as mere intruders into stellar society. Whether or not belonging to it ‘for better for worse,’ they evidently at present form an important part of it, and the problem they present cannot be excluded from a general consideration of sidereal mechanism. Indeed, they furnish a most significant index to the workings of its secret springs. They pursue their careers, so far as observation can yet tell, in right lines, and at a uniform speed. Their high velocities would be otherwise less perplexing; for they might plausibly be supposed due to the powerful attraction of invisible bodies in their neighbourhood, and to represent, by analogy, the rush past the sun of highly eccentric comets. But the evidence is wholly against any such hypothesis. All proper motions known to us—whether of single stars, or of the centres of gravity of multiple stars—are sensibly rectilinear. The centres of curvature, presumably, of the imaginary lines traced out by them are inconceivably remote. A straight line is only part of the circumference of a circle of infinite radius.

The fact then confronts us that not a few of the stars possess velocities transcending the power of government of the visible sidereal system. Is that system then threatened with dissolution, or must we suppose the chief part of its attractive energy to reside in bodies unseen, because destitute of the faculty of luminous radiation? No answer is possible; conjecture is futile. We are only sure that what we can feebly trace is but a part of a mighty whole, and that on every side our imperfect knowledge is compassed about by the mystery of the Infinite.

Physical peculiarities are not, in any obvious way, related to these excessively rapid movements. Arcturus, although, as regards the character of its spectrum, a 'solar' star, combines, we have found reason to believe, a higher temperature with stronger absorption than are present in the sun. Its mass is probably enormous; its light-power certainly is. It proves, in fact, by a moderate estimate, to be equivalent to that of eight thousand suns! The next swiftest star, μ Cassiopeiæ, appears to be about nineteen times more luminous, 1830 Groombridge to be slightly less luminous than the sun. Neither of these stars shows a very distinctive spectrum. Stars with banded and gaseous spectra, and variables of all classes, are, as a rule, almost destitute of proper motion; but this may be an effect of remoteness, rather than of genuine inertness.

An unmistakable connection, however, exists between proper motion and sidereal locality. The late Mr. Proctor drew attention to the prevalence in certain regions of the sky of what he termed 'star-drift.' Here and there, unanimity of movement is, to some extent, substituted for the caprice which is the most striking superficial feature of stellar displacements, when deduction has been made of their common perspective element. Amid seeming confusion, order and purpose by glimpses reveal themselves. Battalions of stars—'flying synods of worlds'—regardless, as it were, of the erratic flittings of the casual surrounding crowd, march in widely extended ranks, by a concerted plan along a prescribed track, under orders sealed perhaps for ever to human intelligence.

Among the stars situated between Aldebaran and the Pleiades, there is next to no relative movement. They all drift in company towards the east, by about 10″ in a century. All, we mean, that have been investigated. It has not yet been ascertained to what extent the fainter neighbouring stars share a tendency largely, if not wholly, due to the sun's oppositely directed progress.

Five of the 'Seven Stars' (*septem triones*) forming the Plough (those excluded being the 'Pointer' next the pole, and η at the extremity of the handle) were regarded by Mr. Proctor as members of a vast united group advancing, but at a higher speed, towards the same point with the sun. But his inference has been in a measure overthrown by Dr. Auwers's exact determinations of the very small proper motions affecting these brilliant objects. The concert assumed to exist between them is thus so gravely disturbed that only two out of the five can now be safely bracketed as companions. These are ε and ζ in the Plough-handle, the separation of which by 4° 22′ of a great circle implies a gap so vast as to be measured by many—we cannot tell how many—years of light-travel. The system thus constituted is, at the very least, a quintuple one, since ζ Ursæ, as our readers are aware,[1] carries with it three dependent stars, one (Alcor) visible to the naked eye, one a telescopic attendant, the third revealed only by the spectroscopic effects of orbital motion.

It is scarcely likely, however, that the system is complete in itself. Common proper motion does not necessarily imply the mutual revolution of the objects to which it belongs. What it does imply is their systemic connection. But that connection need not be of the kind exemplified close at hand by the earth and moon. It may rather be such as prevails between the earth and Venus, or between Jupiter and Saturn. The group in Ursa Major, it is safe to assert, includes examples of both kinds of relationship. Of the movements of two satellites, Mizar (ζ Ursæ) is the undoubted mainspring. The status of Alcor is dubious. Its path at present appears strictly rectilinear; but curvature relative to the large adjacent star *may* be rendered sensible in it by future observations. About what might be called the personal independence, however, of the distant ε, there is little room for doubt. Although dominated by the same influence, it advances on its own account; and since the probabilities are strong against

[1] See *ante*, p. 212.

its keeping strict pace with its travelling companion, their relationship will perhaps eventually cease to be traceable. Slight inequalities betraying differences in the period of revolution round the same remote centre may easily co-exist with what is known as common proper motion. Such discrepancies can alone hold the stars affected by them aloof from binary combination. While travelling along parallel lines, they have still a relative velocity exceeding, at their distance apart, the power of their mutual gravitation to sway into an ellipse. One must hence fall very slowly behind the other, as Saturn falls behind Jupiter after conjunction. Evidence of their affinity is then only temporarily accessible to us. After many ages it will cease to come within the sphere of our possible recognition. There may be, probably are, in distant parts of the sky, stars revolving in boundlessly spacious orbits round the same focus of attraction with ε and ζ Ursæ; but we have no means of identifying them.

'Partial systems,' governed presumably from without, are of tolerably frequent occurrence. The first to become known was discovered by Bessel in 1818.[1] It is composed of a fifth and a seventh magnitude star known respectively as 36 A Ophiuchi and 30 Scorpii, more than twelve minutes of arc apart, yet endowed with an accordant movement of $1''\cdot22$ yearly. The former star has a close attendant; and an intermediate minute object also forms part of the company.[2] Another interesting quadruple group was detected by Flammarion in 1877.[3] Two couples in the Swan, one revolving, the other in appearance fixed, separated by an interval of $15'$, drift slowly southward together in a direction nearly perpendicular to the line of march of the sun. Their movement is hence 'proper' to themselves, perspective effects being unconcerned with it. The stationary pair is the fifth-magnitude yellow star, 17 Cygni, with its bluish satellite at $26''$; the circulating pair consists of two eighth-magnitude stars at $3''$, numbered 2576 in Struve's great Catalogue.

[1] *Fundamenta Astronomiæ*, p. 311.

[2] Flammarion, *Comptes Rendus*, t. lxxxv. p. 783; *Cat. des Étoiles Doubles*, p 105.

[3] *C. Rendus* t. lxxxv. p. 510.

The most curious instance of concerted movement yet brought to light is afforded by two ninth-magnitude stars in Libra, discovered by Schönfeld in 1881 to progress across the sphere at the exceptionally quick rate of 3″·7 annually.[1] Notwithstanding the wide interval (5′) separating them, their advance seems perfectly harmonious. They flit side by side, as if rigidly connected across a chasm probably some thousands of millions of miles in width. The measures of their parallax, now progressing at the Cape, will soon decide whether they are as near the earth as might be supposed, and supply the means of determining their true velocity, so far as it lies square to the line of sight.

To the question—Has the sun any associates in his journey through space? only a provisional answer can as yet be given. None are known, but investigations on the point are barely nascent. The peculiarities which we should expect beforehand to attend such companion-stars are comparative proximity and relative immobility. They should have sensible parallaxes, and be devoid both of radial and tangential velocity. Neither spectroscopic nor telescopic evidence of motion should be derivable from them. It is certain, indeed, that no star up to this completely investigated combines these characters; but then they could not possibly be found in the 'proper motion' stars chosen by preference as the subjects of parallactic observations. When more has been done in photographically registering line-of-sight movements, stars may perhaps be discovered sensibly fixed as regards the sun, because borne along with him at the same translatory speed. The construction of such a group, and the investigation of the peculiarities of its members, might open up a fascinating branch of inquiry. But its existence is purely imaginary; not a single star can be pointed to as at all likely to be coupled with the sun in his advance. They should be diligently searched for, since their occurrence, or non-occurrence, must be of essential importance to any theory of sidereal construction; yet without undue confidence as to the result of the search.

[1] *Sitzungsberichte Niederrheinischer Ges.* Bonn 1881, p. 172.

If the system formed by the stars be destined to permanence in its present shape, some general law of movement must be obeyed by them. Even if its state be one of progressive modification, a prevalent method of change ought to make itself felt. Local irregularities, however, so effectually disguise the fundamental harmony that its presence may long continue a matter of speculative belief.

The assumption is indeed indispensable, as Dr. Schönfeld pointed out in 1883,[1] that the motions of the stars are somehow related to the plane in which the vast majority of them are disposed. For otherwise their actual configuration would be a wildly improbable accident of the time in which we live. The Milky Way, to put it otherwise, should be regarded as an evanescent phenomenon, unsustained by any persistently acting forces, the outcome of a hundred millions of casual conjunctions. If this be incredible (as it surely is), then we are constrained to admit a preference, in the long run, among stellar displacements for the grand level of stellar aggregation. The Milky Way must be, in some true sense, what Lambert called it in the last century—the 'ecliptic of the stars.'

Sir John Herschel imagined the law of harmony to consist in a general parallelism of stellar motions, involving a kind of systematic circulation, as of a solid body round an axis perpendicular to the galactic plane. Innumerable exceptions to any such rule are of course to be found, but they were assumed, in the upshot, to be mutually destructive, the main 'stream of tendency' flowing on irrespective of them. But it is difficult to conceive a physical basis for a quasi-rotational system wholly without warrant from experience. More plausible is M. Ludwig Struve's view that the main part of the revolutions of the stars round their common centre of gravity situated *in* the Milky Way, are performed in planes slightly inclined to that of the zone towards which they are concentrated.[2] His attempt, indeed, to elicit a 'rotation-component' from the secular movements of Bradley's stars, proved unavailing. Yet this is not decisive against the truth

[1] *V. J. S. Astr. Ges.* Jahrg. xvii. p. 255.

[2] *Mémoires de St. Pétersbourg*, t. xxxv. No. 3 pp. 5, 19.

of an hypothesis which leaves open the possibility of a balanced stellar circulation pursued in opposite senses.

M. Rancken was in 1882 more fortunate with a selected list of stars.[1] He admitted only those within thirty degrees on either side of the Milky Way, and possessed of annual proper motions not exceeding a quarter of a second. The solution of his equations showed these movements to include a common element of very slow progressive increase of galactic longitude. That is to say, the 106 stars considered were found subject to a drift along the Milky Way in the direction from Aquila upward towards Cygnus and Cassiopeia, and down past Capella through the Club of Orion towards the Ship. The reality and extent of this drift will be a matter for future investigation. Should the one be confirmed, and the other ascertained, something like a clue to the labyrinth of stellar movements will have been provided. Even the suggestion of its presence is useful as an indication of what may be looked for with some chance of success.

[1] *Astr. Nach.* No. 2482.

CHAPTER XXIII.

THE MILKY WAY.

The Milky Way shows to the naked eye as a vast, zone-shaped nebula; but is resolved, with very slight optical assistance, into innumerable small stars. Its stellar constitution, already conjectured by Democritus, was, in fact, one of Galileo's earliest telescopic discoveries. The general course of the formation, however, can only be traced through the perception of the cloudy effect impaired by the application even of an opera-glass. Rendered the more arduous by this very circumstance, its detailed study demands exceptional eyesight, improved by assiduous practice in catching fine gradations of light. Our situation, too, *in* the galactic plane is the most disadvantageous possible for purposes of survey. Groups behind groups, systems upon systems, streams, sheets, lines, knots of stars, indefinitely far apart in space, may all be projected without distinction upon the same sky-ground. Unawares, our visual ray sounds endless depths, and brings back only simultaneous information about the successive objects met with. We are thus presented with a flat picture totally devoid of perspective-indications. Only by a long series of inductions (if at all) can we hope to arrange the features of the landscape according to their proper relations.

To the uncritical imagination, the Milky Way represents a sort of glorified track through the skies —

> A broad and ample road, whose dust is gold
> And pavement stars, as stars to thee appear,
> Seen in the galaxy, that milky way,
> Which nightly as a circling zone thou seest,
> Powdered with stars.

In American-Indian fancy a mysterious 'path of souls,' its popular German name, 'die Jakobsstrasse,' recalls the time when it stood as a celestial figure of the way of pilgrimage to Compostella. This superficial impression is, however, effaced by closer inspection, and an aspect is assumed resembling rather that of a rugged trunk marked by strange cavities and excrescences, and sending out branches in all directions.

The medial line of the Galaxy is sensibly a great circle.[1] This shows that the formation extends symmetrically on all sides of us. We are placed somewhere in its main level, though not necessarily very near its centre. The superior brilliancy of the Milky Way in its southern sections has, indeed, suggested that the sun lies much nearer to the edge of the apparent ring there than elsewhere; but a positive conclusion on the point would still be premature. None of the various movements affecting the earth bears any obvious relation to the starry collection around it. Neither the equator, the ecliptic, nor the line of the sun's way, shows any trace of conformity to its plane. The great circle of the Galaxy is inclined about sixty-three degrees to the celestial equator, which it intersects in the constellations Monoceros and Aquila. It passes in Cassiopeia within twenty-seven degrees of the north pole of the heavens, in Crux, as near to the south pole, while its own poles are located respectively in Coma Berenices [2] and Cetus. Over two-thirds of the celestial circuit, the general unity of this stupendous structure is preserved. Broken, however, near α Centauri by the interposition of a great fissure, it is only regained, after an interval of some 120°, through the reunion, in the neighbourhood of ϵ Cygni, of the separated portions. Involuntarily, the image presents itself of a great river, forced by an encounter with a powerful obstacle to throw its waters into a double channel, lower down merged again into one. The intervening long strip of islanded rock and gravel might stand for the great rift between the branches of the sidereal stratum, which, although to the eye,

[1] Gould, *Uranometria Argentina*, p. 370.

[2] R.A. 12h. 41m. 20s., Dec. + 227°21′, according to Gould.

owing to the effect of contrast with the 'candid way' on either side, darker than the general sky, is in reality nowhere quite free from nebulous glimmerings. It is encroached upon by fringes, effusions, and filaments, spanned by bridges of light, and here and there it is half filled up by long, narrow, disconnected masses, or *pools* of nebulæ, lying parallel to the general flow of the stream. One such 'brilliant and tortuous streak'[1] extends, in almost complete isolation, over nearly 20°, from the tail of Serpens across a corner of the Shield of Sobieski. Moreover, only one of the two principal branches —that traversing Aquila and the bow of Sagittarius—is continuous. The other, after covering the tail of Scorpio 'with a complicated system of interlaced streaks and masses,'[2] dies out in Ophiuchus, about fifteen degrees south-west of the termination, just at the equator, of the arm sent out to meet it through Cygnus. The gap is, however, partially veiled by a faint luminous extension from the south, and thus shows as absolute only over some five degrees of the sphere.

This is not the only interruption to the course of the Milky Way. And that occurring in Argo is the more remarkable that it cuts sheer across the entire, undivided stream. Here, at south declination 33°, the formation, Sir John Herschel says, 'opens out into a wide, fan-like expanse, nearly 20° in breadth, formed of interlacing branches, all which terminate abruptly in a line drawn nearly through λ and γ Argûs.'[3] On the opposite, or eastern side of a moderately broad blank space, a similar assemblage of branches converges upon the variable star η Argûs. There is an obvious correlation of structure on either side of the chasm; subdivisions mutually correspond; the broken series on one margin is resumed on the other.

The impression is strongly conveyed that star-strata once united, have here yielded to the influence of some unknown dispersive force or forces, perhaps still in operation. Yet we

[1] *Uran. Argent.* p. 381. [2] Herschel, *Outlines*, art. 789.
[3] *Ibid.* art. 787.

can scarcely hope ever to command the means of testing the conjecture. For the proper motions of the faint telescopic stars near the edges of the gap are no doubt of such excessive seeming minuteness that centuries, nay, millenniums may pass before they can become perceptible.

The representation of the Milky Way as a uniform starry stream is purely conventional. Its real texture is of a curdled or flaky description.[1] Between Perseus and Sagittarius, Sir William Herschel counted eighteen luminous patches, 're-sembling the telescopic appearance of large, easily resolved nebulæ;'[2] and his son perceived the lucid ramifications in Sagittarius to be made up of 'great cirrous masses and streaks,' the appearance, as his telescope moved, being 'that of clouds passing in a *scud*, as the sailors call it.' Further on, he remarks: 'the Milky Way is like sand, not strewed evenly as with a sieve, but as if flung down by handfuls (and both hands at once), leaving dark intervals, and all consisting of 14th, 16th, and 20th magnitudes,[3] down to nebulosity, in a most astonishing manner.'[4]

The bright spaces of the galactic zone are commonly surrounded and set off by dark winding channels, and the rapid alternation of amazingly rich with poor, or almost vacant patches of sky, is a constantly recurring phenomenon.[5] The most remarkable instance occurs in the Southern Cross, the brilliant gems of which emblazon a broad galactic mass very singularly interrupted by a pear-shaped black opening eight degrees long by five wide, named by early navigators the 'Coal-sack.' This yawning excavation, which is, however, not absolutely denuded of stars, figures in Australian folk-lore as the embodiment of evil in the shape of an Emu, who lies in wait at the foot of a tree, represented by the stars of the Cross, for an opossum driven by his persecutions to take refuge

[1] Houzeau, *Uranométrie Générale*, p. 16, 1878; Klein, *Wochenschrift für Astr.* 1867, p. 288.

[2] *Phil. Trans.* vol. civ. p. 282.

[3] Herschel's '20th magnitude' corresponds approximately with the 14th on the photometric scale.

[4] *Cape Observations*, p. 388.

[5] Herschel, *Outlines*, arts. 790, 797.

among its branches.[1] The legend reads almost like a Christian parable.

Partial galactic vacuities, evidently of the same nature with the southern Coal-sack, though less perfectly developed, occur elsewhere, notably in Cygnus; but they are inconspicuous to casual observers. A telescopic perforation has been adverted to by Mr. Barnard.[2] 'A most remarkable small, inky black hole in a crowded part of the Milky Way' in Sagittarius, is described by him as 'about 2′ in diameter, slightly triangular, with a bright orange star on its north preceding (north-westerly) border, and a beautiful little cluster following.' This singularly abrupt interruption to the clouds and 'breaking sprays' of stars around it, appears clearly on a negative taken in 3h. 7m., August 1, 1889, with highly suggestive result as regards the same observer's project for photographically charting the Milky Way.

Meantime, incomparably the best delineation of the sections fo it visible in the northern hemisphere, has recently been completed at Parsonstown after five years of labour, by Dr. Otto Boeddicker. The amount of unmistakably genuine detail recorded in it, despite climatic conditions of the least propitious sort, is simply astonishing; and the peculiarities thus brought to light are in many cases of a novel and significant character. The general effect may be best described as that of a thick *stem* of light, closely set with curvilinear ramifications; the stem itself being riddled with dusky convolutions, intricate passages, and 'horse-shoe' or 'key-hole' apertures, separated by lustrous wisps and nebulous 'pointed arches.' The circumstance that 'feelers are thrown out towards nebulæ and clusters,'[3] is of profound interest. The Andromeda nebula, for instance, terminates a feeble branch starting from α Cassiopeiæ. The Pleiades stand at the peaked summit of a dim arch springing, on one side, from near β Tauri, on the other, from ϵ Aurigæ. The Hyades are separately involved; Præsepe is all but reached by a long streamer issuing from the vicinity of β Canis Minoris; while a thin sinuous effusion, perhaps of

[1] MacPherson, *Journ. R. Soc. N. S. Wales*, 1881, p. 72.
[2] *Monthly Notices*, vol. l. p. 311. [3] *Ibid.* p. 12.

a spiral nature, includes the great nebula in its sweep through Orion. A parallel streak of barely traceable nebulosity (also noticed by Mr. Gore in India) obviously guides or is guided by a flow of small, but still lucid stars, marking the lion's skin supposed thrown over the left shoulder of the hero; but other instances of apparent relationship between galactic wisps and bright stars are not improbably of purely physiological origin. A broad effusion is depicted by Dr. Boeddicker as just enveloping the pole-star, and all the cavities and interstices of the formation appeared to him filled with dim luminosity.

The Milky Way is throughout subject to great and sudden variations of width. A brilliant stream, no more than three or four degrees in extent where it enters the Cross, it spreads, exclusive of faint borders, to twenty-two degrees in its bifurcated section between Ophiuchus and Antinous. At some spots, too, the nebulous effect to the eye fades away imperceptibly along the margins;[1] at others, the line of demarcation is so sharp that a telescope may have one half of its field crowded with galactic stars, while the other half is well-nigh blank.[2] A definite semi-circular boundary, for instance, limits the formation near ζ Aquilæ; its southern edge in Ophiuchus was remarked by Sir John Herschel, as 'terminated by an irregular nebulous fringe as if lacerated;'[3] and marginal projections, knobs, and bristling outliers can, in other parts, easily be traced. The prevalent rule seems to be that the smaller the stars considered, the more abrupt is the commencement of the Milky Way; while a more and more gradual condensation accompanies each step upward in brightness.[4]

Sir William Herschel was perfectly satisfied that, with his twenty-foot reflector (equivalent to a modern refractor about fourteen inches in aperture), the Milky Way was, in general, 'fathomable.' The stars composing it, that is to say, were of definite numbers, and appeared projected upon a perfectly black sky. But this was not so everywhere; certain points completely baffled the penetrative faculty of his instrument.

[1] W. Herschel, *Phil. Trans.* vol. civ. p. 283.

[2] Proctor, *Universe of Stars*, p. 86.

[3] Klein, *Wochenschrift*, 1867, p. 285; J. Herschel, *Cape Results*, p. 389.

[4] Celoria, *Memorie del R. Istituto Lombardo*, t. xiv. p. 837.

One such was met with in Cepheus, where he found the small stars to become 'gradually less till they escape the eye, so that appearances here favour the idea of a succeeding, more distant clustering part.' And he remarked, in exploring between Sagitta and Aquila, that 'the end of the stratum cannot be seen.'[1] Again, in the galactic branch traversing Ophiuchus, Sir John Herschel encountered 'large milky nebulous irregular patches and banks, with few stars of visible magnitudes;' described a 'very large space' of the Milky Way in Sagittarius as 'completely nebulous like the diffused nebulosity of the Magellanic Cloud;'[2] and observed a similar spot in Scorpio, 'where, through the hollows and deep recesses of its complicated structure, we behold what has all the appearance of a wide and indefinitely prolonged area strewed over with discontinuous masses and clouds of stars which the telescope at length refuses to analyse.'[3]

Even with the best telescopes of recent construction, this perplexing and indeterminate aspect cannot be altogether got rid of. Professor Holden tells us that the thirty-six inch Lick achromatic shows in the Milky Way 'no final resolution of its finer parts into stars. There is always the background of unresolved nebulosity on which hundreds and thousands of stars are studded—each a bright, sharp, separate point.'[4] The stellar nature of the lingering nebulosity is strongly indicated; but it cannot be pronounced off-hand whether it is due to the presence of innumerable small stars mixed up in the same region with larger ones, or to the indefinite extension outward into space of galactic agglomerations.

The explanations attempted of these complicated phenomena may be divided into disc-theories, ring-theories, and spiral-theories. The 'disc-theory' of the Milky Way was first propounded by Thomas Wright, of Durham, in 1750. He supposed all the stars to be distributed in a comparatively shallow layer, producing, by its enormous lateral extension, the effect of annular accumulation. Irregularities, he thought, were partly

[1] *Phil. Trans.* vol. cvii. p. 326.
[2] *Cape Observations*, p. 389.
[3] *Outlines of Astr* art. 798.
[4] *Sid. Mess.* Aug. 1888, p. 298.

due to our eccentric position within the stratum, partly to 'the diversity of motion that may naturally be conceived amongst the stars themselves, which may, here and there, in different parts of the heavens, occasion a cloudy knot of stars.'[1]

To this view Sir William Herschel gave wide currency and apparent stability by the application to its support of his ingenious method of 'star-gauges.' By counting the stars simultaneously visible in his great reflector in various portions of the sky, he showed that their paucity or abundance depended upon the situation of the gauge-fields relative to the Milky Way. In its neighbourhood, stars were copiously—far from it, they were sparsely—distributed. And this by a regular progression of density from the galactic poles to the galactic equator, the latter region being on an average thirty times richer than the former. Now, if we were to admit, as Herschel did, a nearly equable scattering of stars, there would be no alternative but to suppose the sidereal system extended in any direction proportionately to the number of stars seen in that direction. Their crowding should, *on that hypothesis*, be purely optical—the effect of the indefinite spreading out in the line of sight of their evenly serried ranks. Sounding the star-depths upon this principle, Herschel measured the length of his line by their seeming populousness, and constructed, from the numerical data thus obtained, the 'cloven disc' model, long accepted as representing the true form of the stellar universe.

But his own observations at the very moment of enouncing this theory, fatally undermined it. Already in 1785, he remarked that two or three hundred 'beginning, or gathering clusters,' might be pointed out in the galactic system, and he surmised its eventual separation, 'after numbers of ages,' into so many distinct 'nebulæ.'[2] 'Equable scattering,' then, was an ideal state of things long since abolished by the 'ravages of time.' The conviction that such was the case grew with his experience. 'The immense starry aggregation of the

[1] *An Original Theory of the Universe*, p. 63.
[2] *Phil. Trans.* vol. lxxv. p. 255.

Milky Way, he wrote in 1802,[1] 'is by no means uniform. The stars of which it is composed are very unequally scattered, and show evident marks of clustering together into many separate allotments.' Nor did he fail to perceive, from the gradual increase of brightness towards the centres of these 'allotments,' that they tended to assume a spherical form, and thus suggested 'the breaking-up of the Milky Way, in all its minute parts, as the unavoidable consequence of the clustering power arising out of those preponderating attractions which have been shown to be everywhere existing in its compass.'[2] The formal announcement of his conviction 'that the Milky Way itself consists of stars very differently scattered from those which are immediately about us,'[3] amounted to a recantation of the principle of star-gauging.

With it disappeared from Herschel's mind the conception of an optically-produced galaxy. In his ultimate opinion, the actual corresponded very closely with the apparent structure: it was composed, that is to say, mainly, if not wholly, of real clouds of stars. Credit was thus restored to the early impression of Galileo, who in 1610 described the Milky Way as 'nothing else but a mssa of innumerable stars planted together in clusters.'[4]

Wilhelm Struve's[5] effort towards the re-organisation of the stratum-theory, though aided by all the resources of his great ability and address, could scarcely be counted as a step in advance. Substituting for the hypothesis of equable distribution that of concentration in parallel planes, he imagined the average interval of space between the stars to diminish regularly with approach to the central horizon of the system. The swarming aspect of the Milky Way was hence regarded as agreeing with fact, but the annular appearance as being illusory. Of illimitable dimensions, the system was conceived to stretch away, still preserving its specific character, to an infinite, or at least unimaginable remoteness, comparatively

[1] *Phil. Trans.* vol. xcii. p. 495.

[2] *Ibid.* vol. civ. p. 282.

[3] *Ibid.* vol. xcii. p. 480.

[4] *Sidereus Nuncius*, trans. by E. S. Carlos, p. 42.

[5] *Études d'Astronomie Stellaire*, 1847.

narrow *visual* bounds being, however, set to it by a supposed extinction of light.

But the quasi-geometrical regularity of Struve's galaxy is belied by innumerable details of the original. The swell of the tide of stars towards the galactic plane is neither uniformly progressive,[1] nor does it proceed without conspicuous interruptions. Thus, the region near the horns of Taurus, although close to the Milky Way, is absolutely the poorest in the northern hemisphere; [2] and an almost clean-swept space in Scorpio, on meeting which Sir William Herschel exclaimed in amazement, 'Hier ist wahrhaftig ein Loch im Himmel!' lies on the verge of the galactic stream. But it is the openings in the formation itself which afford the most decisive evidence against any modification of the stratum-theory. Is it credible that a boundlessly extended layer of stars should be pierced, in many of its densest portions, by *tunnels* converging directly upon our situation within it! [3] No sane mind, we venture to say, realising all that such an assertion implies, can assent to it. But, indeed, the entire conformation of the Milky Way,—its streaming offsets, convoluted windings, promontories, and sharply bounded inlets no less than the breaches of its continuity—all flatly contradict the view of its being the optical creation of any universally valid law of star distribution.

We seem then led to the alternative belief that it is a definite structure, at a definite distance from ourselves—a belief forced upon Sir John Herschel by his Cape experiences, notwithstanding his natural reluctance to drift far away from the position originally taken up by his father. The shape suggested by him for the galaxy was that of 'a flat ring, or some other re-entering form of immense and irregular breadth and thickness.' [4] Expanded indefinitely along the central plane, the new model scarcely differed from the old except in so far as the idea of homogeneous construction was given up. The disc remained, but with its centre scooped out. The solar system was located in an enormous space of relative vacuity.

[1] C. S. Peirce, *Harvard Annals*, vol. ix. p. 174.

[2] Argelander, *Bonner Beob.* Bd. v. *Einleitung*; Proctor, *Universe of Stars*, p. 32.

[3] Proctor, *loc. cit.* p. 15.

[4] *Outlines*, art. 788.

The Milky Way was then supposed to consist of an indefinite number of stellar collections 'brought by projection into nearly the same visual line'—to represent the foreshortened effect (more especially at a particular spot in Sagittarius) of 'a vast and illimitable area scattered over with discontinuous masses and aggregates of stars in the manner of the cumuli of a mackerel-sky.'[1] But in an assemblage of this nature—seen *edgewise*—a 'Coal-sack' would be a phenomenon as anomalous as in a uniform stratum; nor could it, without violent improbability, be conceived of as rent by the colossal fractures dividing the actual Milky Way in Argo and Ophiuchus.

To remedy these inconveniences, Professor Stephen Alexander devised in 1852,[2] upon the model of the wheel-shaped nebula in Virgo (M 99), a spiral galaxy with four curvilinear branches diverging from a central cluster formed by the sun and lucid stars. By properly adjusting the mode of projection of these radiating star-streams, the effects of rifts and coal-sacks were duly produced; but the arrangement, however admired for ingenuity, gave no persuasion of reality, and quickly dropped out of remembrance. Essentially different, although with some features in common, was that by which Mr. Proctor replaced it in 1869.[3] Rather than a 'spiral,' indeed, the new design resembled a bent and broken ring, with long, riband-like ends, looped back on either side of an opening, accommodated to the shape of the gap in the visible structure in Argo. One of these loops, by the apparent inter-crossing of its near with its remoter branch, was supposed to generate the Coal-sack in Crux; while the other end, trailing lengthily backward, afforded a deceptive effect of bifurcation. Excessive distance was brought in, as in Professor Alexander's scheme, to explain the cessation of nebulous light in Ophiuchus.

Of the manifold objections to which this hypothesis is liable,[4] only two need here be mentioned. In the first place, it involves a wholly inadmissible rationale of the openings seen

[1] *Cape Observations*, p. 389. [2] *Astr. Jour.* vol. ii. p. 101.

[3] *Monthly Notices*, vol. xxx. p. 50.

[4] See Mr. J. R. Sutton's remarks, *Illustrated Science Monthly*, vol. ii. pp. 63, 199.

in the Milky Way. If these were due to the interlacing by perspective of branches really far apart in space, the enclosing nebula should be markedly fainter on one side than on the other. But this is not so. The borders of the southern Coal-sack, for instance, are approximately of the same brightness all round. A single vivid mass has obviously been the scene of what, in the absence of better knowledge, may be described as an excavatory process.

Again, on the spiral theory, the great rift in the Milky Way should be ordinary sky-background, the branches on either hand being mutually disconnected except through the optical effect of projection. But there can be no doubt that the rift forms, in a certain sense, a part of the galactic structure. Stars are much more thickly strewn upon it than in the external heavens. Argelander, in fact, showed that, down to 9·5 magnitude, they are only $\frac{1}{21}$ less plentiful than in the adjacent branches of the Milky Way itself, and are actually more plentiful than in the section of the undivided stream passing from Perseus to Auriga.[1] Nor can the fissured parts be regarded as truly independent. Their separation is gradually effected. Premonitory cavities seem to announce it beforehand, and even after it has become definitive, abortive efforts towards reunion are indicated by the correspondence of opposite projections. The bifurcation is beyond question a physical reality.

Over-subtlety has been the besetting snare of theorists on this important subject. Perhaps by adopting the simple view that the Milky Way *is* very much what it *seems*, we shall get nearer to the truth than by indulging in more recondite speculations. What it seems to be—especially as delineated by Dr. Boeddicker—is a ring with streaming appendages—outliers extending from the main body in all possible directions, some nearly straight towards, or away from us, others at every imaginable angle with our line of sight. The results in perspective foreshortening must evidently, under these circumstances, be highly complex; the eye being presented with groups and streams of stars, at immensely different real dis-

[1] *Bonner Beob.* Bd. v. *Einleitung.*

tances, but all projected indiscriminately upon the same zone of the heavens. Thus, while some branches, pursued along their outward course, fade at last into dim nebulosity, other Milky Way groups may be distinguished as bright separate stars, because much nearer to us than the generality of their associates.

The internal organisation of the Galaxy must be in the last degree intricate. It collects within its ample round, there is every reason to suppose, an absolutely endless variety of separate systems. A multitudinous aggregate of individual clusters, it exhibits, moreover, as a whole, the structure of one single cluster on a prodigious scale. Its fringed edges, its rifts and vacuities, are, as we have seen, reproduced in miniature in innumerable star-groups. 'Rings,' and 'sprays,' and 'streams' of stars are unmistakably common to the two orders of formation; and the stellar constituents of both are frequently involved with gaseous nebulæ in a way showing most intimate association by origin and development. The laws then governing stellar aggregation in the one case govern it also in the other; and so, from this direction independently, we again reach Herschel's conclusion that the Milky Way 'consists of stars very differently scattered from those which are immediately about us.'

But are these stars *suns*, co-ordinate with our own? or must we regard them as comparatively insignificant bodies, sharing a sun-like nature, indeed, but on a far lower level of power and splendour? The question is equivalent to this other perpetually recurring one, What is their average distance from ourselves? In what portion of space do the true galactic condensations occur? How far outward should we have to travel before finding ourselves actually in the midst of the crowded objects producing, to terrestrial observers, the 'milky' effect of a nebulous stratum?

Now we are not so completely destitute of knowledge on this point as is commonly supposed. Our readers, we believe, will, when the nature of the evidence already at hand has been explained to them, readily admit its all but demonstrative significance to the effect that the star-clouds of the galaxy lie *beyond* the average distance of tenth-magnitude stars.

The numbers of each order of stars down to the ninth magnitude, derived with approximate precision from Argelander's 'Durchmusterung,' are found, on the whole (apart from an excess of bright stars to be separately accounted for), to increase at about the rate to be expected on the hypothesis of an equable distribution through space of bodies on the same general level of luminous power. This kind of uniformity does not of course preclude any extent of individual variety; it only means that individual varieties, however wide, balance each other in the long run, and on a sufficiently extended average. The stars then, down to the ninth magnitude, are distributed roughly according to the space they occupy; there are more of each magnitude in proportion to the increased cubical contents of the successive spheres, the radii of which are the theoretical mean distances corresponding to each successive magnitude. This theoretical increase of numbers concurrently with available space, is measured by the *cube of the distance-ratio*; it is nearly four-fold. That is to say, there should be (supposing equable distribution) close upon four times as many stars of a given magnitude as of the magnitude immediately superior to it. And, speaking roughly, there are. This law of increase, moreover, seems to be conformed to, as closely as could be expected, over the galactic zone itself no less than over the rest of the sky.

It is then perfectly evident that, down at least to the ninth magnitude, the progression outward is unbroken by any great or systematic accumulation of stars. *The Milky Way swarms collect further off.* But there is more. The stars of the 'Durchmusterung,' including a multitude of nominally 9·5, but really tenth magnitude, are, as already mentioned, nearly as thickly strewn over the dark fissure of the Milky Way between Cygnus and Centaur, as over the bright enclosing branches. The nebulous effect of these latter to the eye is then presumably due to more remote collections. As to the further limits of these, we know as yet nothing, except that Herschel's gauge-numbers left it to be inferred that 'thinning out' had become marked before the attainment of fourteenth-magnitude distance. It is worth mentioning that the rift was well marked by deficiency in Herschel's counts as well as among the stars to the eleventh magnitude given by M. Celoria's 'soundings' at Milan in 1876; but it is only from enumerations by order of magnitude that anything can be learned relative to progressive distances in celestial space. Such enumerations, executed by photographic means,[1] will perhaps before long afford solutions to cosmical problems with which it would be vain to attempt otherwise to grapple. Each of these should form the subject of a specially directed inquiry, since thought moves faster when it is not allowed an over-wide range. There is, indeed, some danger lest the facility of modern methods should lead to the unmethodical accumulation of data too numerous and too indiscriminate for the thinking faculty ever properly to master. In such extended fields, observation should put on, to some extent, the nature of experiment by becoming determinate and prearranged.

Enough has been said, it would seem, to justify the conclusion that the main part of the annular structure we call the Milky Way lies at a distance from us intermediate between the distances belonging respectively to the tenth and fourteenth stellar magnitudes. But if this be so, the orbs clustering in its nebulous convolutions are by no means of subordinate importance. At the average distance we are for the present (certainly from imperfect indications) compelled to ascribe to tenth-magnitude stars, our sun would appear many times fainter than the great majority of galactic stars; it would scarcely in fact attain the fifteenth magnitude. And there can at least be very little doubt that galactic stars are in general fully its equals in real lustre.

From a most careful study of the Milky Way at Cordoba, where it was seen to peculiar advantage, Dr. Gould inclined to regard it as the product of two or more superposed galaxies.[1] The fact ndeed of the two narrowest and brightest, and the

[1] See 'Photographic Star Gauging,' *Nature*, vol. xl. p. 344.

[2] *Uram Argentina*, p. 381.

two most diffused parts lying in pairs opposite to each other is a sufficiently remarkable one, and lends some countenance to the surmise that the 'necks' in Cassiopeia and Crux really represent the intersections of the two crossed rings visibly diverging in Ophiuchus. M. Celoria, too, adopted the hypothesis of a compound Milky Way, but of such a form as to allow the possibility of one of its constituent annuli being comprehended by the other.[1] What is unmistakable is that the entire formation, whether single or compound, is no isolated phenomenon. All the contents of the firmament are arranged with reference to it. It is a large part of a larger scheme exceeding the compass of finite minds to grasp in its entirety.

[1] *Memorie dell' Istituto Lombardo*, t. xiv. p. 82.

CHAPTER XXIV.

STATUS OF THE NEBULÆ.

The question whether nebulæ are external galaxies hardly any longer needs discussion. It has been answered by the progress of discovery. No competent thinker, with the whole of the available evidence before him, can now, it is safe to say, maintain any single nebula to be a star system of co-ordinate rank with the Milky Way. A practical certainty has been attained that the entire contents, stellar and nebular, of the sphere belong to one mighty aggregation, and stand in ordered mutual relations within the limits of one all-embracing scheme—all-embracing, that is to say, so far as our capacities of knowledge extend. With the infinite possibilities beyond, science has no concern.

The chief reasons justifying the assertion that the status of the nebulæ is *intra-galactic*, are of three kinds. They depend, first, upon the nature of the bodies themselves; secondly, upon the stellar associations of many of them; thirdly, upon their systematic arrangement as compared with the systematic arrangement of the stars.

The detection of gaseous nebulæ not only directly demonstrated the non-stellar nature of a large number of these objects, but afforded a rational presumption that the others, however composed, were on a commensurate scale of size, and situated at commensurable distances. It may indeed turn out that gaseous and non-gaseous nebulæ form an unbroken series, rather than two distinct classes separated by an impassable barrier. For the bright lines indicating gaseity are accompanied by more or less of continuous light, and the continuous spectra significant of advanced condensation *perhaps* include bright lines, while the proportions of these continuous and discontinuous ingredients differ considerably in different individual nebulæ. But before any settled opinion can be formed as to whether these differences have really the transitional or 'evolutionary' meaning we might be inclined to attribute to them, nebular spectroscopy must be a good deal further advanced than it is at present. Apart from this question, however, there is such strong evidence of relationship between the various orders of nebulæ that to admit some to membership of the sidereal system while excluding others would be a palpable absurdity. And since those of a gaseous constitution must be so admitted, the rest follow as a matter of course.

Of the physical connection of nebulæ with particular stars, fresh and incontrovertible proofs accumulate day by day. Nothing can be more certain than that objects of each kind co-exist in the same parts of space, and are bound together by most intimate mutual ties. To argue the matter seems, as the French say, like 'battering in an open door.' We need only recall the stars of the Pleiades, photographically shown to be intermixed with nebulæ, and those in Orion still bearing in their spectra traces of their recent origin from the curdling masses around. The nuclear positions so frequently occupied in nebulæ by stars single and multiple, reiterate the same assertion of kinship, emphasised still further by the phenomena of stellar outbursts *in* nebulæ. The scenes of these *must* lie within the circuit of the Milky Way, unless we are prepared to assume the occurrence, in extra-sidereal space, of conflagrations on a scale outraging all probability.[1] It has been calculated that if the Andromeda nebula were a universe apart of the same real extent as the Galaxy, it should be situated, in order to reduce it to its present apparent dimensions, at a minimum distance of twenty-five galactic diameters.[2] And a galactic diameter being estimated by the same authority at thirteen thousand light-years, it follows that, on the supposition in question, light would require 325,000 years to reach us from the nebula. The star then which suddenly shone out

[1] This point was frequently insisted upon by the late Mr. Proctor.

[2] Weisse, *Schriften Wiener Vereins*, Bd. v. p. 318.

in the midst of it in August 1885, should have been at 564 times the distance inferrible from its effective brightness. In real light it should have been equivalent to 318,000 stars like Regulus, or to nearly *fifty million* such suns as our own! But even this extravagant result inadequately represents the real improbability of the hypothesis it depends upon; since the Andromeda nebula, if an external galaxy, would almost certainly be at a far greater remoteness from a sister-galaxy than would be represented by twenty-five of its own diameters.

Just as the Milky Way might be described as a great compound cluster made up of innumerable subordinate clusters, so the greater Magellanic Cloud seems to be a gigantic nebula embracing, and bringing into some kind of correlation, multitudes of separate nebulæ. To the naked eye it shows vaguely a brighter axis spreading at the extremities so as to produce a resemblance to the 'Dumb-bell' nebula; it shows, that is to say, signs of definite organisation as a united whole; and it includes, strangely enough, among its inmates a miniature of itself (N. G. C., 1978), but of much greater intensity and distinctness. Sir John Herschel's enumeration in 1847 of the contents of the 'Cloud' gave conclusive evidence of the interstellar situation of nebulæ—evidence the full import of which Dr. Whewell was the first to perceive. Over an area of forty-two square degrees, 278 nebular objects (stars being copiously interspersed) are distributed with the elsewhere unparalleled density of 6½ to the square degree. 'The Nubecula Major,' Herschel wrote, 'like the Minor, consists partly of large tracts and ill-defined patches of irresolvable nebula and of nebulosity in every stage of resolution, up to perfectly resolved stars like the Milky Way, as also of regular and irregular nebulæ properly so called, of globular clusters in every stage of resolvability, and of clustering groups sufficiently insulated and condensed to come under the designation of clusters of stars.' [1]

Here then we find—in a system certainly, as Herschel said, '*sui generis*,' yet none the less, on that account, instructive as to cosmical relationships—undoubted stars and undoubted nebulæ at the same general distance from the earth. Some

[1] *Cape Results*, p. 146.

of the nebulæ may indeed very well be placed actually nearer to us than some of the stars; and the extreme possible difference of their remoteness cannot in any case exceed one-tenth of the interval between the hither edge of the Cloud and ourselves. We learn too the plain lesson that distance is only one factor in the production of 'irresolvability.' For stars in every stage of crowding, from loose groups to the veriest dust-streaks, globular clusters coarse and fine, nebulæ of all kinds and species, range side by side in this extraordinary collection, proving beyond question that differences of aggregation are real and enormous, and need no additional abysses of space to account for them.

Even, however, if all these mutually confirmatory arguments could be dismissed as invalid, the mode of scattering of nebulæ on the sky-surface would alone suffice to demonstrate their association with the sidereal system. Sir William Herschel was early struck with the occurrence of beds of these objects, preceded and followed by spaces void of stars. His assistant was indeed sometimes warned by him, not without good cause, 'to prepare, since he expected in a few minutes to come at a stratum of the nebulæ, finding himself already on nebulous ground.'[1] He attained, too, a partial comprehension of the larger plan of their distribution, as being the *inverse* to that of stars; but the younger Herschel first brought into clear view the distinct and striking division of the nebulæ into 'two chief strata, separated by the Galaxy.' Taking the circle of the Milky Way as a horizon, he remarked that the mass of them gathered together in Virgo and Coma Berenices 'forms, as it were, a canopy occupying the zenith, and descending thence to a considerable distance on all sides, but chiefly on that towards which the (celestial) north pole lies.'[2]

This characteristic of accumulation about the galactic pole is less marked in the southern hemisphere, though here too there is a 'chief nebular region' approximately corresponding to that in Virgo. The distribution is, however, on the whole, much more uniform than in the northern hemisphere, or rather, more uniformly *patchy*, rich districts alternating with

[1] *Phil. Trans.* vol. lxxiv. p. 449. [2] *Cape Results*, p. 137.

more or less ample vacuities. One of these extends about fifteen degrees all round the south pole, the Lesser Cloud marking its edge. The remarkable fact, too, was noticed by Sir John Herschel that the larger nubecula seems 'to terminate something approaching to a zone of connected patches of nebulæ,' reaching across Dorado, Eridanus, and Cetus to the equator, where it unites with the 'nebular region of Pisces.' A similar line of communication is less conspicuously kept open with the minor nubecula, and this feature of 'streams' of nebulæ with terminal aggregations was considered by Mr. Proctor to be distinctive of southern skies.[1] He adverted besides to the coincidence of two of them with stellar 'streams' in Eridanus and Aquarius, and was struck with a significant deficiency of bright stars over the intervals between nebular groups.[2]

The general facts of nebular distribution were correctly described by Mr. Herbert Spencer in 1854. 'In that zone,' he wrote, 'of celestial space where stars are excessively abundant, nebulæ are rare; while in the two opposite celestial spaces that are furthest removed from this zone nebulæ are abundant. Scarcely any nebulæ lie near the galactic circle; and the great mass of them lie round the galactic poles. Can this be mere coincidence? When to the fact that the general mass of nebulæ are antithetical in position to the general mass of stars, we add the fact that local regions of nebulæ are regions where stars are scarce, and the further fact that single nebulæ are habitually found in comparatively starless spots, does not the proof of a physical connection become overwhelming?'[3]

Accompanying, but considerably overlapping the Milky Way along its entire round, is a 'zone of nebular dispersion' (as Mr. Proctor called it)—a wide track of denudation, so far as these objects are concerned. The nebular multitude shrinks, as it were, from association with the congregated galactic stars. A relation of avoidance is strongly accentuated. But withdrawal implies recognition. It implies the subordination

[1] *Monthly Notices*, vol. xxix. p. 340. [2] *Ibid.* p. 344.
[3] *The Nebular Hypothesis* (with Addenda), p. 112.

of stars and nebulæ alike to a single idea embodied in a single scheme. Of our possible acquaintance, then, there is but one 'island universe'—that within whose boundaries our temporal lot is cast, and from whose shores we gaze wistfully into infinitude.

Dismissing, then, the grandiose but misleading notion that nebulæ are systems of equal rank with the Galaxy, we may turn our attention to the problems presented by their situation within it. When the facts connected with it are looked at in detail, distinctions become evident between the different classes of nebulæ—distinctions so marked as to lead almost to their separation.

The 'relation of avoidance' to the Milky Way just adverted to prevails *only* among the 'unresolved' nebulæ. These, it is true, are the great majority of the entire, so that the conclusion of nebular crowding away from that zone remains unimpeachable. For certain classes of minor numerical, but high cosmical importance, the relation is precisely inverted. Over gaseous nebulæ and clusters, the Milky Way seems to exercise an attractive influence equally strong with its repulsive effect upon nebulæ of all other kinds.

Forty out of one hundred and two globular clusters belong to the galactic zone,[1] which is hence thirteen times more richly furnished than the rest of the sky with this peculiar description of objects. And the excess rises to forty-two times for irregular or nondescript clusters, 434 out of 535 of which—that is, eighty-one per cent.—are located in, or close to, the Milky Way. Many clusters, indeed, obviously form an integral part of the formation itself; of others, it is difficult to decide whether they should be ranked as distinct, or simply as intensifications of ordinary galactic star-groupings. To the latter category almost certainly belongs a collection (M 24) visible to the naked eye as a dim cloudlet near μ Sagittarii, and named by Father Secchi, 'Delle Caustiche,' from the

[1] Taken as of the uniform width of thirty degrees, and covering $\frac{1}{4 \cdot 5}$ of the sphere. Major Markwick (*Jour. Liv. Astr. Soc.* vol. vii. p. 182) finds the proportionate area of the Milky Way in the northern hemisphere to be $\frac{1}{4 \cdot 37}$, in the southern $\frac{1}{3}$.

peculiar arrangement of its stars in rays, arches, caustic curves, and intertwined spirals. This again is included in the great oval condensation of galactic stars, shown (from a Lick photograph) in fig. 42, and obviously endowed with some degree of structural independence.

Gaseous nebulæ, like gaseous stars, are nearly exclusive in their galactic affinities.[1] Very few planetaries can be found at any considerable distance from the favoured zone; the spectroscopic search for stellar nebulæ is fruitless unless within its borders; and they embrace—with one exception—*all* the irregular nebulæ. This single exception is a most significant one. It is that of the 'great looped nebula,' one of the numerous constituents of the greater Magellanic Cloud. Plainly, then, the same peculiar conditions which have allowed primitive cosmical matter to remain uncondensed in galactic regions prevails also in the nubecula, although here they are found consistent with the presence, in large numbers, of those species of nebulæ in great measure banished from the Milky Way.

Within its precincts only one in sixteen of those dim, often fantastically-shaped, objects is met with, the analysed light of which gives no indication of gaseity, while their even texture, under the highest telescopic powers, suggests no approach to the stage of breaking up into stars. What then is their nature? Is the difference separating them in appearance from the resolvable nebulæ, i.e. clusters, crowding the Milky Way, a difference of distance solely? Are they, too, clusters of a further degree of remoteness, and therefore inaccessible to effective scrutiny? There is nothing in their aspect to preclude this supposition. So far as observation can tell, they may be of stellar composition. Only it is not easy to understand why the average distance of nebulæ situated near the galactic poles should be many times that of nebulæ thronging the vicinity of the galactic equator.

Mr. Cleveland Abbe[2] imagines the nebulæ to be equably distributed over the surface of a 'prolate ellipsoid,' its longer

[1] Bauschinger, *V. J. S. Astr. Ges.* Jahrg. xxiv. p. 43.
[2] *Monthly Notices*, vol. xxvii. p. 262.

axis coinciding approximately with the axis of the Milky Way; and this arrangement would undoubtedly give an appearance of crowding in the observed directions, since to an eye placed near the centre of such an oval figure, objects uniformly scattered over its surface would produce, by perspective, the effect of running together near its pointed ends. But it is scarcely credible that things should in reality be disposed in this highly artificial manner. It is not enough barely to 'cover the phenomena' with a theory, unless the theory be in itself congruous with the general plan of operations upon which we can see that nature works. Besides, the local distribution of nebulæ is so far from uniform, that antecedent probability is in favour of their general distribution, too, being marked by striking irregularities. The 'canopy' of nebulæ in Virgo is, then, we may rest assured, as genuine an accumulation in its own way as the spherical assemblage in the Magellanic Cloud.

But if there be no systematic difference of distance between the nebular classes occupying contrasted situations as regards the lines of galactic structure, there must be a systematic difference of constitution.[1] The parts of those objects crowding towards the poles must be comparatively small and close together. We have indeed already found reason to believe that clusters do, in point of fact, merge insensibly into nebulæ —that groups of genuine suns at wide intervals stand at the summit of an unbroken gradation of systems with continually smaller and closer constituents, down to accumulations of what is almost literally 'star-dust.' Resolvability is then, we repeat, a question of constitution quite as much as of distance, and we are brought to the conclusion that, while galactic nebulæ are of what we may roughly describe as stellar composition, non-galactic nebulæ are more or less pulverulent. We cannot of course pretend to account for this remarkable distinction. All that can be said is that it *appears* to be actually existent. The irresolvable 'polar' nebulæ perhaps escaped influences powerful over the 'equatorial' ones. Their development, at any rate, seems to have taken a differen course.

[1] Proctor, *Monthly Notices*, vol. xxix. p. 342

No direct proof of motion in nebulæ has—as we have seen in an earlier chapter [1]—yet been obtained. In no single case has either visual change of place or spectroscopic alteration due to recession or approach been ascertained. But the mode of nebular distribution affords indirect evidence of movement. 'Streaminess,' if it mean anything, implies that the bodies affected by it advance in common towards a common goal. Aggregation at the end of a stream prompts the conjecture that a motion of advance was at a certain point, by some supervening attraction, swayed into a motion of revolution. A hint as to the origin of the Magellanic Clouds may hence be derived. They represent in some sort vessels filled through long pipes from a vast reservoir. And since the pipes are still there, the flow may be conceived to be still in progress. Were it to cease, the connection of the nubeculæ with the main nebular body would eventually be interrupted, and their insulation would become complete.

The fidelity with which gaseous nebulæ and clusters adhere to the Milky Way as seen projected upon the sphere warrants the inference that their distribution in space is of a similar character. It would be unreasonable to disconnect them from a formation of which they so closely follow the lines. We can scarcely err in supposing that they lie in general *within*, not behind or in front of it. Thus, the globular clusters richly strewn over the branch of the Milky Way from Scorpio to Ophiuchus, but withdrawn from the conterminous dark rift, plainly belong to the aggregations of minute stars, owing to the absence of which the fissure seems black, although copiously stocked with stars to the tenth magnitude. Other condensed groups stand out from a curtain of apparently still more remote stars, representing possibly a divergent galactic ramification.

Such ramifications must in many cases be greatly foreshortened as viewed from our nearly central position; in some, they may appear only as brilliant knots upon the 'trunk' of the Milky Way. Possibly, the double cluster in Perseus may partake of this optical character. It may be

[1] See *ante*, p. 290.

the termination of a branch spreading inward, and seen nearly end-on. Its constituent stars *appear* to be at vastly different distances from our eyes. The appearance *perhaps* corresponds with the reality. But this is a mere conjecture, and one which there is no immediate prospect of testing by comparison with facts.

Like the Perseus clusters, the Orion nebula gives indications of greater proximity than the main galactic accumulation, to which it is nevertheless beyond doubt structurally related. For a winding nebulous extension from the Milky Way can be traced past α Orionis in the right shoulder through the belt and sword, the bright stars marking which are demonstrably associated with the nebula. The inference then presents itself that the whole mixed system, or series of systems, is placed upon an obliquely directed offset from the galactic zone. Reasoning of the same kind may perhaps apply to the combined nebula and cluster M 8. It occurs as a premonitory outlier of the leading division of the fissured Milky Way, from which it lies a little apart; and it seemed to Sir John Herschel only an 'intense exaggeration' of the stellar collections in its neighbourhood.[1]

Summarising our conclusions, we find the unity of the stellar and nebular systems to be fully ascertained. They are bound together by relations of agreement and contrast scarcely less visibly intimate than those severally connecting individual members of each order. The general plan of nebular distribution is into two vast assemblages, one on either side of the galactic zone; but while this is, comparatively speaking, avoided by the unresolved crowd, it is densely thronged with clusters and gaseous nebulæ. The conditions of aggregation within the zone are hence inferred to differ from those prevailing outside it; but in what respect they are different cannot readily be surmised, far less determined. As to the distances of the nebulæ, we know nothing positive; they no doubt vary extensively; nor can either fineness of grain or faintness of light (both of which may be inherent qualities) serve to distinguish between those nearest

[1] *Cape Results*, p. 387.

to, and those further away from us. We may, however, plausibly conjecture that the hood-like accumulations of the nebulæ are of about the same order of remoteness with eleventh or twelfth magnitude stars, thus constituting, as it were, polar caps on a sphere of which the annular formation of the Milky Way marks the equator.

CHAPTER XXV.

THE CONSTRUCTION OF THE HEAVENS.

SIR WILLIAM HERSCHEL conceived it to be the supreme object of astronomy 'to obtain a knowledge of the construction of the heavens;' and this, in his view, would be accomplished by the 'determination of the real place of every celestial body in space.'[1] Thus limited, the problem would be completely solved could the absolute distance be ascertained of every object telescopically or photographically discernible in the sky. But even the attainment of this unattainable point would never have satisfied Herschel's restless spirit. The real scope of his inquiries went far beyond it. They had an historical, as well as a statistical aim. 'Looking before and after,' they embraced the past and future, no less than the present of the Cosmos.

Modern investigators are of the same mind. The heavens are regarded by them from a *physiological*, rather than from a purely *anatomical* point of view. Mere knowledge of structure, however accurate, will not content them. The vital functions of the organism, the mutual dependence of its parts, the balance of internal forces tending towards destruction and preservation, the dimly-apprehended aim of its divinely sustained activity, engage their eager attention. The heavens live and move, and the laws of their life and motion involve the material destiny of man. It is impossible that he should be indifferent to them.

Even, however, if our instinctive interest in the working of the machine were less keen, we should be driven to search out the dynamical relations of its parts by the impossibility of otherwise arriving at a true knowledge of their geometrical relations. Not only are these variable from one moment to

[1] *Phil. Trans.* vol. cvii. p. 302.

another, but acquaintance with them at any single moment is not *conceivably* accessible to us apart from previous acquaintance with modes and laws of motion. For our view of sidereal objects is not simultaneous. Communication with them by means of light takes time, and postdates the sensible impressions by which we are informed of their whereabouts, in the direct proportion of their distances. We see the stars not where they are—not even where they were, at any one instant, but where they were on a sliding scale of instants. The epoch corresponding to the apparent position of each is different, and the range of difference extends over many thousands of years. The reduction of those positions to a common epoch so as to get a survey of the genuinely contemporary relations in space of all sidereal objects—ideally feasible at best—could not so much as be thought of as possible without a preliminary knowledge of their displacements during the centuries, or millenniums, elapsed since the ethereal vibrations they originate started on their several journeys hither. Thus the study of configurations blends with the study of movements and forces; the restrictions placed upon thought by the effort to exclude all but a single aspect of phenomena fall away of themselves, and we are confronted, whether we will or no, by the stupendous problem of the universe as a vital whole.

As a whole; but not necessarily as the whole. The sidereal world presents us, to all appearance, with a finite system. Human reason would, indeed, otherwise be totally incompetent to deal with the subject of its organisation. There would be nothing for it but to lay down the arms of our understanding before its transcendental and appalling magnitude. But the probability amounts almost to certainty that star-strewn space is of measurable dimensions. For from innumerable stars a limitless sum-total of radiations should be derived, by which darkness would be banished from our skies; and the 'intense inane,' glowing with the mingled beams of suns individually indistinguishable, would bewilder our feeble senses with its monotonous splendour. This laying bare, so to speak, of the empyrean would be the simple and certain result of the continuance *ad infinitum* of any arrangement of sidereal objects comparable with that prevailing in our neighbourhood. Unless, that is to say, light suffer some degree of enfeeblement in space. If this be the case, then our reasonings are put to silence, and a veil is drawn impenetrable to scrutiny. But there is not a particle of evidence that any such toll is exacted; contrary indications are strong; and the assertion that its payment is inevitable depends upon analogies which may be wholly visionary.[1] We are then, for the present, entitled to disregard the problematical effect of a more than dubious cause. The sidereal system cannot accordingly be regarded as in any true sense infinite. The scale upon which it is constructed baffles, it is true, the utmost strain of the imagination to conceive; in the multitudinous splendour of its components, in the number and variety of the subordinate groups constituted by them, in the magnificent play of forces it unfolds, in the dim processes of development it suggests, it bears glorious witness to the power and wisdom of the Almighty Designer; yet it has limits, and for that reason it is a fit subject for the exercise of limited understandings. With further systems, 'pinnacled deep' out of our sight for ever, we have, properly speaking, no scientific concern; we only know that 'when a man hath done, then shall he begin' to declare the wonderful works of God.

Regarding the visible world of stars and nebulæ as an isolated, though excessively complex, system, we may try to give the best order we can to our ideas respecting its constitution. Let us see what are the available data. The number of stars actually registered, as stated in the opening chapter of this work, is about 650,000,[2] of which three-quarters, or thereabouts, are of magnitudes between the ninth and tenth, and the rest are brighter. Beyond the limits of this great census, minute stars abound; but to how many millions they would sum up if completely enumerated, cannot even be guessed with a show of probability. Sir John Herschel esti-

[1] Hirn, *Constitution de l'Espace Céleste*, p. 297.

[2] Including those in Dr. Gill's southern 'Durchmusterung,' the photographic work for which was completed in 1889, while the resulting catalogue may be looked for in a couple of years.

mated at five and a half millions the stars (to the fourteenth photometric magnitude) perceptible over the entire sky with his twenty-foot reflector; Struve calculated them at twenty millions; and it has been vaguely surmised that a hundred millions could be shown by the most powerful modern telescopes. The truth is, that we are still almost wholly ignorant on the point. Different parts of the sky vary enormously in richness. In some, telescopic stars literally swarm; in others, they occur by comparison scantily. It has been computed by Mr. Gore[1] that if the whole heavens were as thickly strewn as the region of the Pleiades, the number of stars to the seventeenth (nominal) magnitude, would be about thirty-three millions. But the method of distribution within a definite cluster evidently gives no clue to that prevailing outside it. A fair specimen-field is, indeed, all but impossible to choose. Counts in the Milky Way, extended in the same proportion over the sphere, would enormously exaggerate the crowding of the stars; which would, in an equal degree, be underrated by counts executed apart from it.

Reliable data on the subject can only, it would seem, be collected with practical usefulness by the method of 'photographic star-gauging.' Reckonings of the stars in their light-ranks, upon plates exposed for various lengths of time, ought to tell with certainty how far the ideal law of augmenting numbers holds good, and where 'thinning-out' becomes apparent. In an equable stratum (as our readers are aware),[2] the stars must nearly quadruple at each descent of a magnitude, simply because the cubical space holding them is quadrupled. Should this rule be overthrown by excess, a real aggregation is indicated, *at the distance corresponding to the altered rate of increase*; if by defect, then obviously the supply of stars in the region examined is becoming exhausted, their scattering is sparser than in our nearer vicinity, and the termination of the series is at hand, if not already reached.

Now even Herschel's Milky Way gauges afforded such indistinct evidence of a terminating series as alone could be derived from aggregate counts, in the fact that the numbers of

[1] *Jour. Liv. Astr. Soc.* vol. vii. p. 180. [2] See *ante*, p. 365.

stars recorded by him amounted to only one-third of what might have been anticipated from the penetrating power of his instrument applied to an indefinitely extended system. And for a 'mean sounding,' at the northern galactic pole, M. Celoria, with a refractor showing, at the utmost, eleventh-magnitude stars, obtained a number almost identical with that given by Herschel's great reflector. The larger instrument, then, here revealed no additional stars. Similar symptoms of

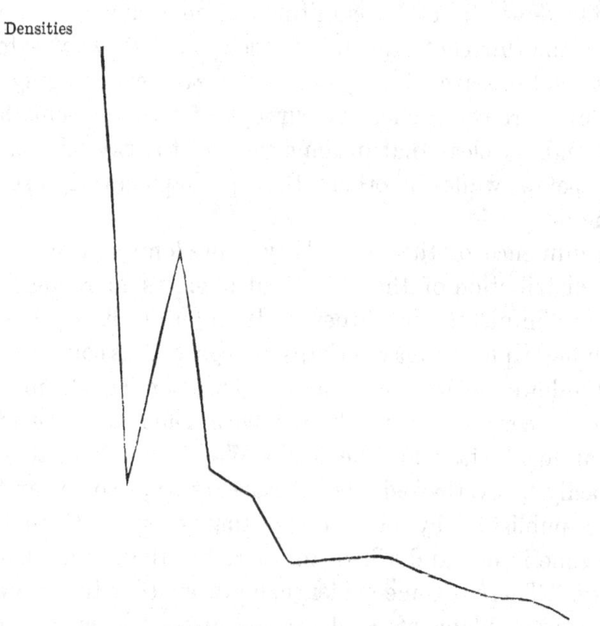

Fig. 50.—Distribution of 934 stars within 1° of the pole, showing the ratio of numbers to space for each half-magnitude.

exhaustion in the star-supplies may be found in Professor Pickering's photographic catalogue of 947 stars within one degree of the celestial north pole.[1] A single glance at the synoptical table giving the numbers for each half-magnitude suffices to show that the numerical representation of the lower ranks is inadequate. The small stars are overwhelmingly too few for the space they must occupy if of average brightness;

[1] *Harvard Annals*, vol. xviii. p. 202.

and they are too few in a constantly increasing ratio. The accompanying diagram (fig. 50) represents graphically the decrease outward of density (or the proportion of numbers to space), deducible from Professor Pickering's enumeration on the sole supposition of the equal average lustre of each class of stars. Those of the ninth are the most thickly strewn; the intervals between star and star widen rapidly and continuously (for the sudden dip at 9·5 magnitude is evidently accidental) down to 11·5 magnitude, when a slight recovery, lasting to the thirteenth magnitude, sets in. To what extent these fluctuations are of a systematic character, can only be decided by more comprehensive surveys; for the present they serve to make it clear that in some parts of the sky faint stars fall far below, while in others they perhaps largely exceed their due proportions.

The influence of the Milky Way is predominant over the general distribution of the stars, but it grows more marked with their diminishing brightness. Its inferior efficacy in the southern hemisphere [1] may perhaps be regarded as corresponding to the lower degree of polar condensation visible in the southern, as compared with the northern nebulæ. The preference of lucid stars for the Milky Way is so slight that it might easily be overlooked; but it appears from some careful statistics published by Mr. Gore [2] that even in them the galactic zone is one and a half times richer than other parts of the sky. There is some evidence, however, that this crowding is towards a plane of condensation distinct from, though very close to, that of the galaxy.

A zone of large stars traversing the southern hemisphere was thought by Sir John Herschel to be the projection of 'a subordinate sheet, or stratum, deviating some twenty degrees from parallelism to the Milky Way.' [3] The hint was further developed by Dr. Gould. 'Few celestial phenomena,' he considered to be 'more palpable than the existence of a stream or belt of bright stars,' traceable 'with tolerable distinctness through the entire circuit of the heavens, and forming a great

[1] Seeliger, *Sitzungsberichte*, Munich, 1884, p. 521, 1886, p. 220.

[2] *Jour. Liv. Astr. Soc.* vol. vii. pp. 175, 182. [3] *Cape Results*, p. 385.

circle as well defined as that of the galaxy itself,' [1] which it crosses at an angle of about 20° in Crux and Cassiopeia. Traversing in the southern hemisphere Orion, Canis Major, Argo, the Centaur, Lupus, and Scorpio, it pursues its way in the northern through Taurus, Perseus, Cassiopeia, Cepheus, Cygnus, and Lyra, its line being less obviously continued by the stars of Hercules and Ophiuchus.[2] Like the Milky Way, it seems to bifurcate near α Centauri, the branch there thrown off reuniting with the parent stem in Andromeda. That the stars thus marked out, to the number of about five hundred, constitute with the sun a cluster 'of a flattened and somewhat bifid form,' 'distinct from the vast organisation of the Milky Way,' grew into a conviction with the progress of Dr. Gould's observations. Since, moreover, they were found, on fuller inquiry, to trace on the sphere only a small circle, to the south of the parallel great circle, the sun's place within the group must be removed to the north of its medial plane, towards Leo, and is supposed to be nearly marginal in the direction of Hercules and Ophiuchus, where the track of the 'belt' is almost effaced by the wide scattering of the constituent stars.

Two circumstances appear to show that this 'solar cluster' is in some way a reality. In the first place, an enumeration of the stars in photometric order discloses a systematic excess of objects brighter than the fourth magnitude, making it certain that there is actual condensation in the neighbourhood of the sun—that the average allowance of cubical space per star is smaller within a sphere enclosing him with a radius, say, of 140 light-years, than further away. Again, when deduction has been made of some five hundred stars of the higher orders, the remainder, to the ninth magnitude, form a tolerably regular series, increasing in numbers nearly in the theoretical proportion of their diminishing light.[3] And the conclusion that these five

[1] *Amer. Jour. of Science*, vol. viii. p. 333 (1874). [2] *Uran. Argentina*, p. 355.

[3] The 'empirical ratio' (that resulting from actual enumeration) of multiplication of numbers per magnitude, is 3·912; the theoretical ratio is 3·987. —*Uran. Argentina*, p. 367.

hundred superfluous objects do in fact compose a group apart, is strengthened by the symmetrical arrangement, with regard to the 'belt,' of the bright stars outside it. Their tendency to collect towards its central plane is, according to Dr. Gould, irrespective of the Milky Way, except in so far as the two formations coincide by the projection of one upon the other.

It is worth notice, too, that the present direction of the solar movement agrees with the general 'lie' of the belt; although the other stars it is supposed to include give no sign of conformity to any fundamental plane of revolution. The truth of the hypothesis can, however, be tested only by a detailed study of motions, presumably of extreme complexity. The problem indeed of separating the individual displacements of the clustered stars from those common to all and shared by ourselves, as well as from the independent shiftings of extraneous objects, baffles all the present resources of science. The ascertained translation of the sun must, if the 'solar cluster' be really organised on a separate dynamical basis, be mainly 'interstitial'; it must represent the revolution of the solar system round the centre of gravity of the collection; for it emerged, essentially as we know it now, from Herschel's interpretation of the apparent displacements of a few undoubted members of the group. The application of the same method (as recommended by Professor Holden)[1] to stars unequivocally external to the group, would, if practicable, be of extreme interest. Thus, if at all, the bodily transport through space of the entire solar cluster might be determined, and Herschel's forecast realised of a still higher grade in the hierarchy of motions than that corresponding to the sun's journey towards λ Herculis. It is even conceivable that such a movement of the entire stellar system of the sun with reference to other similar systems, or to what Herschel called 'intersystematical stars,' might account for an otherwise inexplicable excess of movement among the smaller stars,[2] the apparent displacements of which would be increased by a perspective element absent from those of brighter and nearer objects.

[1] *Century Magazine*, Sept. 1889.

[2] See *ante*, p. 338.

The attempt, however, ideally to compose the solar cluster encounters enormous difficulties. Should, for instance, the stars of Orion and the Pleiades be admitted to, or excluded from it? Visually, they belong to the 'belt;' but they are associated with the Milky Way as well; so that distinctness of constitution cannot well be granted to any stellar collection of which they form part. But distinctness of constitution is an essential feature of Dr. Gould's scheme. The adoption into it, besides, of a subordinate system on the scale of the Pleiades would imply a vaster organisation than it seems feasible to admit; while its rejection, together with that of the Orion group, would seriously impair the integrity of the starry girdle forming the *sensible* groundwork of the theory under consideration.

The velocities, moreover, of the stars belonging to the supposed cluster far exceed its possible gravitative power to produce. Estimating the total mass of the system at 2,500 times that of the sun, and supposing it all concentrated in one point, the utmost speed which it could generate at the distance of one 'sidereal unit' (the interval of space corresponding to a parallax of one second) would be three miles a second. But in the hypothetical cluster, this rate could practically never be attained, far less surpassed; while among the stars as we know them, it is commonly sextupled, decupled, nay exceeded scores of times. It seems to follow, then, that the movements of the sun and stars are *not* interstitial movements, and that Gould's cluster is *not* organised on an independent dynamical footing.

Its problematical existence was entirely disregarded in Mr. Maxwell Hall's elaborate effort to fix the elements of the sidereal system. Contemplating it as an undivided whole, he nevertheless restricted the scope of his immediate endeavours to the establishment of a bond of union between the motions of the sun and the nearer stars. These, by his fundamental hypothesis, are governed by the same attractive force, and pursue immense orbits round a centre excessively remote even on the sidereal scale. Having thus laid down the broad outlines his scheme, he was enabled to set up certain formulæ, from which, with the aid of tolerably secure data respecting a few stars, he arrived, by long processes of calculation, at some remarkable conclusions, corrected from time to time as advancing research improved the quality of the materials available.[1]

The first step was to decide upon a centre for the sun's movement (taken to be known in direction), and to assign the law of force ruling it. Motion in a circle is of course always aimed along a line perpendicular to the line joining the moving object with the centre; hence, if the solar orbit be nearly circular, its centre must be removed by ninety degrees from the solar apex—it must be situated, that is to say, somewhere on a great circle of which the apex is one of the poles. Two diametrically opposite points on this great circle, one in Andromeda, the other in Hydra, gave satisfactory results on preliminary trial; but the choice between them depended upon the form assumed for the law dominant over sidereal revolutions. If, like those of the planets, they are swayed by a preponderating central mass, then the law is the familiar one of inverse squares, and the linear velocity, or speed in miles per second, of each member of the system decreases with increasing distance from the attractive centre. If, on the other hand, the movements of the stars are governed solely by their mutual attractions, then the sum-total of the gravitative power acting upon each is greater in the direct ratio of its remoteness from the common centre. Swiftness in a system thus constructed increases *outwards*, because angular velocity remains the same. All its members, in other words, circulate in an identical period, and will be restored, at the end of their *annus magnus*, to their precise original positions.

Rightly judging the latter to be incomparably the more probable plan of stellar organisation, Mr. Hall chose for its centre a spot a little to the east of ε Andromedæ, from which the velocities of certain well-known stars seemed to increase outward. The supposed real interval between this point and the sun—that is, the radius of the sun's orbit—is of no less than 150 sidereal units, implying a light-journey of 490 years;

[1] *Memoirs R. Astr. Soc.* vol. xliii. p. 157; *Monthly Notices*, vols. xxxix. p. 126, xlvii. p. 521.

and the common period ascribed to the innumerable bodies revolving round it somewhat exceeds thirteen million years. Their revolutions, it may be remarked, although conducted (according to Mr. Hall's hypothesis) on a fundamental plan of almost monotonous simplicity, would present a far from uniform appearance. The orbits of the various stars might lie in any planes, be inclined at all possible angles, and be traversed in either direction. The sole elements of necessary agreement between them would be those of being described about a common centre in a common period. Among the stars coming within our range of measurement, however, there could, if their paths approximated to circularity, be little variety in real speed, since the differences of their distances from the supposed enormously remote centre would be insignificant; and for this reason highly eccentric ellipses had to be assigned to exceptionally swift stars. Thus, for 1830 Groombridge a cometary orbit (eccentricity = 0·87) was laid down; moderately oval ones being appropriated to ζ Toucani, Oeltzen 17415, Oeltzen 11677, Lacaille 9352, σ Draconis, and o^2 Eridani. The conditions of movement, however, assumed for these bodies are totally diverse from those under which comets sweep through the solar system. For the stellar centre of force is placed, not in one focus, but at the middle point of ellipses traversed with velocities accelerated not *towards*, but *away from* that centre; and two equal maxima of speed are accordingly attained at opposite extremities of the major axis. Stars moving much more rapidly than their fellows were then located by Mr. Hall near one or other of the far ends of an elongated track; nor could the observed inequalities be otherwise accounted for on the basis adopted for his calculations. The combined mass brought out by them for the stellar system about equalled that of seventy-eight million suns, a result not in itself improbable, since Professor Newcomb, as we have seen, concluded for a maximum value of a hundred millions.

Two circumstances, however, tend to undermine confidence in Mr. Hall's method. In the first place, all we seem to know of the sun's motion in space depends upon the preliminary assumption that the stellar shiftings from which it has been elicited, include no systematic element apart from the effects of perspective—that deviations from the parallactic line occur indifferently in all directions, and are in a sense casual. But acquaintance with the sun's motion was an indispensable preliminary to the construction of Mr. Hall's scheme. The insecurity then of the data with which he worked appears to follow necessarily from the admission of the genuine character of an edifice reared upon self-destructive principles. If it exists, we are in ignorance as to the solar movement; and if we are ignorant of the solar movement, we cannot tell whether it exists or not. A second objection is that the ascription of approximately circular or only moderately elliptic paths to the majority of the stars, is scarcely warranted by the analogy of binary stars. But if highly elliptical orbits preponderate, the problem of sidereal construction, as attacked by Mr. Hall, is practically indeterminate.

The proffered solution of it, however, can be tried by a direct appeal to facts. The equations established by Mr. Hall gave him the means of calculating, from the known proper motions, the parallaxes and velocities in the line of sight of the stars considered. Subsequent determinations, then, in many cases supply a test as to the correspondence of the conditions embodied in the formulæ with those existing in nature. The following parallaxes have been selected as among the best and most recently observed:—

Star	Computed	Observed	C–O
Arcturus	0″·188	0″·018	+0″·170
Capella	0·029	0·107	−0·078
Procyon	0·074	0·266	−0·192
α Orionis	0·018	negative	
Pollux	0·06	0·068	−0·008
Regulus	0·02	0·093	−0·073
α Cygni	0·004	negative	
Polaris	0·008	0·052	−0·044
γ Cassiopeiæ	0·022	0·007	+0·015
β Cassiopeiæ	0·065	0·167	−0·102
70 Ophiuchi	0·111	0·155	−0·045
e Eridani	0·328	0·143	+0·185
μ Cassiopeiæ	0·199	0·036	+0·163

Large discrepancies are, it will be seen, redeemed by only one telling coincidence; and what traces of general agreement are discernible may be attributed to the effects of the inevitable correlation of distance to proper motion. Radial velocities similarly computed show an equally distant relation to facts. But since one tolerably exact comparison outweighs a dozen loose ones, we have included in the following little table only the stars photographically measured at Potsdam:—

Star	Computed [1]	Observed	C–O
Capella	+ 9	+ 17	− 8
Aldebaran . . .	+ 14	+ 30	
Polaris	− 13	− 16	+
Procyon	+ 9	− 7	+
α Persei	No result	− 7	
Rigel	No result	+ 39	

Here the sum of the errors exceeds half the sum of the observed quantities, to say nothing of the failure of the theory in two cases out of six.

Mr. Hall cannot then be said to have overcome the difficulties encountered in his arduous enterprise; but it must not be concluded thence that his labours were thrown away. The subject is one which, for the present, can only be approached tentatively; it would be highly undesirable that investigators should, for that reason, be discouraged from approaching it at all. The life of a science is in the thought that binds together its facts; decadence has already set in when they come to be regarded as an end in themselves. 'Man is the interpreter of nature;' to draw up an inventory, however, is not to interpret. It is true that speculation is prone to wander into devious ways; but then 'truth emerges more easily from error than from confusion.' And in sidereal science especially, there is danger lest investigators, seduced by the wonderful facilities of novel methods, should exhaust their energies upon the accumulation of data, and leave none for the higher work

[1] The sign + signifies that the star in question is receding from the earth, the sign − that it is approaching it. The figures given above indicate English miles per second.

of marshalling them along the expanding lines of adequate theory.

It is scarcely probable, indeed, that indications as to the general plan of the sidereal world sufficiently definite for purposes of numerical calculation can be gathered during the present era of human knowledge. A limitless field of fruitful research, however, lies open even now in the systems of various degrees of subordination, the federated combination of which we may reasonably suppose to constitute the supreme unity of the cosmos. From double, triple, multiple groupings to knots, drifts, clusters, clouds of stars, an ascending scale of complex arrangement leads upwards to the unknown—perhaps beyond it to the unknowable.

It may be that no star in the entire stellar scheme escapes this law of particular aggregation. Even our own sun, solitary in appearance though it be, has, we are led to suppose, closer ties with some stars than with others, though the significance of the resulting movement at present escapes us. Nor can we be sure that a dynamical equilibrium, such as prevails in the planetary system, has been established in the solar or any other such star-group. Association of a transitory nature —on the cosmical time-scale—is even conceivable. One set of combinations may be dissolved to give place to others; a single star may pass from one vast confederacy to the next, seeking its fortune, as it were, through space; or, breaking away from the entire congeries of systems, rush out into the ethereal desert, to find itself, after milliards of ages, within the precincts of a strange galaxy beyond terrestrial ken with telescope or camera.

The more attentively the configuration of the stars is studied, the more clearly do special phenomena of grouping come into view. Among the minute stars of the Milky Way, above all, the tendency towards arrangement in typical *patterns* is, in certain parts of the sky, almost as unmistakable as it would be in a ball-room crowded with dancers suddenly arrested in threading their way through the figures of a quadrille or a minuet. Yet in the heavens, methodical distribution must always be to some extent masked by the projection upon the same surface, of objects at totally different distances. That, under these circumstances, it should often be effaced is less remarkable than that it should occasionally become apparent.

One of the 'typical forms' in which stars seem to collect is that of an ellipse, or circle seen in perspective.[1] Radiated structures also occur; and Father Secchi, who early drew attention to this curious subject, regarded the presence of a large red star in a commanding situation among minor objects, as a common trait of physical groupings.[2] As specimens of a class of objects to which the 'persevering student' could make large additions 'with an increasing conviction as to the mutual interdependence' of their constituents, Mr. Webb singled out in the constellation Hercules, a 'wreath,' a 'double chain,' and a 'recurved line of small stars proceeding from one of 7·5 magnitude.'[3] The predominance of the 'spray' form of arrangement was dwelt upon by Mr. Proctor; while some regions, Professor Holden tells us, are characterised by 'streams' of stars, defiling in rows of a dozen or two, across the telescopic field. 'Others,' he continues, 'are rich in small definite ellipses of stars, often all of the same size. In some cases, stars are surrounded by circles of other fainter stars. In other instances, the ellipses become tolerably regular ovals, often of large size. These interlace in the most intricate manner. . . . We can frequently trace new and highly interesting features of the kind by paying attention to stars of one magnitude only. If we regard the eleventh-magnitude stars alone, for example, we may find rings and ovals of these stars forming a regular pattern in the sky. Interlacing with these may be found another pattern of similar ovals (usually smaller) of stars of the twelfth magnitude, and so on.'[4]

It must, indeed, be admitted that the search for 'star-patterns' is over somewhat treacherous ground. Imagination may here easily play the part of will-o'-the-wisp. Among a

[1] *Century Magazine*, Sept. 1889, p. 787.

[2] *Atti dell' Accad. Pont.* t. vii. p. 67.

[3] *Cel. Objects*, p. 323.

[4] *Monthly Notices*, vol. l. p. 62. Cf. Barnard, *Ibid.* p. 314, Backhouse, *Ibid.* p. 374.

multitude of scattered objects, counterfeit groupings of any given design can be put together by a very slight stretch of fancy. It has been shown experimentally by Mrs. Huggins, that the random distribution of dots of Indian ink (shaken from a brush over the paper) does not exclude the possibility, to a predisposed eye, of forming them into almost any desired figures. But the illustration should not, nor was it designed to lead us to set down as purely imaginary the visible peculiarities of stellar arrangement. This would be an extreme quite as mischievous as that of unqualified credulity, since there is cumulative evidence that the peculiarities in question are, in many cases, real and significant. Only their recognition should be pursued with caution.

The late Mr. Proctor expressed his conviction that 'star streams' will eventually prove themselves genuine by the unanimity of their proper motions.[1] And it is obvious that they can only subsist upon this condition. Their order would otherwise be only a passing coincidence. Stars thus marching 'in Indian file' are presumably swayed by an identical force acting upon them from a very great distance. The group that they form is not self-centred, but makes only a part of a larger organisation. Segregation, on the other hand, is the distinguishing 'note' of true clusters. They might be called autonomous democracies, each of their members obeying the united commands of all, while outside influences, although exerted upon them collectively, are without effect upon their internal regimen.

A 'streaming' can then be discriminated (at least ideally) from a 'clustering' collection of stars, by the circumstance that the centre of movement of the one is external, of the other internal. It may possibly be found that the two plans of organisation prevail respectively in different sections of the Milky Way; there is some appearance that they not unfrequently compete or combine within the same cluster, the streaming tendency working towards the dissolution, the clustering tendency for the preservation of the system. This is not the only feature of sidereal construction which conveys a hint to us that the

[1] *Universe of Stars*, p. 84.

world of stars and nebulæ is in a state of transition. We see it in only one phase of its long development. To regard its condition as settled upon an unalterable basis, would be to misconstrue signs everywhere legible to attentive scrutiny.

Since stars and nebulæ are undeniably united into a single scheme, our view of the universe must embrace both classes. The distribution of nebulæ is in fact complementary to the distribution of stars. Groupings of the one kind fill in the outlines left blank by the groupings of the other. The Milky Way, so far as can be discerned, is a rifted and irregular ring, furnished with innumerable *tentacular* appendages, and composed of stars in every stage of aggregation. This ring, however, has obvious geometrical relations with the rest of the sidereal structure. It marks the equator of a vast globe, of which the poles are canopied by the nebulæ. Necessarily, too, of a rotating globe, since axial movement alone gives rise to the distinction between poles and equator.

The opinion that the *shape* of the visible universe is spherical, or spheroidal, rather than lenticular, has been lately expressed by Radau,[1] Klein, and Falb.[2] The polar relations of the nebulæ to the plane of the Milky Way admit indeed of no other interpretation. And these relations can scarcely have been determined otherwise than by the rotation on an axis of the enormous, undivided volume. The condition thus indicated as primitively existing may have become modified with time; we could, at any rate, know nothing of its prevalence, since the relative situations of the heavenly bodies would, as M. Falb has remarked, be absolutely unaffected by it.

All that we can see clearly is that an universal movement of rotation had much to do with the present distribution of matter in sidereal space. Whether the forces which have brought it about are still active must remain an open question. The opposite tendencies of stars to gather in the equatorial plane, and of nebulæ to stream towards the poles of the system, may not even yet be exhausted; but the point can

[1] *Bull. Astr.* t. ii. p. 88.

[2] *Handbuch der Himmelserscheinungen*, Th. ii. p. 312; *Sirius*, Bände v i. p. 10, viii. p. 198.

only be investigated by long and arduous methods. Their activity (if ascertained) would apparently imply an inherent difference in the qualities of the objects respectively displaying them; while, on the other hand, the hypothesis of stellar development from nebulæ is indispensable, and wears the aspect of truth.

We can indeed hesitate to admit neither the fundamental identity of the material elements of the universe, nor the nebulous origin of stars. The transition from one to the other of the two great families of the sidereal kingdom is so gradual as to afford a rational conviction that what we see contemporaneously in different objects has been exhibited successively by the same objects. Planetary nebulæ pass into gaseous stars on one side, into nebulous stars on the other; the greater nebulæ into clusters. The present state of the Pleiades refers us inevitably to an antecedent condition closely resembling that of the Orion nebula; the Andromeda nebula may represent the nascent stage of a splendid collection of suns. But even though stars without exception have sprung from nebulæ, it does not follow that nebulæ without exception grow into stars. The requisite conditions need not invariably have been present. Other ends than that of star production are perhaps subserved by the chief part of the present nebulous contents of the heavens. The contrast between stellar and nebular distribution is intelligible only as expressing a definitive separation of the life-histories of the two classes—a divergence destined to be perpetual between their lines of growth.

The view that stars arise from a condensation of nebulous matter, encounters one singular difficulty. Nebulæ, as we have seen, are devoid of sensible motion; stars usually possess high, sometimes enormous velocities; and measures in the line of sight give it to be understood that the distinction is real and well-marked. It is one, however, which no theory of evolution hitherto broached is competent to explain; and only one of two alternatives accordingly seems open to us—either to reject all such theories, or to assent to the inference, anomalous though it appear, that stellar movements are not *innate*, but are gradually acquired.

Progress, then, is the law of the universe. From its present state we can obscurely argue a 'has been' and a 'shall be.' The face of the skies is not cast in stereotype. 'As a vesture Thou shalt change them, and they shall be changed.' They shall change, by no caprice of hazard, but in subjection to laws unalterable in their essence although infinitely various in their applications, divinely directed towards the continually more perfect embodiment of the unfolding Eternal Thought.

But the glory of the heavenly bodies, it is asserted, must come to an end. It results from a merely transitory state of things. The radiations, by virtue of which they shine, are the outcome of what may be figuratively termed the effort of nature to establish a universal thermal equilibrium. This condition will be attained when the frigid 'temperature of space' reigns in all the millions of bodies which once were suns, and will thenceforward revolve, amid 'darkness that may be felt,' the mechanism of their movements unimpaired, but inert, lifeless, and invisible. Is this then the predestined end? Science replies in the affirmative. That is to say, it knows no better. Yet there is much as to which it is ignorant. Matter rests upon a subsensible basis, into the arcana of which no inquiry has penetrated. The observation of phenomena leads, it may be said, to the shore of an all-diffusive ocean of force, the existence of which is implicated in their occurrence. That is all we know; at the brink of the ocean we pause, helpless to sound its depths, or number the modes of its manifestations, or predict the tasks of renovation or preservation committed to it. We can only recognise with supreme conviction that He who made the heavens can restore them, and that when the former things have passed away, and the scroll of the skies is taken out of sight, 'like a book folded up,' a 'new heaven and a new earth' shall meet the purified gaze of recreated man.

APPENDIX

TABLE I.—STARS SHOWING BRIGHT

Name of Star	Approximate Place (1890)	
	α	δ
	h. m. s.	° ′
R Andromedæ	0 18 13	+37 58·1
γ Cassiopeiæ	0 50 4	+60 7·3
S Cassiopeiæ	1 11 34	+72 1·9
φ Persei	1 36 46	+50 8·1
o Ceti (Mira)	2 13 47	− 3 28·6
Pleione	3 42 38	+23 48
Anon. Cœli	4 36 44	−38 27·2
θ¹ Orionis	5 29 52	− 5 27·7
θ² Orionis	5 29 59	− 5 29·3
U Orionis	5 49 17	+20 9·3
Lalande 13412	6 49 33	−23 47·3
R Geminorum	7 0 44	+22 52·4
γ Argûs	8 6 19	−47 0·7
Anon. Argûs	8 51 19	+47 10·0
R Carinæ	9 29 29	−62 18·1
R Leonis	9 41 39	+11 56·3
Arg. Gen. Cat. 14626	10 37 5	−58 56·7
Anon. Argûs	10 37 15	−58 10·8
Arg. Gen. Cat. 14684	10 39 39	−59 23
η Argûs	10 40 47	−59 6·3
Arg, Gen. Cat. 15177	11 0 25	−64 58·8
Arg. Gen. Cat. 15220	11 1 56	−64 45·2
Arg. Gen. Cat. 15305	11 5 25	−60 22·9
δ Centauri	12 2 40	−49 56·5
θ Muscæ	13 1 2	−64 33·4
Anon. Centauri	13 11 3	−57 33·2
R Hydræ	13 23 43	−22 42·7
μ Centauri	13 43 0	−41 46·5
Arg. Gen. Cat. 20052	14 41 49	−36 2·7
Arg. Gen. Cat. 22748	16 43 53	−40 59·2
Arg. Gen. Cat. 22763	16 44 36	−41 36·7
Stone 9168	16 46 36	−41 38·2
Arg. Gen. Cat. 23072	16 56 21	−37 56·8
Arg. Gen. Cat. 23073	16 56 29	−37 38·4
Arg. Gen. Cat. 23416	17 11 25	−45 29·3
Argelander-Oeltzen 17681	18 1 46	−21 16·2
β Lyræ	18 46 1	+33 14·1
D M + 30°3639	19 31 30	+30 17·5
R Cygni	19 33 52	+49 57·2
D M + 48° 2940	19 39 50	+48 31·5
χ Cygni	19 46 20	+32 38·2
D M + 35° 3952	20 1 43	+35 21·6
D M + 35° 4001	20 6 4	+35 51·1
D M + 35° 4013	20 7 42	+35 51·1
D M + 37° 3821	20 8 6	+38 1·5
D M + 36° 3956	20 10 25	+36 19·5
D M + 36° 3987	20 12 55	+37 4·9
P Cygni	20 13 44	+37 41·4
D M + 38° 4010	20 15 31	+38 23·0
D M + 43° 3571	20 37 12	+43 31·5
π Aquarii	22 19 40	+ 0 49·1

LINES IN THEIR SPECTRA.

Magnitude	Remarks
5·6 to 12·8	F, D_3, &c., bright near maximum
2·3	Bright lines variable
6·7 to 13	F, &c., bright near maximum
4·2	White star. Hydrogen lines bright
1·7 to 9·5	Bright lines near maximum
6·2	Thin bright hydrogen lines
7·0 to 10·5?	Bright lines near maximum
4·4	Unknown bright lines in photographed spectrum
4·9	Details wanting
6·1 to 12	F, D_3 &c., bright near maximum
7·0	Wolf-Rayet type
6·6 to 12·3	Carbon bands bright near maximum
2·0	Wolf-Rayet type
8·8	Wolf-Rayet type
4·3 to 9·3	*G* and *h* bright
5·2 to 10	Hydrogen lines bright near maximum
7·3	Wolf-Rayet type
9·8	Wolf-Rayet type
6·9	Wolf-Rayet type
−1 to 7·6	Bright hydrogen lines
8·5	Wolf-Rayet type
8·3	Wolf-Rayet type
9·0	Wolf-Rayet type
2·8	Bright hydrogen lines
6·0	Wolf-Rayet type
9·0	Wolf-Rayet type
4 to 10	Hydrogen-lines bright near maximum
3·4	Bright hydrogen lines
6·8	Wolf-Rayet type
5·9	Wolf-Rayet type
7·5	Wolf-Rayet type
7·5	Wolf-Rayet type
6·5	Wolf-Rayet type
7·1	Wolf-Rayet type
7·3	Wolf-Rayet type
7·0	Wolf-Rayet type
3·4 to 4·5	Bright lines variable
9·3	Unknown lines bright
5·9 to 13·0	Hydrogen-lines bright near maximum
7 to 11·5	Hydrogen-lines bright near maximum
4·0 to 12·8	Bright lines near maximum
7·5	Wolf-Rayet type ?
8·0	Wolf-Rayet. Cygnus No. 1
8·5	Wolf-Rayet. Cygnus No. 2
7·0	Wolf-Rayet type
8·0	Wolf-Rayet. Cygnus No. 3
8·0	Wolf-Rayet type
5·0	Thin bright hydrogen lines
9·0	Wolf-Rayet type ?
7·5	Wolf-Rayet type
4·6	Bright hydrogen lines

Table II.—VARIABLE DOUBLE STARS.[1]

ζ Piscium = Σ 100, 6, 8 mags. at 23″·7. Fixed, with common proper motion. Photometric mags. (Harvard), 5·4, 6·4; but Webb thought large star rose at times to 4 mag. A close companion to B detected by Burnham in 1888. Colours, both white. Spectrum, type I.

h 2036. Composed of two white stars never differing by above a mag., but both probably variable. Binary; in retrograde motion.

γ Arietis = Σ 180. Combined mag. in Harvard Photometry 4·3. Slight relative variability recorded by Flammarion. Colours yellowish; different in each component. Spectrum, type I.

α Piscium = Σ 202. Both stars vary in light, smaller in colour as well. Spectrum type I.

Arietis = Σ 333. Struve's estimates of mag. ranged from 4·5 to 6·5 for one, from 5 to 6·5 for the other component. Colours, both white. Spectrum ?

U Tauri. Divided by Knott in 1867 into two 9·7 mag. stars. The combined object fluctuated considerably in light 1865-71, but has been omitted from recent lists of variables.

δ Orionis = Σ 14¹. Chief star varies 2·2 to 2·7. Companion, 6·8 mag. at 52″·7. Colours, greenish-white and white. Spectrum, type I.

Lacaille 2145, 8″·2, 8″·5 mags. Probably binary, certainly variable (Tebbutt, *Observatory*, iv. 211). Of 6 mag. in Lacaille's Catalogue.

η Geminorum varies to the extent of one mag. in a period of 229 d. Divided into a 3 and a 9 mag. star at 0″·96 by Burnham, Nov. 11, 1881. Spectrum, type III.

S (15) Monocerotis = Σ 950. Chief star varies from 4·9 to 5·4 mags. in 3 d. 10 h. 38 m. Colours, pale green and bluish. Spectrum, type I.

38 Geminorum = Σ 982. Differences of brightness from 1·5 to 4 mags. observed by Struve. Colours, light yellow and purple to greenish. Spectrum, type I. Slow binary.

Σ 1058. Companion invisible since 1844 (Burnham, *Memoirs R. Astr. Soc.* xliv. 245).

61 Geminorum. Mags. 5·7 to 7·5, and 9 to below 12. Distance, 60″. Larger star deep yellow.

U Puppis = Lalande 14551, varies 6 to 6·8 mags. in 14 d. Resolved by Burnham, Jan. 28, 1875, into a pair of 6·5 and 8·5 mags. at 0″·8. Colour, yellowish. Spectrum, type I.

Σ 1517. Slight alternate variability confirmed by O. Struve. Binary. Yellowish colour.

γ Virginis = Σ 1670. Each star varies from about 3 to 3·5 mag. Colour, pale yellow. Spectrum, type I.

OΣ 256 = Lalande 24098. Each star varies from 7 to about 8 mag. Spectrum, type II ?

Y Virginis = Lalande 25086. Irregularly variable 5 to 8 mag. Composed of two nearly equal stars at 0″·48 (Burnham). Spectrum, type I.

[1] See *Nature*, vol. xxxix. p. 55.

π Boötis = Σ 1864. Each component varies irregularly to the extent of at least one mag. Colours, white and ashen to yellow and blue. Spectrum, type I.

ζ Boötis = Σ 1865. Relative variability noticed by F. and O. Struve. Colours, both white. Spectrum, type I.

Σ 1875. One compo nt varies from 8·5 to 10 mag. Colours, both white.

β Scorpii. Mags. estimated by Struve at 2 and 4; in Harvard Photometry, 3, 5·2. Webb found a difference of 3·2 mags.; J. Herschel of only 1. Colours, yellowish-white and lilac. Spectrum, type I.

ρ (5) Ophiuchi. Disparities recorded from 0 to 2·5 mags. Colours, yellow and blue.

λ Ophiuchi = Σ 2055. Mags. 4 and 6; but companion varies both in light and colour. Large star, greenish-white. Spectrum, type I.

Σ 2062. Companion at times invisible under the best conditions.

α Herculis = Σ 2140. Large red star varies irregularly from 3·1 to 3·9 mag. Spectrum, type III. Green companion thought by Struve to vary from 5 to 7 mag. Spectrum shows extensive absorption of less refrangible rays.

Σ 2344. Mags. 8·5 and 10 to below 12. Companion frequently invisible.

θ Serpentis = Σ 2417. Both stars seem variable. Differences from 0·34 to 1·69 mags. registered at Harvard College in 1878. Colours, both yellowish-white. Spectrum, type I.

β Cygni. Large star changes very slowly from 3·3 to 3·9 mag. Colours, golden and azure. Spectrum, type II with modifications.

h 1470 = Lalande 38428. Mags. estimated 7 and 8 by Secchi in 1856, but suspected to vary. Continuous observations desirable. Distance apart, 23″·8. Colours, 'superb' red and blue. Spectrum of larger star, type III.

U Cygni = Schjellerup 239*a*. Varies from 7·7 to below 11 mag. in 466 d. Colour, deep ruby. Spectrum, type IV. Companion at 62″ appears to fluctuate from 8 to 8·7 mag. in light, and in colour from blue to white or reddish.

Σ 2718. Each component alternately for a short time superior.

δ Cephei. Large star varies regularly from 3·7 to 4·9 mag. in a period of 5 d. 8 h. 48 m. Colours, orange and blue. Spectrum, type II.

U Cassiopeiæ = OΣ (App.) 254. A similar pair to U Cygni. Red star (= Schj. 280) varies irregularly from 7 to 9; blue star, from 8 to 10 mag. Distance apart, 59″. Spectrum, type III.

TABLE III.—

	Name of Star	Magnitude	Approx. R.A. 1890	Approx. Decl. 1890	Authority
			h. m. s.	° ′	
1	β Cassiopeiæ . .	2·4	0 3 18	+58 32·8	Pritchard
2	Groombridge 34 .	7·9	0 12 6	+43 23·9	Auwers
3	ζ Tucanæ . .	4·1	0 14 20	−65 31·3	Elkin
4	α Cassiopeiæ . .	2·2	0 34 16	+55 56·0	Pritchard
5	η Cassiopeiæ . .	3·6	0 42 26	+57 13·9	O. Struve
6	μ Cassiopeiæ . .	5·2	1 0 57	+54 22·8	Pritchard
7	β Andromedæ . .	2·2	1 3 34	+35 2·2	Pritchard
8	Polaris . . .	2·1	1 18 29	+88 43·3	Pritchard
9	α Arietis . . .	2·0	2 0 58	+22 56·5	Pritchard
10	e Eridani . . .	4·4	3 15 30	−43 29·5	Elkin
11	α Persei . . .	1·9	3 16 28	+49 28·2	Pritchard
12	40 (o²) Eridani . .	4·5	4 10 13	− 7 49·2	Gill
13	Aldebaran . .	1·0	4 29 36	+16 17·3	Elkin
14	Capella . . .	0·2	5 8 34	+45 53·1	Elkin
15	ψ⁵ Aurigæ (comes) .	8·5	6 38 48	+43 41	Schur
16	Sirius . . .	−1·4	6 40 18	−16 34·0	Gill & Elkin
17	51 Hev. Cephei .	5·3	6 48 48	+87 13	Wagner & De Ball
18	Procyon . . .	0·5	7 33 33	+ 5 30·4	Elkin
19	Pollux . . .	1·1	7 38 35	+28 17·5	Elkin
20	10 Ursæ Majoris .	4·2	8 53 30	+42 13·1	Wagner & Belopolsky
21	Lalande 18115 . .	7·7	9 7 12	+53 10	Kapteyn
22	θ Ursæ Majoris .	3·2	9 25 30	+52 10·7	Kapteyn
23	Lalande 19022 .	8·0	9 36 30	+43 14	Kapteyn
24	20 Leonis Minoris .	6·0	9 54 42	+32 29	Kapteyn
25	α Leonis . . .	1·4	10 2 31	+12 30·3	Elkin
26	Groombridge 1618 .	6·5	10 4 46	+50 1·0	Ball
27	Groombridge 1646 .	6·2	10 21 18	+49 24	Kapteyn
28	Lalande 21185 .	6·8	10 57 20	+36 56·4	Winnecke
29	Lalande 21258 . .	8·5	11 0 1	+44 5·5	Auwers
30	Σ 1516 . . .	7·0	11 8 3	+74 4·25	De Ball
31	Arg. Oeltzen 11677 .	9·0	11 14 25	+66 26·5	Geelmuyden
32	Σ 1561 *seq.* . .	6·0	11 33 0	+45 43	Kapteyn
33	Groombridge 1830 .	6·5	11 46 38	+38 30·5	Brünnow
34	Lalande 22810 . .	7·5	12 4 6	+41 51	Kapteyn
35	α Centauri . .	−0·5	14 32 8	−60 22·7	Gill & Elkin
36	η Herculis . .	3·7	16 39 6	+39 8	Wagner & Belopolsky
37	π Herculis . .	3·4	17 11 12	+36 56	Wagner & Belopolsky
38	ν¹ Draconis . .	4·9	17 30 1	+55 15·6	Wagner & Belopolsky
39	ν² Draconis . .	4·8	17 30 6	+55 14 9	Wagner & Belopolsky
40	Arg. Oeltzen 17415 .	9·0	17 34 4	+68 27·4	Krüger
41	70 Ophiuchi . .	4·1	17 59 54	+2 31·5	Krüger
42	α Lyræ . . .	0·2	18 33 13	+38 40·9	Elkin
43	Σ 2398 . . .	8·2	18 41 33	+59 27·7	Lamp
44	α Draconis . .	4·7	19 32 34	+69 28·4	Brünnow
45	α Aquilæ . . .	1·0	19 45 25	+8 34·7	Elkin
46	ε Cygni . . .	2·7	20 41 45	+33 33·5	Pritchard
47	61 Cygni . . .	5·1	21 1 58	+38 12·5	Ball
	61 Cygni . . .	5·1	21 1 58	+38 12·5	Pritchard
48	α Cephei . . .	2·6	21 16 0	+62 7	Pritchard
49	ε Indi . . .	5·2	21 54 56	−57 14·3	Gill & Elkin
50	Lacaille 9352 . .	7·5	22 58 46	−36 29·2	Gill
51	Bradley 3077 . .	5·5	23 7 56	+56 33·6	Gyldén
52	85 Pegasi . . .	5·8	23 56 26	+26 30·1	Brünnow

STELLAR PARALLAXES.

Method	Date	Parallax	Distance in Light-years	Proper Motion	In Miles per Second
Photography	1888	0″·16	20	0″·55	10·8
Micrometer	1867	0·29	11	2·80	28·3
Heliometer	1883	0·057	57	2·05	101
Photography	1888	0·07	47	0·05	2
Micrometer	1855	0·15	22	1·20	22·8
Photography	1888	0·04	82	3·25	305
Photography	1890	0·092	35·4	0·196	6·2
Photography	1888	0·05	63	0·045	2·5
Photography	1890	0·08	40·5	0·22	8·1
Heliometer	1883	0·14	24	3·03	63
Photography	1890	0·074	44	0·035	1·4
Heliometer	1883	0·166	19·6	4·05	71·5
Heliometer	1888	0·116	28·1	0·19	4·8
Heliometer	1888	0·107	30·5	0·43	11·8
Micrometer	1886	0·111	29·4	?	?
Heliometer	1883	0·39	8	1·31	9·8
Meridian circle	1885	0·027	121	0·059	6·4
Heliometer	1888	0·266	12·3	1·26	13·9
Heliometer	1888	0·068	48	0·64	27·6
Meridian circle	1889	0·20	16	0·51	7·5
Meridian circle	1889	0·087	37·5	1·69	47·3
Meridian circle	1889	0·046	70·9	1·11	70·7
Meridian circle	1889	0·072	45·3	0·79	32
Meridian circle	1889	0·071	46	0·69	28·5
Heliometer	1888	0·093	35·1	0·27	8·5
Micrometer	1881	0·32	10·2	1·43	13
Meridian circle	1889	0·109	30	0·89	24
Heliometer	1872	0·501	6·5	4·75	27·8
Heliometer	1863	0·262	12·4	4·4	49·2
Micrometer	1887	0·104	31·3	0·42	11·8
Micrometer	1880	0·26	12·5	3·04	34·3
Meridian circle	1889	0·047	70	0·64	39·9
Micrometer	1870	0·089	36·6	7·05	232
Meridian circle	1889	0·067	48·7	0·33	14·4
Heliometer	1883	0·75	4·35	3·67	14·4
Meridian circle	1889	0·40	8·15	0·08	0·6
Meridian circle	1889	0·11	29·6	0·04	1·1
Meridian circle	1889	0·32	10·2	0·16	1·4
Meridian circle	1889	0·28	11·6	0·16	1·6
Heliometer	1863	0·247	13·6	1·27	15
Heliometer	1863	0·15	21·7	1·13	22
Heliometer	1888	0·034?	96	0·36	31
Micrometer	1887	0·353	9·2	2·40	20
Micrometer	1873	0·250	13	1·84	21·6
Heliometer	1888	0·199	16	0·65	9·6
Photography	1890	0·115	28·3	0·48	12·2
Micrometer	1878	0·467	7	5·16	32·4
Photography	1887	0·432	7·5	5·16	35]
Photography	1889	0·061	53·5	0·15	72
Heliometer	1883	0·20	16·3	4·6	67·8
Heliometer	1883	0·285	11·4	6·96	71·6
Micrometer	1882	0·283	11·5	2·09	21·6
Micrometer	1873	0·054	60·4	1·29	71·7

Table IV.—ANNUAL PROPER MOTIONS OF 1″ AND UPWARDS.

(*Kindly communicated by Professor E. Schönfeld of Bonn.*)

Name of Star	Mag.	R. A. (1855)	Decl. (1855)	Motion	Remarks
		h. m. s.	° ′	″	
Groombridge 1830	6·5	11 44 36	+38 45·5	7·0	B.B. vii., 112 = Ll. 22369
Lacaille 9352 .	7·2	22 56 29	−36 41·2	6·9	
Cordoba 32,416 .	8·2	23 56 51	−38 4·2	6·1	
61 Cygni . .	5·1	21 0 24	+38 2·3	5·2	dupl. 17′
Lalande 21185 .	7·3	10 55 24	+36 56·4	4·7	B.B. vii., 104
ε Indi . . .	5·2	21 52 14	−57 22·5	4·6	
Lalande 21258 .	8·6	10 58 15	+44 16·2	4·4	B.B. vii., 105
40 Eridani . .	4·4	4 8 36	− 7 52·9	4·1	tripl. (comes dupl. 4″, dist. 82″)
μ Cassiopeiæ .	5·2	0 58 39	+54 12·7	3·7	
α Centauri . .	−0·5	14 29 48	−60 13·9	3·7	dupl. 15″
Arg. Oeltz. 14318 .	9·2	15 2 17	−15 45·9	} 3·6	dist. = 5′·0
Arg. Oeltz. 14320 .	9·0	15 2 17	−15 40·9		
Lacaille 8760 .	7·6	21 8 44	−39 25·5	3·4	
e Eridani . .	4·4	3 14 7	−43 37·6	3·0	
Arg. Oeltz. 11677 .	9·0	11 12 27	+66 37·9	3·0	
Groombridge 34 .	7·9	0 10 7	+43 12·0	2·8	dupl. 40″
Piazzi II., 123 .	6·3	2 28 8	+ 6 11·5	2·4	
Lalande 25372 .	8·0	13 38 23	+15 40·7	2·3	
α Boötis . .	0·2	14 9 3	+19 56·3	2·3	
Σ 2398 . . .	8·5	18 41 10	+59 24·5	2·3	dupl. 16″. Struve p.m. 2164
β Hydri . .	3·0	0 18 4	−78 4·3	2·2	
Lalande 7443 .	8·5	3 53 31	+34 55·1	2·2	
Weisse V. 592 .	9·5	5 24 7	− 3 42·3	2·2	
Bradley 3077 .	5·9	23 6 19	+56 22·1	2·1	
ζ Toucanæ . .	4·1	0 12 28	−65 43·7	2·0	
Lalande 15290 .	8·3	7 44 17	+31 2·9	2·0	
Piazzi XIV., 212 .	5·9	14 49 0	−20 45·5	2·0	dupl. 13″. B. A. C. 4923
τ Ceti . . .	4·3	1 37 20	−16 42·6	1·9	
Lacaille 661 . .	6·5	2 4 35	−51 32·2	1·9	
σ Draconis . .	4·7	19 32 38	+69 24·8	1·9	
Weisse IX. 954 .	9·2	9 43 55	−11 35·7	1·8	
Fedorenko 1457, 1458 . . .	7·7, 7·9	9 4 29	+53 18·3	1·7	dupl. 20″. Ll. 18115
Lalande 30694 .	6·9	16 45 41	+ 0 16·7	1·6	
δ Pavonis . .	5·5	19 54 26	−66 32·6	1·6	
Lacaille 8362 .	6·0	20 1 40	−36 27·8	1·6	
Lacaille 2957 .	6·0	7 40 11	−33 53·3	1·5	
61 Virginis . .	4·8	13 10 49	−17 30·2	1·5	
Lalande 31055 .	7·5	16 57 31	− 4 49·0	1·5	
Piazzi 0 h. 130 .	5·6	0 29 54	−25 33·9	1·4	
20 Mayer . .	5·7	0 40 47	+ 4 32·0	1·4	B. A. C. 221
Lalande 6888 .	8·2	3 37 9	+41 1·2	1·4	
Fedorenko 1384 .	8·8	8 41 40	+71 21·0	1·4	
Groombridge 1618	6·5	10 2 28	+50 11·1	1·4	
Lalande 30044 .	7·5	16 23 21	+ 4 33·2	1·4	
Lalande 38383 .	7·3	19 57 49	+22 58·3	1·4	

TABLE IV.—ANNUAL PROPER MOTIONS OF 1″ AND UPWARDS—*continued.*

Name of Star	Mag.	R. A. (1855)	Decl. (1855)	Motion	Remarks
		h. m. s.	° ′	″	
ν Indi . . .	6·5	22 12 5	−72 57·8	1·4	
Lalande 46650 .	8·7	23 41 38	+ 1 38·0	1·4	
ι Persei . .	4·1	2 58 38	+49 3·7	1·3	
Weisse IV. 1189 .	6·5	4 53 37	− 5 55·6	1·3	
Sirius . . .	−1·4	6 38 46	−16 31·3	1·3	dupl. 10″
Procyon . .	0·5	7 31 43	+ 5 35·6	1·3	
Lacaille 3386 .	6·5	8 27 12	−31 2·4	1·3	
Lacaille 4887 .	5·5	11 39 36	−39 42·6	1·3	
Lalande 27744 .	7·0	15 6 35	− 0 47·2	1·3	
γ Serpentis . .	3·5	15 49 46	+16 8·3	1·3	
Arg. Oeltzen 17415	9.0	17 37 17	+68 28·3	1·3	
Piazzi XX. 29 .	5·3	20 6 14	−27 27·7	1·3	
Weisse XXIII. 175	8·2	23 9 34	−14 35·6	1·3	dupl. 1″ (Burnham 182)
85 Pegasi . .	5·8	23 54 43	+26 18·9	1·3	dupl. 0·″5 (Burnham 733)
η Cassiopeiæ .	3·6	0 40 21	+57 2·8	1·2	dupl. 7″
δ Trianguli . .	5·0	2 8 14	+33 33·5	1·2	
Lalande 15565 .	7·5	7 51 33	+29 39·0	1·2	
Lacaille 4955 .	6·7	11 50 45	−26 52·5	1·2	
43 Comæ . .	4·4	13 5 7	+28 36·9	1·2	
Lalande 28607 .	7·0	15 35 18	−10 27·3	1·2	
Weisse XVI. 906 .	8·8	16 47 45	− 8 3·8	1·2	
36 Ophiuchi . .	6·0	17 6 27	−26 22·1	1·2	dupl. 4″ } dist. = 12′·1
30 Scorpii . .	7·0	17 7 18	−26 19·0		
Weisse XVII. 322	8·2	17 18 33	+ 2 17·5	1·2	
Lamont₂ 3744 .	9·0	18 50 55	+ 5 46·0	1·2	
Arg. Oeltzen 20452	8·2	20 15 2	−21 47·4	1·2	
ζ¹ Reticuli . .	6·0	3 14 35	−63 7·9	1·1	dist. = 5′·1
ζ² Reticuli . .	5·7	3 15 1	−63 3·7		
Lacaille 2138 .	5·6	5 48 48	−80 34·3	1·1	
θ Ursæ Majoris .	3·2	9 23 8	+52 20·1	1·1	
20 Crateris . .	6·0	11 27 27	−32 3·8	1·1	
Lalande 27298 .	7·6	14 51 4	+54 14·9	1·1	
70 Ophiuchi .	4·1	17 58 8	+ 2 32·3	1·1	dupl. 4″
Lalande 16304 .	5·7	8 11 31	−12 8·6	1·0	
Lalande 27026 .	7·7	14 43 27	−23 41·1	1·0	
w Herculis . .	5·4	17 15 8	+32 39·4	1·0	
b Aquilæ . .	5·3	19 18 14	+11 38·3	1·0	
Lacaille 8620 .	6·0	20 48 2	−44 38·6	1·0	

TABLE V.—MASSES OF BINARY STARS.

Name of Star	Parallax	Period	Mean distance in seconds of arc	Mean distance in radii of earth's orbit	Mass of system Solar mass = 1	Combined light. Solar emission = 1	Remarks
Sirius . . .	0″·38	58·5 years (Gore)	8″·58	22·31	3·25	69	Mass of bright star = 2·15 ; of satellite = 1·1
α Centauri . .	0·75	88 ,, (Doberck)	18·45	11·26	1·9	4·7	Components of equal mass
61 Cygni . .	0·45	782 ,, (Peters)	29·48	65·5	0·459	0·12	Orbit known only approximately
70 Ophiuchi .	0·162	88 ,, (Gore)	4·5	27·8	2·776	2·3	
η Cassiopeiæ .	0·154	149 ,, (L. Struve)	8·79	57	8·33	4	Mass of brighter star = 6·57 ; of satellite = 1·76
o² Eridani (close pair) . . .	0·166	139 ,, (Gore)	5·99	36	2·4	0·03	
85 Pegasi . .	0·054	22·3 ,, (Schaeberle)	0·96	17·8	11·34	3·6	Parallax doubtful
Algol . . .	?	2·867 days	?	0·051	0·67	?	Spectroscopic. Equal mean density of components assumed
ζ Ursæ Majoris .	?	104 ,,	?	1·537	40	?	Spectroscopic. Values of distance and mass minima
β Aurigae . .	?	4 ,,	?	0·085	2·3	?	Spectroscopic. Values of distance and mass minima
α Virginis . .	?	4 ,,	?	0·033	2·4	?	Mass of dark companion assumed equal to that of bright star

Table VI.—MOVEMENTS IN LINE OF SIGHT

(The signs + and – signify respectively recession from, and approach to, the earth).

Name of Star	Photographically measured at Potsdam	Visually measured at Greenwich
Aldebaran	+ 30 English miles a second	+ 31
Capella	+ 17 " " "	+ 23
Rigel	+ 39 " " "	+ 18
ε Orionis	+ 34·6 " " "	+ 15
Polaris	– 16·3	
α Virginis	– 14 " " "	– 17
α Persei	– 7 " " "	– 22
Procyon	– 7 " " "	+ 3
Algol	– 2·3 " " "	
Arcturus	. . . .	– 45
Vega	. . . .	– 34
α Cygni	. . . .	– 36
Pollux	. . . .	– 33
α Aquilae	. . . .	– 27
α Orionis	. . . .	+ 28

INDEX

Printed in the United States
By Bookmasters